Mechanical Properties of Advanced Metallic Materials

Mechanical Properties of Advanced Metallic Materials

Editors

Yang Zhang
Yuqiang Chen

MDPI • Basel • Beijing • Wuhan • Barcelona • Belgrade • Manchester • Tokyo • Cluj • Tianjin

Editors
Yang Zhang
Harbin Engineering University
China

Yuqiang Chen
Hunan University of Science and Technology
China

Editorial Office
MDPI
St. Alban-Anlage 66
4052 Basel, Switzerland

This is a reprint of articles from the Special Issue published online in the open access journal *Crystals* (ISSN 2073-4352) (available at: https://www.mdpi.com/journal/crystals/special_issues/mechanical_metallic).

For citation purposes, cite each article independently as indicated on the article page online and as indicated below:

LastName, A.A.; LastName, B.B.; LastName, C.C. Article Title. *Journal Name* **Year**, *Volume Number*, Page Range.

ISBN 978-3-0365-6956-7 (Hbk)
ISBN 978-3-0365-6957-4 (PDF)

© 2023 by the authors. Articles in this book are Open Access and distributed under the Creative Commons Attribution (CC BY) license, which allows users to download, copy and build upon published articles, as long as the author and publisher are properly credited, which ensures maximum dissemination and a wider impact of our publications.

The book as a whole is distributed by MDPI under the terms and conditions of the Creative Commons license CC BY-NC-ND.

Contents

About the Editors vii

Preface to "Mechanical Properties of Advanced Metallic Materials" ix

Zhenxin Li, Yang Zhang, Kai Dong and Zhongwu Zhang
Research Progress of Fe-Based Superelastic Alloys
Reprinted from: *Crystals* **2022**, *12*, 602, doi:10.3390/cryst12050602 1

Songlin Shen, Mei Zhan, Pengfei Gao, Wenshuo Hao, Fionn P. E. Dunne and Zebang Zheng
Microstructural Effects on Thermal-Mechanical Alleviation of Cold Dwell Fatigue in Titanium Alloys
Reprinted from: *Crystals* **2022**, *12*, 208, doi:10.3390/cryst12020208 27

Juan Pu, Peng Xie, Weimin Long, Mingfang Wu, Yongwang Sheng and Jie Sheng
Effect of Current on Corrosion Resistance of Duplex Stainless Steel Layer Obtained by Plasma Arc Cladding
Reprinted from: *Crystals* **2022**, *12*, 341, doi:10.3390/cryst12030341 37

Andrei Mostovshchikov, Fedor Gubarev, Pavel Chumerin, Vladimir Arkhipov, Valery Kuznetsov and Yana Dubkova
Solid Energetic Material Based on Aluminum Micropowder Modified by Microwave Radiation
Reprinted from: *Crystals* **2022**, *12*, 446, doi:10.3390/cryst12040446 51

ZF Liu, N Tian, YG Tong, YL Hu, DY Deng, MJ Zhang, ZH Cai and J Liu
Mechanical Performance and Deformation Behavior of CoCrNi Medium-Entropy Alloy at the Atomic Scale
Reprinted from: *Crystals* **2022**, *12*, 753, doi:10.3390/cryst12060753 63

Mingzhu Fu, Suping Pan, Huiqun Liu and Yuqiang Chen
Initial Microstructure Effects on Hot Tensile Deformation and Fracture Mechanisms of Ti-5Al-5Mo-5V-1Cr-1Fe Alloy Using In Situ Observation
Reprinted from: *Crystals* **2022**, *12*, 934, doi:10.3390/cryst12070934 75

Yuqiang Chen, Hailiang Wu, Xiangdong Wang, Xianghao Zeng, Liang Huang, Hongyu Gu and Heng Li
Corrosion Mechanism and the Effect of Corrosion Time on Mechanical Behavior of 5083/6005A Welded Joints in a NaCl and $NaHSO_3$ Mixed Solution
Reprinted from: *Crystals* **2022**, *12*, 1150, doi:10.3390/cryst12081150 91

Parinaz Seifollahzadeh, Morteza Alizadeh, Ábel Szabó, Jenő Gubicza and Moustafa El-Tahawy
Microstructure and Mechanical Behavior of Cu–Al–Ag Shape Memory Alloys Processed by Accumulative Roll Bonding and Subsequent Annealing
Reprinted from: *Crystals* **2022**, *12*, 1167, doi:10.3390/cryst12081167 107

Sunder Jebarose Juliyana, Jayavelu Udaya Prakash, Sachin Salunkhe, Hussein Mohamed Abdelmoneam Hussein and Sharad Ramdas Gawade
Mechanical Characterization and Microstructural Analysis of Hybrid Composites ($LM5/ZrO_2/Gr$)
Reprinted from: *Crystals* **2022**, *12*, 1207, doi:10.3390/cryst12091207 129

Jing Xu, Zichun Wu, Jianpeng Niu, Yufeng Song, Chaoping Liang, Kai Yang, Yuqiang Chen and Yang Liu
Effect of Laser Energy Density on the Microstructure and Microhardness of Inconel 718 Alloy Fabricated by Selective Laser Melting
Reprinted from: *Crystals* **2022**, *12*, 1243, doi:10.3390/cryst12091243 . . . . **151**

Yang Liu, Zichun Wu, Qing Wang, Lizhong Zhao, Xichen Zhang, Wei Gao, Jing Xu, et al.
Optimization of Parameters in Laser Powder Bed Fusion TA15 Titanium Alloy Using Taguchi Method
Reprinted from: *Crystals* **2022**, *12*, 1385, doi:10.3390/cryst12101385 . . . . **165**

Qingyou Han, Yanfei Liu, Cheng Peng and Zhiwei Liu
Engulfment of a Particle by a Growing Crystal in Binary Alloys
Reprinted from: *Crystals* **2022**, *12*, 1421, doi:10.3390/ cryst12101421 . . . . **179**

Raj Soni, Sarang Pande, Santosh Kumar, Sachin Salunkhe, Harshad Natu and Hussein Mohammed Abdel Moneam Hussein
Wear Characterization of Laser Cladded Ti-Nb-Ta Alloy for Biomedical Applications
Reprinted from: *Crystals* **2022**, *12*, 1716, doi:10.3390/cryst12121716 . . . . **193**

Maria-Rosa Ardigo-Besnard, Aurélien Besnard, Galy Nkou Bouala, Pascal Boulet, Yoann Pinot and Quentin Ostorero
Effect of Pre-Oxidation on a Ti PVD Coated Ferritic Steel Substrate during High-Temperature Aging
Reprinted from: *Crystals* **2022**, *12*, 1732, doi:10.3390/cryst12121732 . . . . **205**

Vlastimil Novák, Lenka Řeháčková, Silvie Rosypalová and Dalibor Matýsek
Wetting of Refractory Ceramics with High-Manganese and Structural Steel and Description of Interfacial Interaction
Reprinted from: *Crystals* **2022**, *12*, 1782, doi:10.3390/cryst12121782 . . . . **219**

About the Editors

Yang Zhang

Yang Zhang is an Associate Professor of the Harbin Engineering University, head of Institute of New Materials and Advanced Manufacturing Technology. He obtained his Ph.D. degree from Shandong University, China. Then, he worked as a postdoctoral researcher at the Pennsylvania State University, USA. His research focuses on the mechanical properties and irradiation effect of nanoprecipitate and nanotwin strengthened alloys. He was awarded the IAAM Innovation Award in 2022. He was selected as part of the young scientific and technological elite talents of CNNC. He won the first prize of the Heilongjiang Science and Technology Award and the first prize of the Heilongjiang University Science and Technology Award (both ranked second). As the project leader, he has undertaken more than 10 national and provincial projects, such as the NSFC funding and the China–Ukraine intergovernmental exchange program. In recent years, he has published more than 40 SCI/EI papers in journals such as *Int. J. Plasticity*, *Adv. Sci.*, and *Scripta Mater*. He has applied for more than 20 invention patents, and 15 patents have been authorized. He was selected as a member of IAAM Fellow and Vebleo Fellow. He has attended more than 10 academic conferences, and delivered more than 10 keynote, invited, and oral lectures.

Yuqiang Chen

Yuqiang Chen is a Professor of the Hunan University of Science and Technology, vice dean of the graduate school, and vice dean of the Hunan Engineering Research Center of Forming Technology and Damage Resistance Evaluation for High Efficiency Light Alloy Components. He obtained his Ph.D degree from the Central South University, China. His research focuses on microstructure characterization, forming mechanism and fatigue behavior of light alloys. He was selected as "Hunan Young Talents", "the Young Core Instructor from the Education Commission of Hunan Province", and "Training Program for Hunan Science and Technology Innovation and Entrepreneurship Elites to Germany in 2018". As the project leader, he has undertaken more than 30 scientific research projects, such as NSFS funding and key research project of the Hunan Province. He has published more than 70 journal papers, applied more than 10 national invention patents. He is a reviewer of the journals such as *Journal of Materials Science & Technology*, *Journal of Alloys and Compounds*, *Journal of Materials Research and Materials Science and Engineering A*, etc.

Preface to "Mechanical Properties of Advanced Metallic Materials"

With many of today's emerging technologies, the primary emphasis is on the mechanical properties of the metallic materials used in the fields of ocean, air and aerospace, bridge and nuclear engineering. Because the mechanical property is an extremely important indicator to evaluate whether the materials can be applied in the above fields. Strength is the main indicator of the mechanical property. Different strengthening mechanisms, such as phase transformation strengthening, solid-solution strengthening, dislocation strengthening, grain-boundary strengthening, precipitation strengthening and load transfer via the introduction of strong phases, can be used to achieve high strength/hardness. These strengthening methods are accompanied by various deformation mechanisms, such as Transformation-Induced Plasticity (TRIP), Twinning-Induced Plasticity (TWIP), etc. Achieving high strength–ductility synergy is a long-time challenge and becomes a topic of general interest. To promote the further progress of this field, a Special Issue focusing on the "Mechanical Properties of Advanced Metallic Materials" is now published.

This Special Issue includes 1 review paper and 14 original research papers devoted to the mechanical properties of the superelastic alloy, titanium alloy, shape memory alloy, various kinds of steels, medium-entropy alloy, high-temperature alloy, etc. The review article of Prof. Y. Zhang and Prof. Z.W. Zhang summarized the research progress of Fe-based superelastic alloys, mainly including the superelastic mechanical properties and its influence factors. Prof. Y.Q. Chen, Prof. J. Liu, Prof. Z.B. Zheng, Prof. Y. Liu, Prof. M. Alizadeh, etc., investigated the mechanical performance and deformation behavior of various kinds of alloys and steels, including titanium alloys, CoCrNi medium-entropy alloy, 5083/6005A alloy, Cu–Al–Ag shape memory alloys, Inconel 718 alloy and hybrid composites (LM5/ZrO2/Gr). Besides, wear characterization, corrosion resistance, pre-oxidation microwave radiation, interfacial interaction and engulfment mechanisms in titanium and various kind of steel and binary alloys were studied by Prof. S. Pande, Prof. J. Pu, Prof. F. Gubarev, Prof. V. Novák, Prof. M. Ardigo-Besnard, Prof. Q.Y. Han, etc.

We kindly wish that the readers will find this Special Issue to be informative and the published papers will bring forth new ideas on different related aspects. This should be of great significance for promoting the further development of this field.

Yang Zhang and Yuqiang Chen
Editors

Review

Research Progress of Fe-Based Superelastic Alloys

Zhenxin Li, Yang Zhang *, Kai Dong and Zhongwu Zhang *

Key Laboratory of Superlight Materials and Surface Technology, Ministry of Education, College of Materials Science and Chemical Engineering, Harbin Engineering University, Harbin 150001, China; lizhenxin@hrbeu.edu.cn (Z.L.); dongkai1006@hrbeu.edu.cn (K.D.)
* Correspondence: zhangyang0115@hrbeu.edu.cn (Y.Z.); zwzhang@hrbeu.edu.cn (Z.Z.)

Abstract: In recent years, superelastic alloys have become a current research hotspot due to the large recoverable deformation, which far exceeds the elastic recovery. This will create more possibilities in practical applications. At present, superelastic alloys are widely used in the fields of machinery, aerospace, transmission, medicine, etc., and become smart materials with great potential. Among superelastic alloys, Fe-based superelastic alloys are widely used due to the advantages of low cost, easy processing, good plasticity and toughness, and wide applicable temperature range. The research progress of Fe-based superelastic alloys are reviewed in this paper. The mechanism of thermoelastic martensitic transformation and its relation to superelasticity are summarized. The effects of the precipitate, grain size, grain orientation, and texture on the superelasticity of Fe-based superelastic alloys are discussed in detail. It is expected to provide a guide on the development and understanding of Fe-based superelastic alloys. The future development of Fe-based superelastic alloys are prospected.

Keywords: Fe-based alloy; superelasticity; martensitic transformation; precipitate

1. Introduction

The essence of superelastic alloys is actually a kind of shape memory alloys, which has two characteristics of the shape memory effect and superelasticity. The superelasticity of the shape memory alloys is generally produced by temperature-induced reverse martensitic transformation to recover deformation. However, the superelastic alloy is a special type of shape memory alloy. The superelasticity in superelastic alloys relies on the thermoelastic martensitic transformation, that is, stress-induced martensitic transformation is used to recover deformation. Therefore, different with the shape memory alloys, superelastic alloys can exhibit superelasticity at a constant temperature without heating and cooling [1]. Superelasticity means that when the alloy undergoes a limited amount of plastic deformation (non-linear elastic deformation) under the action of higher than the transformation temperature and stress, the stress can be directly released to recover to its original shape. Superelastic alloys are favored by scholars due to their special mechanical behaviors, and thus have a wide range of applications in the automotive machinery, aerospace, telecommunications conduction, smart sensors, and other fields due to their extremely recoverable deformation [2–5]. Obvious stress hysteresis can be observed in the tensile curves of the superelastic alloys, which makes them possess hysteretic energy dissipation characteristics, absorbing energy, and reducing vibration [6]. This feature broadens the applications of superelastic alloys. The achievement on the hysteretic energy dissipation characteristic of superelastic alloys is expected to be applied in the fields of non-destructive testing, vibration and shock protection, biomedical imaging, and soft machinery [7–10]. In the decades of development of superelastic alloys, according to their composition classification, superelastic alloys can be divided into NiTi-based, Cu-based, and Fe-based superelastic alloys [1]. NiTi-based superelastic alloy is currently the most widely used due to the high superelastic strain of up to 8%, good mechanical properties, excellent corrosion resistance, and good biocompatibility. However, it also has some disadvantages, such as high cost and processing difficulty [11,12]. The resistivity of Cu-based superelastic alloy is relatively small, which is about an order of magnitude smaller than that of NiTi-based alloy [13]. It is not suitable for the applications of heating and charging. In addition, its low strength and high sensitivity to temperature changes further limit its application [13]. Fe-based superelastic alloy has the advantages of low cost, good ductility, high strength, easy processing, and weldability, etc., making it a substitute for NiTi-based superelastic alloy, and becoming a hot spot in current researches [14–16].

In this paper, the research progress of Fe-based superelastic alloys is reviewed. The effect of thermoelastic martensitic transformation on the superelasticity of Fe-based superelastic alloys is analyzed and discussed. The factors on superelasticity in Fe-based superelastic alloy are summarized. Finally, the future development directions of Fe-based superelastic alloys are prospected.

2. Superelastic Mechanisms of Fe-Based Superelastic Alloys

2.1. Thermoelastic Martensitic Transformation

Martensitic transformation is a very important means to strengthen the alloys [17–19]. In Fe-based superelastic alloys, the superelasticity is extremely dependent on the thermoelastic martensitic transformation. When quick quenching is conducted from the austenite zone, there is no time for the decomposition process of eutectoid diffusion to occur, generating martensites [20–22]. During the martensitic transformation, the movement of a single atom is less than the distance between two atoms [23,24]. According to the morphology, martensite can be divided into lath martensite and plate martensite. Figure 1 shows the morphologies of lath martensite and plate martensite in alloys [25]. Figure 1a shows the body centered cubic (BCC) martensite in Fe-14Mo-11.5Cr-9Ni-7Co-2Cu-0.6Ti-0.4Al (weight %) alloy austenitized at 1150 °C, and then quenched to room temperature in water. Figure 1a illustrates the lath martensite composed of clusters of laths. An austenite grain can form several lath groups with different orientations. The lath group is composed of lath bundles, which contains many slender martensite laths arranged almost in parallel. Figure 1b presents plate martensite with the body centered tetragonal (BCT) structure in Fe-0.47Cr-0.27Si-0.22Mn-1.67C (weight %) alloy, possessing needle-like or bamboo-leaf morphology [25], while the martensite plates with different sizes are not parallel to each other. In fact, the spatial morphology of plate martensite is convex lens. Lath martensite usually possesses good toughness. The dislocations are not uniformly distributed in cellular dislocation substructure, resulting in the formation of low-density dislocation areas, which provide room for dislocation motion [26]. Dislocation movement can alleviate local stress concentration, delay crack nucleation, and weaken the stress peak at the crack tip, which is beneficial to toughness [26]. The twin substructure of plate martensite will reduce the effective slip system, decreasing the toughness, but increasing the hardness and strength. With the in-depth studies of martensite and martensitic transformation by scholars, it is found that martensite and martensitic transformation are not unique in steels [27–30]. As long as certain conditions are met, the transformation can be called martensitic transformation. Xu summarized the characteristics of martensitic transformation and put forward a simple definition: martensitic transformation is a first-order and nucleation-length phase transformation characterized by invariant plane strain, in which the replacement atoms undergo non-diffusive shear displacement, resulting in shape change and surface projection [31]. The basic characteristics of martensitic transformation are: (1) non-diffusivity; (2) occur mainly through shearing with surface relief feature; (3) the martensite and matrix possess a certain orientation relationship; (4) the habit plane of martensite does not produce distortion, and does not rotate during the process of phase transformation [32–36]. All phase transformations that meet these characteristics can be called martensitic transformation.

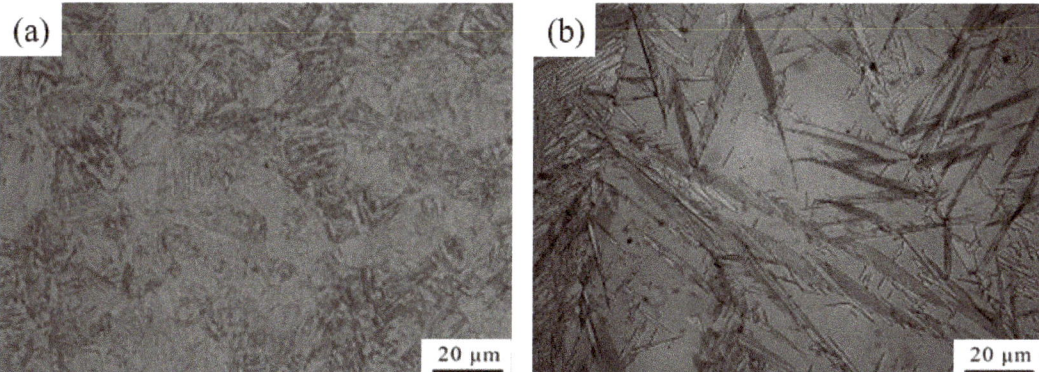

Figure 1. Martensites in alloys [25]: (**a**) Lath martensite in Fe-14Mo-11.5Cr-9Ni-7Co-2Cu-0.6Ti-0.4Al(weight %) alloy austenitized at 1150 °C, and then quenched to room temperature in water, (**b**) Plate martensite in Fe-0.47Cr-0.27Si-0.22Mn-1.67C(weight %) alloy austenitized at 1100 °C, and then quenched in brine. Reprinted with permission from Ref. [25]. Copyright 2013 Elsevier.

Most Fe-based shape memory alloys undergo non-thermoelastic martensitic transformation, and thus do not exhibit superelasticity [37–39]. The martensitic transformation in Fe-based superelastic alloy must be thermoelastic so that the alloy can possess superelasticity. The thermoelastic martensitic transformation is a sufficient and necessary condition for the alloy to possess superelasticity [40]. When the shape change of martensitic transformation is coordinated with elastic deformation, this transformation is called thermoelastic martensitic transformation. During the cooling process, the thermoelastic martensite will expand; when the temperature increases, the thermoelastic martensite will shrink. In Fe-based superelastic alloys, stress can also induce thermoelastic martensitic transformation [41]. A schematic diagram of stress-induced martensitic transformation is shown in Figure 2. During loading, the martensitic transformation occurs. After unloading, the martensite undergoes a reverse phase transformation and transforms into the parent phase, recovering the strain. In case of thermoelastic martensitic transformation, the driving force of thermoelastic martensite is mechanical, and comes from the interaction between the applied stress field and the transformation strain. In the thermoelastic martensitic transformation, the volume change is small, because the deformation caused by the transformation is basically elastic. In addition, during the reverse phase transformation of thermoelastic martensite, the thermal hysteresis is relatively small. According to the characteristics of thermoelastic martensitic transformation, Xu et al. gave the judgment basis for thermoelastic martensitic transformation: (1) Both the critical driving force of phase transformation, and the thermal hysteresis during the phase transformation are small; (2) Phase interface can move with the phase transformation or reverse transformation; (3) The strain produced by the phase transformation is elastic, and the elastic strain energy stored in the martensite is the driving force for the reverse phase transformation [42,43]. Studies have shown that the driving force for superelastic martensite transformation in Fe-Mn-Al-Ni superelastic alloy is only 32 J/mol, while the driving force for martensite phase transformation in Fe-based non-superelastic alloy is as high as 1000 J/mol [44]. It can be seen that the driving force for the transformation of non-thermoelastic martensite is about 10 times more than that of thermoelastic martensite.

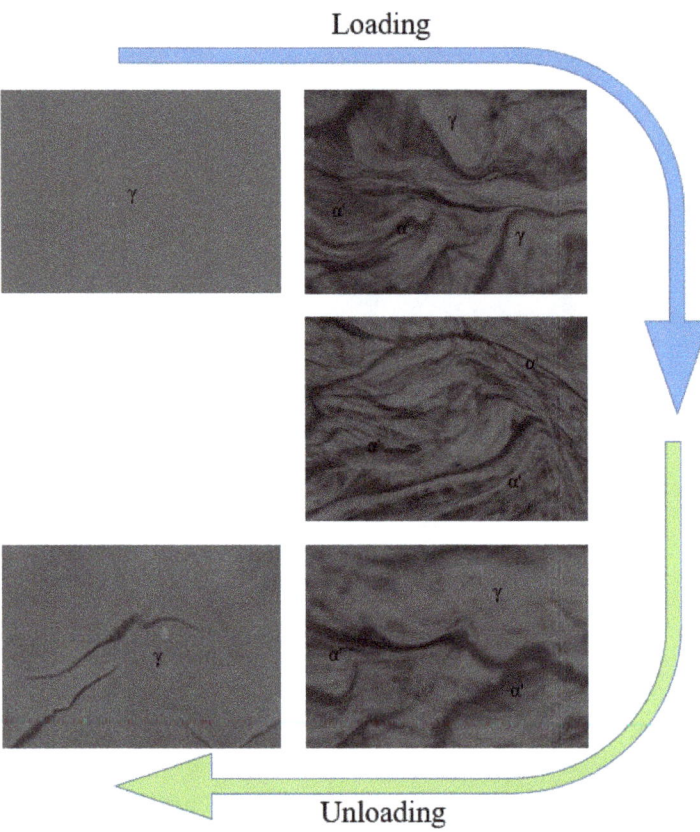

Figure 2. A schematic diagram of stress-induced martensitic transformation (γ refers to the parent phase, and α' refers to martensite).

The superelasticity of Fe-based alloys relies on the occurrence of thermoelastic martensitic transformation, and its reverse transformation to recover deformation [42]. In Fe-Mn-Al-Ni superelastic alloy, martensitic transformation occurs from the BCC matrix to face center cubic (FCC) martensite phase, while the transformation is from the FCC matrix to BCT martensite phase in Fe-Ni-Co-Al and Fe-Ni-Co-Ti superelastic alloys. Here, α refers to the austenite phase and γ refers to the martensite phase. When the thermoelastic martensitic transformation occurs, temperature varies. At high temperature, the α austenite phase has greater entropy, so the Gibbs free energy of α austenite phase is larger than that of the γ martensite phase. Thus, the α phase is more stable than γ phase at high temperature, which leads to γ→α phase transformation at high temperature [44]. The α phase is ferromagnetic, and its entropy can be reduced by magnetic ordering. The low temperature will make the magnetic ordering, thereby reducing the entropy, causing the ferromagnetic α phase to possess a lower entropy than the γ phase, leading to the α→γ phase transformation at low temperature [44]. These two transformations can change back and forth with the difference in temperature to recover the strain, thus possessing superelasticity.

2.2. Clausius–Clapeyron Equation

The critical stress (σ_{cr}) for stress-induced martensitic transformation tends to change with temperature. Clausius–Clapeyron equation can quantitatively represent the critical stress as a function of temperature [45,46]. Clausius–Clapeyron equation is shown as follows [45,46]:

$$\frac{d\sigma_{cr}}{dT} = -\frac{\Delta S}{V_m \varepsilon} = -\frac{\Delta H}{\varepsilon_0 T_0} \qquad (1)$$

where T is the temperature, ΔS and ΔH are the transformation entropy and transformation enthalpy, V_m is the molar volume, ε is the critical strain, ε_0 is the orientation-dependent lattice deformation, and T_0 is the chemical equilibrium temperature [45,46]. The martensitic transformation temperature can be obtained from Equation (2) [47]:

$$M_s(\sigma_d) = T_0^d(\sigma_d) - \Delta T \qquad (2)$$

where $M_s(\sigma_d)$ is the martensitic transformation temperature when the stress reaches σ_d, $T_0^d(\sigma_d)$ is the phase equilibrium temperature of the parent phase and martensite when the stress is σ_d, ΔT is the degree of subcooling required for phase transformation [47]. Equation (3) can be obtained from Equation (2). Substituting Equation (3) into Equation (1) yields Equation (4). The relationship between stress and start temperature of martensitic transformation can be obtained from Equation (4). Based on the Clausius–Clapeyron equation, the temperature sensitivity of the critical stress for the martensitic transformation can be identified. According to Equation (4), one can know the effect of stress on the start temperature of martensitic transformation.

$$\frac{dT_0^\sigma}{d\sigma_d} = \frac{dM_s}{d\sigma_d} \qquad (3)$$

$$\frac{d\sigma_d}{dM_s} = -\frac{\Delta H}{\varepsilon_0 T_0} \qquad (4)$$

2.3. Superelasticity

Superelasticity is achieved by the dynamic changes of martensite and austenite at low temperatures. Temperature and stress can induce martensitic transformation and its recovery. Superelasticity is a dynamic combination of these two stages, and caused by martensitic transformation [48]. When loading temperature is above the austenite transformation temperature, the austenite will undergo martensitic transformation accompanied by a transformation strain. Subsequent to the isothermal unloading, the martensite reverse transformation will occur. In the martensite reverse transformation, the transformation strain generated during the martensitic transformation can be eliminated, exhibiting a hysteresis phenomenon at the same time. The schematic diagram of the thermoelastic martensitic transformation is shown in Figure 3. Above the A_f temperature (end temperature of austenite transformation), the material is in austenite state. After loading, the stress induces martensitic transformation, resulting in the formation of martensite variants with different orientations. At this time, the material undergoes macroscopic deformation, and then changes from austenite to martensite with continuous loading. After unloading, the martensite is reversely transformed into austenite, recovering deformation. The superelasticity caused by martensite reorientation is attributed to the reorientation of martensite variants caused by stress. However, the reoriented martensite variant can be recovered to its original direction by recovery force, which will also cause hysteresis [49,50]. Figure 4 shows the stress–strain curve of Fe-based superelastic alloy during loading and unloading. It can be seen that in the initial stretching stage, the stress–strain curve is a straight line, which belongs to the elastic deformation stage. When the stress continues to increase, the stress–strain curve deviates from the straight line. In this case, martensitic transformation occurs. The stress value at the deviated point, ε_{Ms}, is the critical stress value of martensitic transformation. At this time, there will be a stress plateau. When the curve breaks away from the stress plateau, it marks the end of the martensitic transformation. During unloading, the reverse martensitic transformation will occur, thereby eliminating the existing strain, and recovering to the original state.

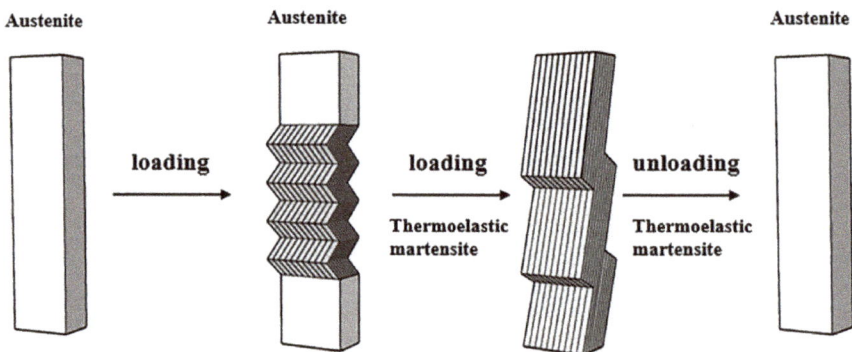

Figure 3. Schematic diagram of the thermoelastic martensitic transformation process in superelastic alloys.

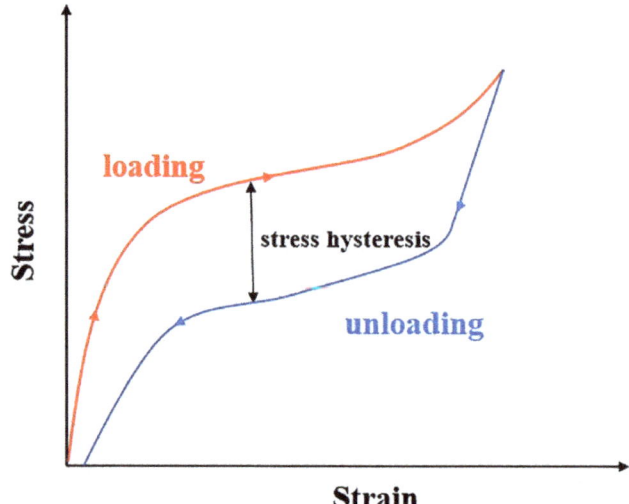

Figure 4. Typical superelastic stress–strain curve.

Although superelastic alloy belongs to shape memory alloy, it is different from shape memory alloys that recover deformation through temperature change. Superelastic alloy can recover deformation at a constant temperature. Both superelasticity and shape memory effect are closely related to temperature, exhibiting various macroscopic mechanical properties at different temperatures [51]. Figure 5 shows the mechanical behaviors of the alloys at different temperatures. In Figure 5, M_d is the critical temperature of stress-induced martensitic transformation, A_s is the start temperature of austenite transformation, A_f is the end temperature of austenite transformation, and M_f is the end temperature of martensitic transformation. When $T > M_d$, the critical stress required to form martensite is very large, so the material undergoes plastic deformation before the occurrence of stress-induced martensitic transformation. Fracture may occur before the occurrence of martensitic transformation. When $A_f < T < M_d$, the alloy is composed of austenite, and stress-induced martensitic transformation occurs in this stage. If the martensitic transformation at this time is thermoelastic, the material possesses superelasticity. When $T < M_f$, after stretching and unloading, only a part of the elastic deformation can be recovered. Residual strain still exists. Only by heating the alloy above the A_s temperature can make the material recover to its original shape, as shown by the red dashed line in Figure 5.

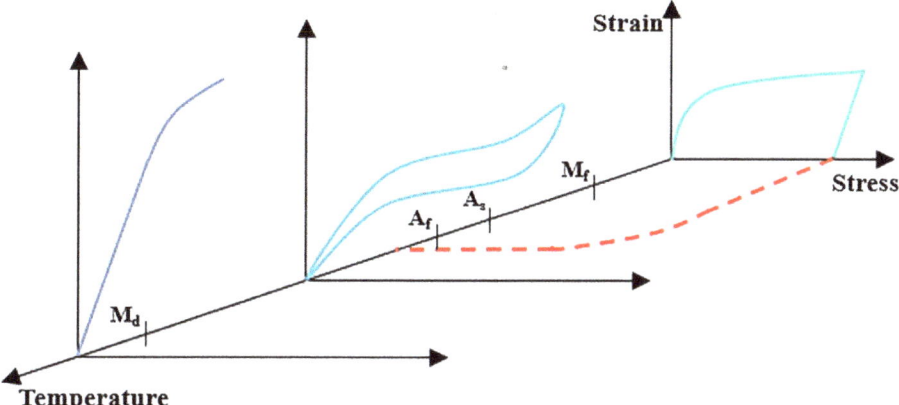

Figure 5. The alloy exhibits different mechanical behaviors at different temperatures.

Superelasticity significantly depends on the thermoelastic martensitic transformation. Only the occurrence of phase transformation and reverse phase transformation from thermoelastic martensite can achieve superelasticity. The reason why thermoelastic martensite can undergo a reverse transformation is that the driving force for the phase transformation is the difference in chemical-free energy between the two phases. The driving force of the phase transformation and elastic energy are in a dynamic equilibrium state [52]. When the thermoelastic martensitic transformation occurs, the driving force causes the phase transformation to occur. The material deforms to store up the elastic energy. The reverse phase transformation of thermoelastic martensite occurs during unloading. Then, the stored elastic energy is released. The thermoelastic balance of these two opposite energy terms makes the alloy stretch or shrink according to the change of external stress [53]. At the same time, because the energy barrier of lattice shear in the process of thermoelastic martensitic transformation is higher than that of non-thermoelastic martensitic transformation, it is difficult for plastic deformation to occur, so that the deformation can be recovered [54–56].

To sum up, in order to obtain good superelasticity, it is necessary to control the phase transformation to be thermoelastic martensitic phase transformation, that is, controlling the thermal balance between the elastic energy and driving force. Among Fe-based superelastic alloys, including Fe-Mn-Al-Ni system, Fe-Ni-Co-Al system, and Fe-Ni-Co-Ti system alloys, the most effective method is to introduce coherent precipitates to change non-thermoelastic martensitic transformation into thermoelastic martensitic transformation [57,58]. On the one hand, the coherent precipitates are dispersedly distributed in the matrix, which can strengthen the matrix, and make the matrix better adapt to the deformation caused by external stress. On the other hand, the disperse distribution of coherent precipitates can also suppress the slip of dislocations. In this case, critical stress of martensitic transformation can be reached before plastic deformation of the dislocation slips. The occurrence of martensitic transformation inhibits plastic deformation. In addition, the existence of the coherent precipitates also provides elastic energy, which can achieve the mutual transformation between the elastic energy and driving force of the phase transformation. All the above three points should be satisfied to ensure the occurrence of thermoelastic martensitic transformation, making the alloy obtain excellent superelasticity [40,59,60]. For example, Fe-Mn-Al-Ni superelastic alloy can obtain superelasticity by strengthening the parent phase and controlling the grain size to obtain the bamboo structure [61]. The superelasticity in Fe-Ni-Co-Al superelastic alloy is limited due to the precipitation of brittle phases at the grain boundary. It is found that the addition of B element can control the precipitation of grain boundaries together with large cold-rolling deformation amount to possess small-angle grain boundaries, obtaining good superelasticity [41]. In Fe-Ni-Co-Ti superelastic alloy, the precipitation of brittle phases at grain boundaries can be suppressed by adding B element or Cu element to reduce thermal hysteresis to obtain superelasticity [62,63].

3. Effects of Precipitates on Superelasticity

3.1. Effects of Coherent Precipitates on Superelasticity

The superelasticity of Fe-based alloys depends on thermoelastic martensitic transformation. The martensitic transformation of many systems is usually non-thermoelastic or semi-thermoelastic, such as Fe-Mn-Al, Fe-Mn-Si, Fe–C, Fe-Mn-C alloys, etc. [64–67]. Ando et al. studied the martensitic transformation of Fe-Mn-Al alloy [64], whose martensitic lath contains high-density twins, resulting in high thermal hysteresis of phase transformation. This leads to the occurrence of the incomplete martensitic reverse transformation, and incomplete restoration of martensitic to original phase. According to the thermoelastic martensitic transformation criterion described above, the martensitic transformation of Fe-Mn-Al alloy is semi-thermoelastic. There is no equilibrium between the thermal driving force and elastic energy in the non-thermoelastic or semi-thermoelastic martensitic transformation. Adjusting the equilibrium between the thermal driving force and elastic energy can make the martensitic transformation become thermoelastic. The introduction of coherent precipitates is the most common method to achieve this goal. Thermoelastic martensitic transformation can be achieved by introducing coherent precipitates in Fe-based alloys, which can be obtained by adding alloying elements or heat treatment [41,61,68–70].

3.1.1. Effects of Coherent Precipitates on Martensitic Transformation

In Fe-based superelastic alloys, thermoelastic martensitic transformation can be achieved by introducing coherent precipitates in order to obtain superelasticity. The types of coherent precipitates are different in different alloy systems. For example, the coherent precipitates in Fe-Mn-Al-Ni superelastic alloys are B2 phase, and $L1_2$-γ' phase in Fe-Ni-Co-Al and Fe-Ni-Co-Ti superelastic alloys.

The effect of coherent precipitates on the martensitic transformation can be directly reflected by M_s point. The effect of precipitation on M_s point can be attributed to the following three aspects: (1) the content of Ni in matrix decreases with the precipitation, which leads to the increase of thermodynamic equilibrium temperature; (2) the precipitates possess strengthening effect, increasing the energy barrier of martensitic transformation; (3) the stress field energy is different between the parent phase/precipitate and the martensite/precipitate [70]. To a certain extent, the decrease of M_s point will reduce the thermal hysteresis. The most obvious characteristic of thermoelastic martensitic transformation is the small thermal hysteresis (generally <100 K) [42].

In Fe-Mn-Al-Ni superelastic alloys, the precipitation of B2 phase makes the martensitic transformation become thermoelastic, because B2 precipitate is coherent with parent phase and martensitic phase [61]. B2 phase can contribute elastic energy, which is conducive to reverse transformation, and makes the phase transformation into thermoelastic, thus obtaining superelasticity [71,72].

The γ' phase in Fe-Ni-Co-Al and Fe-Ni-Co-Ti superelastic alloys can change the deformation mechanism from slip to twinning [51,73]. The critical shear stress required by twinning is larger than slip, and thus the energy required to start the plastic deformation mechanism is higher. This makes it easier for the alloy to satisfy the criterion that the critical stress of thermoelastic martensitic transformation is less than the yield stress. The thermoelastic martensitic transformation occurs before the initiation of plastic deformation mechanism, and superelasticity is obtained. The emergence of precipitate can effectively avoid dislocation slip during martensitic transformation. The precipitate improves the tetragonal degree of martensite, and is conducive to the interface compatibility between parent phase and martensite phase [41]. In addition, the precipitate also reduces the elastic strain energy during martensitic transformation and makes martensitic transformation into thermoelastic [74].

3.1.2. Effects of Alloying on Coherent Precipitates

In both Fe-Mn-Al-Ni and Fe-Ni-Co-Al systems, coherent precipitates are obtained by adding alloying elements. Omori et al. added Ni element into Fe-Mn-Al alloy, making the martensitic transformation become thermoelastic, forming Fe-Mn-Al-Ni superelastic alloy [61]. Tanaka et al. added strong γ' phase elements into Fe-Ni-Co-Al alloy to introduce γ' phase, occurring thermoelastic martensitic transformation and exhibiting superelasticity [41,68,69].

The predecessor of Fe-Mn-Al-Ni superelastic alloy is Fe-Mn-Al alloy. In NiTi-based superelastic alloy, some studies show that introducing precipitate can improve the matrix order degree, strengthen the matrix to make plastic deformation difficult to occur, and further improve the superelasticity [75]. For Fe-Mn-Al alloy, it can be considered to introduce the precipitate to realize the thermoelastic martensitic transformation, and obtain superelasticity. Omori et al. first added Ni element to Fe-Mn-Al alloy to investigate the effect of Ni element on phase transformation [61]. There is no mismatch at the interface between the parent phase and β phase, indicating that the β phase and parent phase are coherent, which is beneficial to the occurrence of thermoelastic martensitic transformation [61]. The internal stress generated by β phase precipitation is regulated by nano-twin, which does not change the interatomic fit, and does not affect the thermal balance of thermoelastic martensitic transformation. However, Omori et al. only stated that the β phase precipitation after the addition of Ni could make the martensitic transformation into thermoelastic, without further explaining the reason [61]. Roca et al. further investigated how the β phase affected the martensitic transformation [71,72], and found that B2 precipitates could store elastic energy. On the one hand, the elastic energy acts as the driving force to carry out the reverse transformation. On the other hand, it is beneficial to maintain the thermoelastic equilibrium, thus producing the thermoelastic martensitic transformation. At the same time, the B2 phase also increases the hardness, thus inhibiting the occurrence of irreversible deformation [71]. In addition, Walnsch et al. developed a thermodynamic model to further illustrate the contribution of B2 phase to the generation of thermoelastic martensitic transformation [72]. The dispersed B2 phase stabilizes the austenite matrix, and transforms into the elastic $L1_0$ phase after martensitic transformation, which can store the energy released during the reverse transformation from martensite to austenite [72].

The thermoelastic martensitic transformation of Fe-Ni-Co-Al system's superelastic alloy is from FCC-γ parent phase to body centered tetragonal (BCT)-α' martensite phase. Thermoelastic martensitic transformation does not occur in Fe-Ni-Co-Al alloys. However, with the addition of strong γ' elements such as Nb, Ti, and Ta, the alloy is superelastic with the occurrence of thermoelastic martensitic transformation [41,68,69]. After adding strong γ' phase elements, the solid-solution temperature of γ' phase increases. More γ' phase will be precipitated after heat treatment [41]. These γ' phases will increase the hardness and the tetragonality of martensite. After adding Nb, Ti, and Ta, the alloys obtained superelasticity of 5%, 4.2%, and 13.5%, respectively [41,68,69].

In addition, Vallejos et al. found that the addition of Al significantly affected the martensitic transformation temperature, and the formation of B2 phase in Fe-Mn-Al-Ni alloy [76]. The martensitic transformation temperature of 17Al is lower than −250 °C, while that of 15Al is about −10 °C [76]. However, calculation shows that the martensitic transformation temperature of the Fe-34Mn-15Al-7.5Ni alloy is 900 °C, which is far higher than the results reported by Vallejos et al. [77]. The reason for this significant difference is due to the thermodynamic contribution of β phase to the formation of martensite. As high Al content makes the parent phase stable, a larger degree of undercooling is needed to provide enough energy for phase transformation, leading to a lower M_s point of 17Al. When the parent phase is stable, the temperature where the two-phase miscibility gap starts will increase. If the two-phase miscibility gap is reached at a higher temperature, the driving force of precipitation will increase, resulting in a large amount of B2 phase precipitation after quenching, with a superelasticity of about 8.4% [76]. Some studies have shown that Al element has a significant effect on obtaining a wide range of single α phase region or γ phase region in the phase diagram [78]. Heat treatment in the region of single phase or uniform solid solution is beneficial to obtain B2 phase [78]. In Fe-Mn-Al-Ni alloy, the addition amount of Al element is limited. The addition of 5 at. % Al requires the compensation of a large amount of Mn and Ni elements [78]. However, in terms of manufacturing, both Mn and Ni elements cannot be added in large amounts. Mn has strong volatility, which will cause the chemical composition of the alloy to be unstable, and also cause the negative influence to production [79]. Although Ni element will not cause production problems, the addition of a large amount of Ni requires extremely high cost [80]. Starting from increasing the single-phase region in the high temperature range, Kaputkina et al. proposed to increase the amount of Al by adding C element [78]. After the addition of C element, a single-phase region of phase diagram appears in the range of 1000 °C to 1200 °C, and thus, the content of Al element is allowed to increase [78].

3.1.3. Effects of Heat Treatments on Coherent Precipitates

In some alloy systems, coherent precipitates can be obtained only by heat treatment without alloying elements. In addition, although the alloy system with alloying elements can precipitate coherent precipitates, their sizes can be further optimized to obtain better superelasticity. Studies have shown that the size of coherent precipitate is too small or too large to ensure its coherence with parent phase, and cannot play a role in strengthening parent phase to obtain good superelasticity [81]. At this time, the size of coherent precipitate needs to be controlled by heat treatment.

The addition of Ni or strong γ′ elements can make the martensitic transformation of Fe-Mn-Al-Ni alloy and Fe-Ni-Co-Al alloy into thermoelastic. However, in order to obtain better superelasticity, coherent precipitates can be regulated by heat treatment (controlling the aging temperature and time), as shown in Figure 6a [81]. Tseng et al. studied the effects of aging temperatures (200 °C and 300 °C) and time on the compression superelastic response of <100> orientation FeMnAlNi single crystal, and found that the size of B2 phase increased with the increase of aging time [81]. The same results were obtained in Fe-Ni-Co-Al superelastic and Fe-Ni-Co-Ti superelastic alloys [82,83]. Evirgen et al. found that the γ′ phase size is 3–4 nm at 700 °C aging and 5 nm at 600 °C in FeNiCoAlTa single crystal [82]. Figure 6b shows the relationship between aging time and superelasticity in FeMnAlNi alloy [81]. When aging at 200 °C for 1–3 h, the superelasticity increases with the increase of aging time, and reaches a peak value of 7.2% at 3 h [81]. FeNiCoAlTa single crystal obtained 4.3% superelasticity after 90 h aging at 600 °C [82]. The longer aging time increases the size and volume fraction of B2 phase, but decreases the number density. This indicates that the precipitate is still growing during the aging process, but the nucleation has ended. As the aging time continues to increase, the superelasticity gradually decreases. When the aging time is not long enough, the B2 phase is too small and insufficient to strengthen the austenite matrix to resist plastic deformation, and thus the superelasticity is poor [81]. When the aging time is too long, the precipitate and the parent phase lose the coherence. The effect of precipitation strengthening is reduced, weakening the superelasticity. The best superelasticity can be obtained for 200 °C/3 h in FeMnAlNi alloy. At this time, the size of the B2 phase is 6–10 nm, which achieves the balance of precipitation strengthening and strong coherence, showing a 7.2% superelasticity [81]. Titenko et al. also used this method to obtain the optimum superelasticity of 4.5% in Fe-Ni-Co-Ti alloy aged at 650 °C for 10 min [84]. With the increase of aging time, the composition of precipitate changes, which will affect the martensitic transformation. In Fe-Mn-Al-Ni superelastic alloy, the content of Al and Ni elements decreases in the matrix, while the content of Fe element increases with the increasing aging time [81]. In Fe-Ni-Co-Al superelastic alloy, the aging process also changes the composition of the coherent γ′ phase [85]. The matrix is rich in Fe and Co elements, while the precipitates are rich in Ni, Al, and Ta after short-time aging [85]. As the aging time increases, the content of Ni, Al, and Ta in the matrix gradually decreases. The shape of the precipitate also changes from plate to granular [85]. The change of the composition of coherent precipitates also affects the composition of the matrix, which affects the transformation temperature. The temperature of martensitic transformation is very sensitive to the Ni content. The decrease of Ni content in matrix leads to the increase of martensitic transformation temperature [86]. At the same time, the Co element has the effect of increasing hardness of the austenite matrix, promoting the formation of the thin-plate martensite phase, and reducing the phase volume change caused by the Invar effect [41,87]. Increasing the critical slip stress can enable the occurrence of martensitic transformation before plastic deformation. In addition, the increase of the martensitic transformation temperature also reduces the thermal hysteresis. At the same time, element changes in the matrix promotes the occurrence of martensitic transformation [87]. These above three factors make the martensitic transformation into thermoelastic.

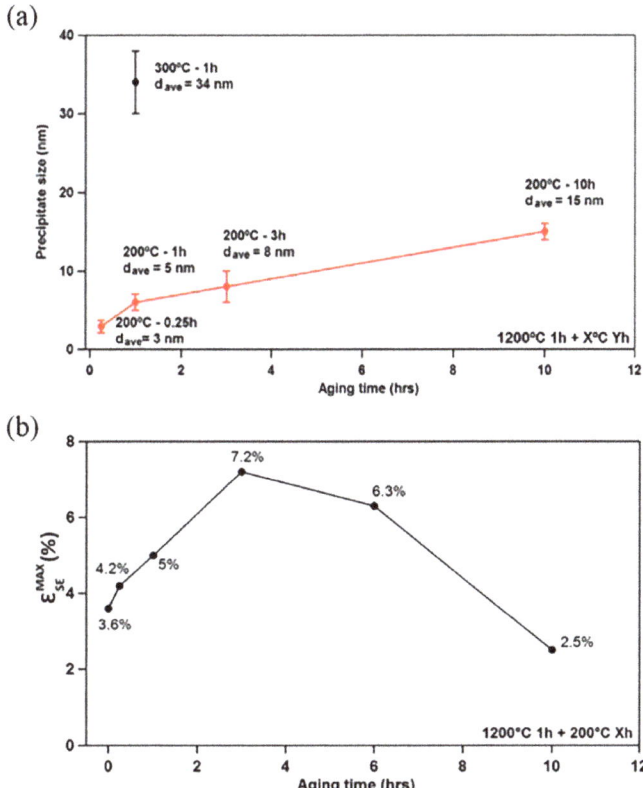

Figure 6. (a) The relationship between aging time and the size of B2 phase; (b) the relationship between aging time and superelasticity in FeMnAlNi single crystal [81]. Reprinted with permission from Ref. [81]. Copyright 2015 Elsevier.

An interesting phenomenon was discovered by Ozcan et al. [88]. At room temperature, the FeMnAlNi alloy has a natural aging phenomenon [88]. No matter after solution treatment, or in samples that have been aged at 200 °C for 3 h, natural aging can occur at room temperature, and the B2 phase can be precipitated. The FeMnAlNi alloy presents no superelasticity after solution treatment. After the precipitation of B2 phase by natural aging, the alloy acquires superelasticity. After 30 days of natural aging, the alloy achieved 5% superelasticity, and the size of the B2 phase also increased from 5 nm to 7 nm [88]. However, the author did not explain the essence of room temperature aging, which requires further investigations. Natural aging is a double-edged sword. If it is well controlled, superelasticity can be improved without additional heat treatment. Otherwise, superelasticity will be unstable and affect the applications.

3.2. Effects of Grain Boundary Precipitates on Superelasticity

Not all precipitates in Fe-based superelastic alloys are conducive to superelasticity. In addition to coherent precipitates, other precipitates may also be generated during thermomechanical treatment. These precipitates tend to adhere to grain boundaries, and have harmful effects on the superelasticity, such as γ phase in Fe-Mn-Al-Ni alloy, B2 phase in Fe-Ni-Co-Al alloy, and η-Ni3Ti phase ($D0_{24}$ structure) in Fe-Ni-Co-Ti alloy, etc. [60,63,89].

These precipitates are apt to precipitate at grain boundaries. In general, B2 phase in Fe-Ni-Co-Al alloy and η-Ni3Ti phase ($D0_{24}$ structure) in Fe-Ni-Co-Ti alloy are brittle, causing the alloy to fracture before exhibiting superelasticity [60,63]. For example, the superelasticity of Fe-28.9Ni-18.2Co-8.3Ti alloy possesses only 0.7% superelasticity due to the precipitation of η phase at grain boundaries [90]. In Fe-Ni-Co-Al superelastic alloy, it is also found that the B2 phase at grain boundary is not conducive to the superelasticity [60]. γ phase in Fe-Mn-Al-Ni alloy will form the serrated interface, which is not conducive to thermoelastic martensitic transformation [89]. The most serious problems are: (1) the precipitate at grain boundaries leads to stress concentration and fracture during deformation; (2) the precipitate pins martensite/matrix interface, hindering reverse martensitic transformation [60,89].

To solve the above problems, alloying elements were added to inhibit the precipitates at grain boundaries. For example, in Fe-Ni-Co-Al and Fe-Ni-Co-Ti superelastic alloys, the addition of B element can inhibit the precipitation of brittle phases at grain boundaries [60,63]. In Fe-Mn-Al-Ni superelastic alloy, Ti element is used to reduce the γ phase precipitation in non-rapid cooling [89]. Some studies have shown that B element can strengthen grain boundary, and reduce the precipitation of brittle phase at grain boundary in Ni-based and Fe-based superalloys [91]. It can be seen clearly in Figure 7 that after the addition of B element inhibits the precipitation at grain boundary [60,63,68]. The addition of B element reduces the grain boundary energy, making the nucleation at grain boundary more difficult, so the precipitation of brittle phase at grain boundary is inhibited [60].

Quenching in cold water will lead to crack formation along the grain boundaries in the Fe-Mn-Al-Ni alloy [89]. In this case, superelasticity cannot be obtained. At present, studies have shown that grain boundary cracking can be effectively prevented by controlling the cooling rate during quenching, that is, quenching in hot water to reduce the cooling rate [92]. In the case of non-rapid cooling, the second phase generates the serrated interface at the grain boundaries, as shown in Figure 8 [89]. This structure is very unfavorable to the superelasticity. As γ phase is not the phase that can occur in phase transformation, it will produce plastic deformation, and dissipate part of the elastic strain energy when martensitic transformation occurs [89]. Moreover, the serrations of the interface also increase the constraint of the reverse martensitic transformation. Studies have shown that the thin-layer γ phase at the grain boundary can prevent intergranular cracking during quenching, but does not significantly affect the superelasticity [92]. However, in the Fe-Mn-Al-Ni alloy, obtaining a thin-layer γ phase requires a fast cooling rate. When the material has a large cross-section, it is difficult to obtain a fast cooling rate, limiting the practical applications of Fe-Mn-Al-Ni alloy. In order to solve this problem, Vollmer et al. added Ti element into Fe-Mn-Al-Ni alloy to control the quenching sensitivity [89]. After the addition of 1.5 at. % Ti element, the volume fraction of precipitates at the grain boundary was significantly reduced. A thin-layer γ phase appeared at the grain boundary, as shown in the inset of Figure 8b [89]. Ti can stabilize the matrix α phase, thus inhibiting the formation of large amounts of γ phase by inhibiting the short-range diffusion of Mn and Al across the boundary of α/γ phase [89]. In a single crystal sample with nearly <102> orientation, the alloy almost completely recovered its deformation under the compression test from $-150\ °C$ to $20\ °C$ and 1.5% strain [89]. In tensile test, the superelasticity at $20\ °C$ to 10.5% strain reaches 4.5% [89].

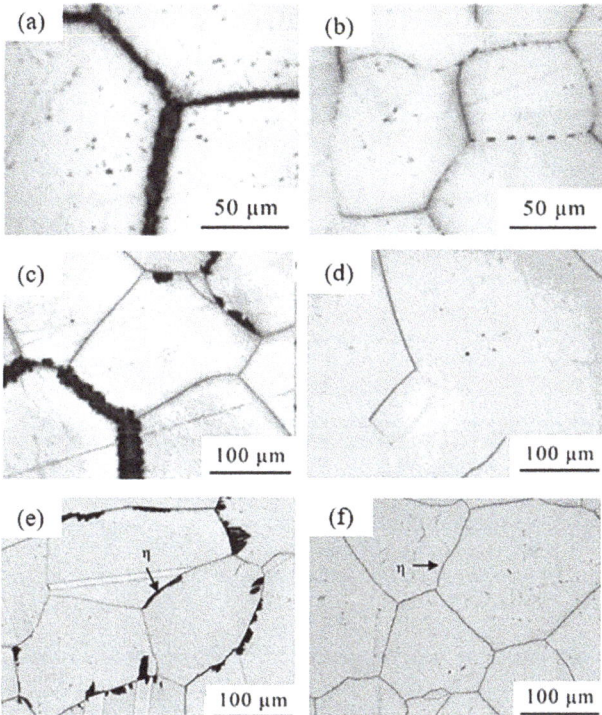

Figure 7. Optical micrographs: (**a**) FeNiCoAlNb; (**b**) FeNiCoAlNb-0.05B; (**c**) FeNiCoAlTa; (**d**) FeNiCoAlTa-0.05B; (**e**) FeNiCoTi; (**f**) FeNiCoTi-0.02B [60,63,68]. Reprinted with permission from Ref. [60], Ref. [63] and Ref. [68]. Copyright 2013 Elsevier and 2015 Elsevier.

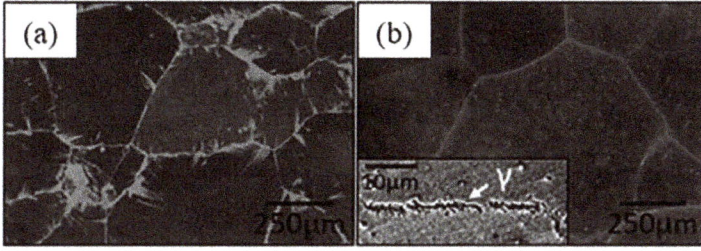

Figure 8. Optical micrographs of air-cooled sample after solution treatment at 1225 °C for 1 h: (**a**) Fe-34.0%Mn-16.5%Al-7.5%Ni (at. %); (**b**) Fe-34.0%Mn-15.0%Al-7.5%Ni-1.5%Ti (at. %) [89]. Reprinted with permission from Ref. [89]. Copyright 2017 Elsevier.

4. Effects of Grain Size on Superelasticity

Grain size is a very important factor affecting superelasticity for the practical applications in many materials [93–96]. In Cu-based superelastic alloys, superelasticity is affected by the grain size due to the phase transformation and the anisotropy of grains [97–99]. The grain boundaries have a restrictive effect on the deformation. In NiTi-based superelastic alloys, the effect of grain size on superelasticity is not significant. There are 24 kinds of martensite variants in NiTi-based superelastic alloys, in which 12 kinds of martensite variants can be activated to coordinate deformation when the alloy undergoes the phase transformation from B2 to B19′ [100]. In Fe-based alloys, there are 12 kinds of martensite variants for Nishiyama orientation and 24 for Kurdjumov–Sachs (K–S) orientation relationships [101]. However, only three kinds of martensite variants are related to the martensitic transformation of Fe-based superelastic alloys [102]. When the number of martensite variants related to the martensitic transformation is small, the phase interface reciprocal migration is difficult [102]. This is not conducive to the occurrence of thermoelastic martensitic transformation. The grain boundaries can be regarded as the phase interface during phase transformation, and the grain size affects the volume fraction of the grain boundary, which further affects the phase interface during phase transformation. Inferred from this, the grain size also has an effect on the superelasticity in Fe-based superelastic alloys. Omori et al. made the alloy into wire to investigate the effect of grain size on the superelasticity [103]. They used d/D to measure the grain size, where d is the average grain size and D is the diameter of the wire. Two models, Taylor and Sachs, are used to calculate the critical stress value and recoverable strain value of martensitic transformation [104,105]. It is found that as the d/D value increases, the recoverable strain value becomes larger, while the critical stress becomes smaller. When $d/D > 1$, the Fe-Mn-Al-Ni wire with bamboo structure possesses ~5% superelasticity due to the decrease of grain constraint effect [103]. When $d/D > 1$, almost every grain can be transformed into martensite independently. The resistance is relatively small during martensitic transformation and recover transformation, which is extremely beneficial to superelasticity [103]. Tseng et al. also proved that the increase in grain size can enhance the plasticity, reduce the critical transformation stress, and increase the recoverable strain [106]. When d/t (t is the width of the sample) is 0.67, the superelasticity is 1.5%; when the d/t is 1.67, the superelasticity is 3% [106]. This is because the increase in grain size reduces the amount of grain boundaries. During the deformation of polycrystalline alloys, the presence of grain boundaries restricts the deformation of the grains, thereby affecting the superelasticity. At present, the effect of grain size on the superelasticity of Fe-Ni-Co-Al and Fe-Ni-Co-Ti superelastic alloys is rarely studied, needing further investigations.

The larger the grain size, the better the superelasticity. Therefore, the method of obtaining large-size grains is particularly important. The principle of obtaining large-size grains is to provide energy for grain growth, which can be divided into normal grain growth and abnormal grain growth. Omori et al. used the normal grain growth method to prepare large α-phase grains [103]. Solid solution at 1200 °C was conducted for different times, so that the grains can obtain enough energy to grow. In fact, the grain size obtained by normal grain growth is not particularly large and thus, abnormal grain growth is usually used to prepare large-size grains. Studies have shown that the presence of texture pinning is a key factor in determining abnormal grain growth [107]. If there is texture pinning after the precipitate is dissolved, abnormal growth of columnar crystals will occur; if there is no texture pinning, only normal growth will occur, limiting the grain size [107]. Cyclic heat treatment is the most commonly used method for abnormal grain growth. Omori et al. first proposed the use of cyclic heat treatment in 2016 to prepare 10 times larger grains than the normal grains [108]. They carried out cyclic heat treatment between the α single-phase zone (1200 °C) and the α + γ dual-phase zone (600–1100 °C). When the α single-phase zone is cooled to the α+γ dual-phase zone, the γ phase precipitates out. When the α + γ dual-phase region is heated to the α single-phase region, the γ phase transforms into the α phase and sub-grains are formed. The grains grow by the gradual annexation of the sub-grains, producing large-sized grains [108]. The cyclic heat treatment method used by Tseng et al. is different from that by Omori et al. They heat-treated the sample at 1200 °C for 0.5 h, and then air-cooled to room temperature, which was used as a cycle for multiple cycles [108]. After 5 cycles, they produced large-sized grains of more than 5 mm. Vollmer et al. investigated the effect of Ti and Cr elements on abnormal grain growth of Fe-Mn-Al-Ni alloy by cyclic heat treatment in single-phase and dual-phase regions [109]. The results show that Ti can promote the abnormal grain growth, while Cr can inhibit the abnormal grain growth. As shown in Figure 9, the grain size of Fe-Mn-Al-Ni-Ti is larger than that of Fe-Mn-Al-Ni in the micrograph after a single- cycle heat treatment [109].

However, the existence of triple junctions can still be seen in Figure 9a,c, which is very unfavorable to superelasticity [109]. After a single-cyclic heat treatment, the average grain size of Fe-Mn-Al-Ni is 2.3 mm, the average grain size of Fe-Mn-Al-Ni-Ti is 7.2 mm, and the average grain size of Fe-Mn-Al-Ni-Cr is 0.66 mm (no difference from the average grain size of the non-cyclic heat treatment) [109]. The addition of Ti element promotes the grain boundary mobility, accelerating the grain growth. The grain boundary mobility of Fe-Mn-Al-Ni-Ti was determined to be 1.84×10^{-5} m/s, which is more than 7 times higher than that of Fe-Mn-Al-Ni (2.5×10^{-6} m/s) [109]. The increase of grain boundary mobility is due to the decrease of the sub-grain size. The small size of sub-grain increases the driving force of abnormal grain growth, and makes the grain grow larger [109]. The reason why Cr element inhibits the abnormal growth of grains is that the part enclosed by the red dashed line in Figure 10 has no sub-crystals near the grain boundaries, while there are a large number of sub-grains in the center area, resulting in the formation of low-density area of sub-grains [109]. The premise of making the grain abnormal growth is that the sub-crystalline low-density region needs to be overcome by the normal growth of grains at large angular grain boundaries, otherwise the abnormal growth of grains will not arise. The abnormal growth of the grains occurs when the grain first contacts the sub-grain. In Figure 10, the large-angle grain boundaries and sub-grains are clearly separated by the low-density region, so the abnormal growth of grains is strongly suppressed [109]. In this study, Vollmer et al. prepared 220 mm long Fe-Mn-Al-Ni-Ti single crystal rod, obtaining a good superelasticity in tensile test under 8% applied strain [109].

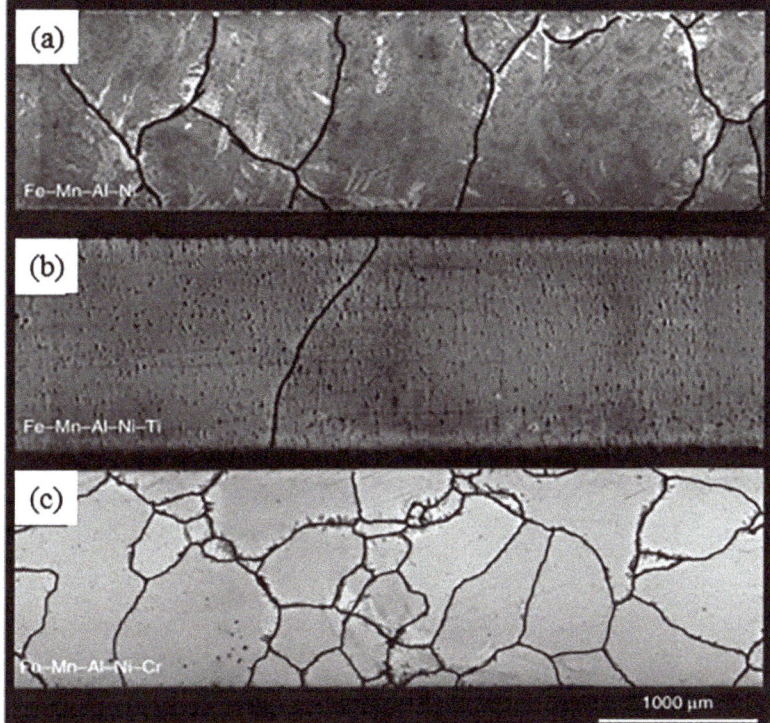

Figure 9. Microstructures after a single cycle of heat treatment: (**a**) Fe-Mn-Al-Ni; (**b**) Fe-Mn-Al-Ni-Ti; (**c**) Fe-Mn-Al-Ni-Cr [109]. Reprinted with permission from Ref. [109]. Copyright 2019 Springer Nature.

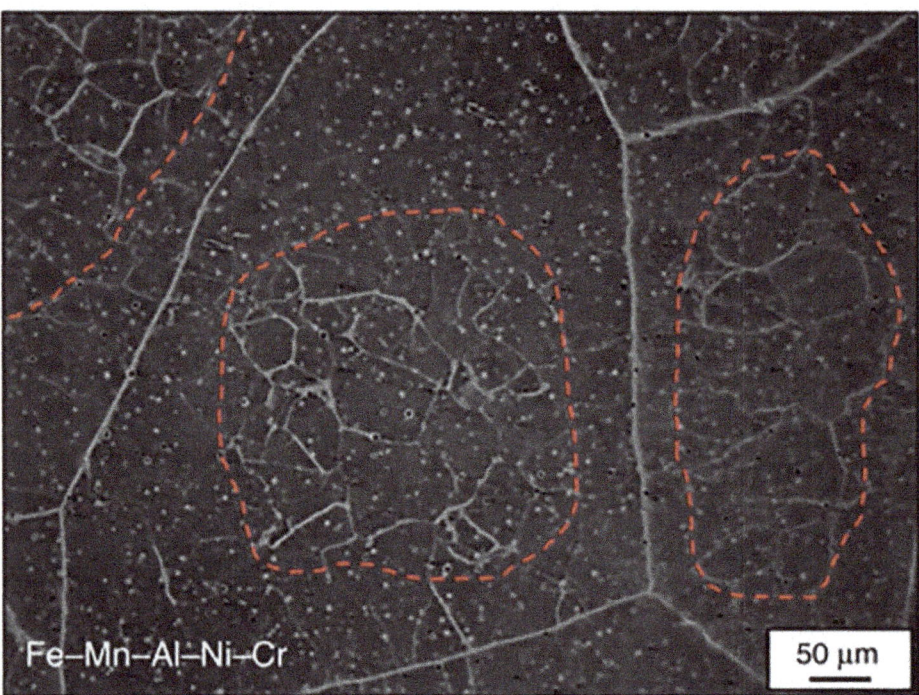

Figure 10. Photomicrograph of Fe-Mn-Al-Ni-Cr after single-cycle heat treatment and quenching [109]. Reprinted with permission from Ref. [109]. Copyright 2019 Springer Nature.

The use of cyclic heat treatment makes it possible to prepare large-sized grains, but this method has a fatal problem, that is, it takes an extremely long time. For cyclic heat treatment, it takes 48.2 h to prepare a 60 mm single crystal [110]. Vallejos et al. pioneered the combination of directional annealing and cyclic heat treatment, which solved the long time-consuming problem [110]. The principle of cyclic directional annealing is a combination of directional annealing to produce strong thermal gradients to cause grain growth and cyclic heat treatment to cause abnormal grain growth. The schematic diagram of cyclic directional annealing is shown in Figure 11. At the beginning of heating, the grains in the hot zone grow equiaxed, as shown in Figure 11a. The grains outside the region cannot grow due to the insufficient energy provided by the temperature to migrate the grain boundaries. When the sample rod moves, the grains that have grown up just after the heating move, passing through the ungrown grains. The difference in grain size provides the driving force, which continuously promotes the migration of grain boundaries and the grain growth, as shown in Figure 11b. α sub-grains will be produced during the thermal cycling, as shown in Figure 11c. These sub-grains can continuously provide driving force, which makes the grain boundaries continue to migrate. The grains grow larger and larger, even becoming single crystals, as shown in Figure 11d. In directional annealing or cyclic heat treatment, the driving force will be dissipated as the grains grow, which hinders the migration of grain boundaries. This is also the reason for the long time-consuming cycle of heat treatment. However, cyclic directional annealing also has some disadvantages. The size of the grains is limited by the hot zone size, but this process also provides a method for preparing large grains in a short time.

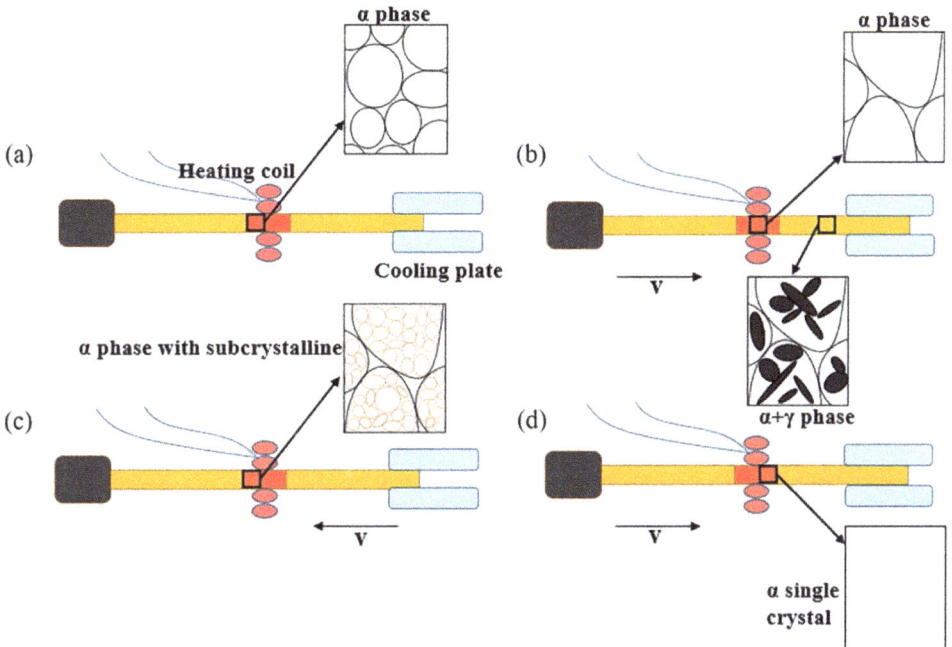

Figure 11. Schematic diagram of cyclic directional annealing: (**a**) the start of cyclic directional annealing; (**b**) after forward movement of the specimen; (**c**) after a cycle is completed; (**d**) after multiple cycles.

5. Effects of Grain Orientation and Texture on Superelasticity

Studies have shown that the superelasticity strongly depends on the grain orientation in NiTi-based superelastic alloys [111]. The orientation of the grains can affect the martensitic variation to adjust the strain in NiTi-based superelastic alloys. It is speculated that grain orientation affects the superelasticity of Fe-based superelastic alloys by affecting the martensite variation. Large-sized grains or single crystals can be obtained by abnormal grain growth by cyclic heat treatment in Fe-Mn-Al-Ni superelastic alloys. Therefore, it is obvious that grain orientation significantly affects the superelasticity of Fe-Mn-Al-Ni superelastic alloys. For polycrystalline alloys, grains are arranged in order along certain directions to produce texture. However, Fe-Ni-Co-Al and Fe-Ni-Co-Ti superelastic alloys are extremely difficult to prepare large grains or single crystals due to the low mobility of grain boundary. In the case of polycrystalline alloys, texture has a significant effect on the superelasticity.

5.1. Effects of Grain Orientation on Superelasticity

Grain orientation has a significant effect on the superelasticity [102]. Tseng et al. calculated the superelasticity of <100> and <123> orientations as 10.5% and 9%, respectively, according to the energy minimization theory and lattice deformation theory in Fe-Mn-Al-Ni superelastic alloy [102,112,113]. In the tensile experiment, the superelasticity of <123> orientation is ~7.8%, which is similar to 9% calculated by the theoretical model. However, the superelasticity of the <100> orientation is only 3.5%, which is quite different from the calculation result [102]. Figure 12a shows the TEM bright field image of Fe-Mn-Al-Ni single crystal in <100> orientation. It can be seen that there is a large amount of parallelism dislocations and hairpin dislocations at the austenite–martensite interface and austenite matrix [102]. These dislocations pin the martensite phase, and suppress the occurrence of the reverse martensitic transformation, resulting in a low recovery strain. Figure 12b shows that high-density dislocations are not observed in <123> orientation, but two martensite variants are found, making it possible for martensite to easily form twins to accommodate the lattice strains, and reducing the likelihood of dislocation formation [102]. Only one martensite variant is available to accommodate tensile strain along the <100> orientation. This makes it difficult to accommodate lattice mismatch between the austenite and martensite, and thus results in the creation of dislocations that pin the martensite [102]. The austenite–martensite phase interface has a higher mobility, and can adapt to the transformation of thermoelastic martensite, thereby obtaining good superelasticity [102]. It is worth mentioning that the phenomenon of different orientation and different superelasticity was also observed in the compression experiment. The superelasticity of the <100>, <111>, and <123> oriented samples are 7.2%, 5.7%, and 1%, respectively [114].

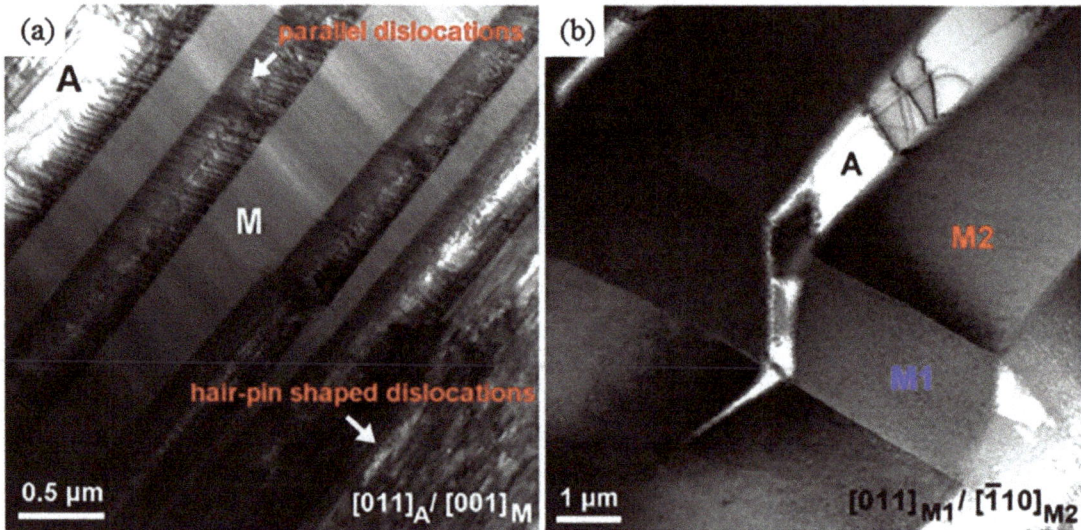

Figure 12. TEM bright field image of Fe-Mn-Al-Ni single crystal: (**a**) <100> orientation; (**b**) <123> orientation (A refers to austenite, M refers to martensite, M1 and M2 refer to two kinds of martensite variants) [102]. Reprinted with permission from Ref. [102]. Copyright 2016 Elsevier.

The mechanism diagram of grain orientation on superelasticity is summarized in Figure 13. There are different kinds of martensite variants in different orientations. If the type of martensite variants is less than 2, a large number of dislocation is easily generated at the interface between austenite and martensite, and the dislocation is deposited at the interface. When martensitic transformation occurs, the phase interface is difficult to reciprocate, which prevents the occurrence of thermoelastic martensitic transformation, making it difficult to obtain superelasticity. The more the martensite variants, the easier the reciprocating migration of phase interface, which is beneficial for the occurrence of thermoelastic martensitic transformation. According to the above mechanism, good superelasticity can be achieved by obtaining more martensite variants in a reasonable orientation to reduce dislocation density, facilitate phase interface migration, and promote the occurrence of thermoelastic martensitic transformation.

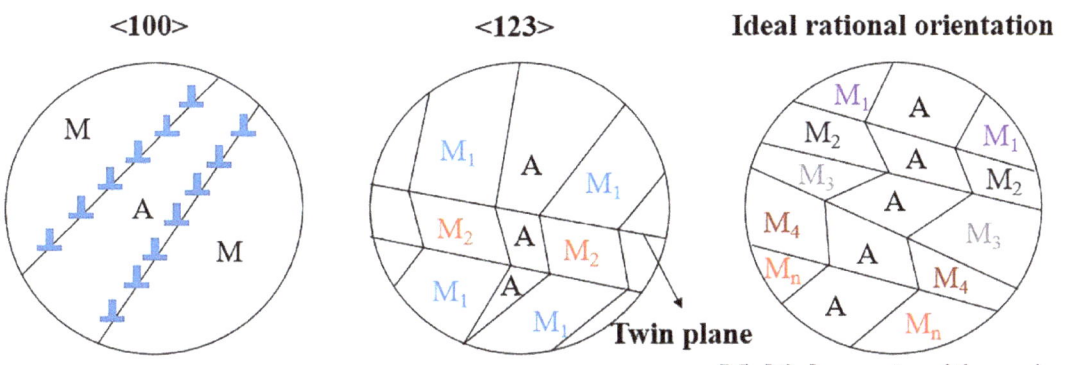

Figure 13. Schematic diagram of the effect of grain orientations on superelasticity.

5.2. Effects of Texture on Superelasticity

In Fe-Ni-Co-Al superelastic alloy, if there is no strong texture, it will fracture before showing superelasticity during deformation [41]. Even if thermoelastic martensitic transformation is obtained by the precipitation of the γ' phase, the sample breaks before the phase transformation occurs. Figure 14 shows that the FeNiCoAlTiB alloy specimen with random grain orientation breaks under 90% cold-rolling deformation without exhibiting superelasticity, as shown in Figure 15a [69]. After 98.5% large cold-rolling deformation, most of the grains are <100> orientation along the rolling direction, as shown in Figure 15b. In Figure 15c, it can be seen that the sample has a strong {012}<100> texture and relatively weak {112}<110> texture, showing 4.2% superelasticity, as shown in Figure 14 [69]. FeNiCoAlNbB alloy also has two kinds of textures of {111}<110> and {112}<110> after 98.5% large deformation cold-rolling, showing 5% superelasticity [68]. FeNiCoAlTaB alloy has a {035}<100> texture after 98.5% cold-rolling, and exhibits 13.5% superelasticity [41], while no superelasticity is presented in the case without strong textures. Although the addition of B element inhibits the precipitation of the B2 phase at the grain boundary, there are still a certain amount of B2 phases observed at the grain boundary without strong textures. In the alloy samples with strong textures, the B2 phases are precipitated only at specific types of grain boundaries [69,115]. The selective precipitation of B2 phase improves the mechanical properties of the alloy and contributes to the excellent superelasticity. The existence of strong textures affects the characters of grain boundaries. In the samples without texture, there are a large number of large-angle grain boundaries, which possess high energy. B2 phase is easy to nucleate and grow up at large-angle grain boundaries, weakening the grain boundaries. Due to the presence of strong texture, many small-angle grain boundaries and coincidence site lattice boundaries appear. The energy of these grain boundaries is extremely small, which inhibits the precipitation of B2 phase [69,115]. In addition, the small-angle grain boundary and coincidence site lattice boundaries also reduce the constraint of the grain boundary during deformation. These factors provide sufficient conditions for obtaining superelasticity.

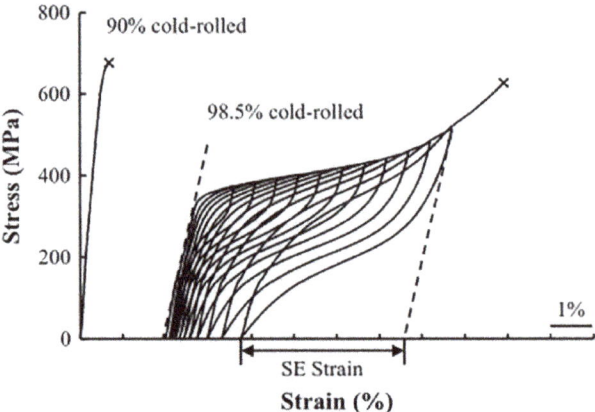

Figure 14. Tensile curves of FeNiCoAlTiB alloy after 90% and 98.5% cold-rolling [69]. Reprinted with permission from Ref. [69]. Copyright 2014 Elsevier.

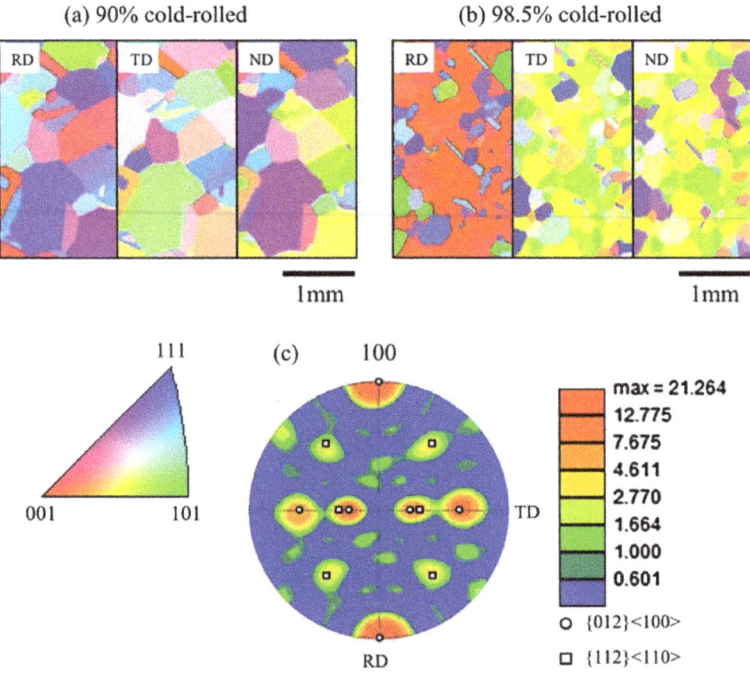

Figure 15. FeNiCoAlTiB alloy quasi-colored orientation maps in rolling direction (RD), transverse direction (TD) and normal direction (ND): (**a**) 90% cold-rolling; (**b**) 98.5% cold-rolling; (**c**) (100) pole figure in 98.5% cold-rolling specimen [69]. Reprinted with permission from Ref. [69]. Copyright 2014 Elsevier.

6. Other Influencing Factors on Superelasticity

The thermal hysteresis also has a significant effect on the thermoelastic martensitic transformation. Reducing thermal hysteresis is beneficial to superelasticity. Reducing the shear modulus of austenite is an effective method to reduce the thermal hysteresis of martensitic transformation in Fe-Ni-Co-Ti alloy, while this will lead to a reduction in the elastic properties of the martensite grown together with austenite [116]. In Fe-Ni alloys, Cu alloying can reduce the elastic modulus of austenite [117]. Kokorin et al. added Cu element into Fe-Ni-Co-Ti alloy to investigate the martensitic transformation characteristics [62]. After the addition of Cu, a relatively small thermal hysteresis of about 60 K is obtained [62]. This is firstly due to the transformation of thermoelastic martensite. Secondly, the alloying of Cu reduces the elastic modulus of austenite, which reduces the elastic energy of growing martensite grains. In the cyclic tensile unloading experiment, the Fe-Ni-Co-Ti-Cu alloy exhibited a superelasticity of 4.5% [118]. This is much higher than the 0.7% superelasticity obtained by Kokorin et al. in the Fe-Ni-Co-Ti alloy, which is the same as obtained by Titenko et al. in the Fe-Ni-Co-Ti alloy through heat treatment [84,90]. The Fe-Ni-Co-Al superelastic alloy with strong γ' phase element also possesses low thermal hysteresis, as shown in Figure 16. The resistivity curves of FeNiCoAlNbB, FeNiCoAlTiB, and FeNiCoAlTaB alloys are closed with small thermal hysteresis of 20 K, 31 K, and 24 K, respectively, corresponding to 5%, 4.2%, and 13.5% superelasticity [41,68,69].

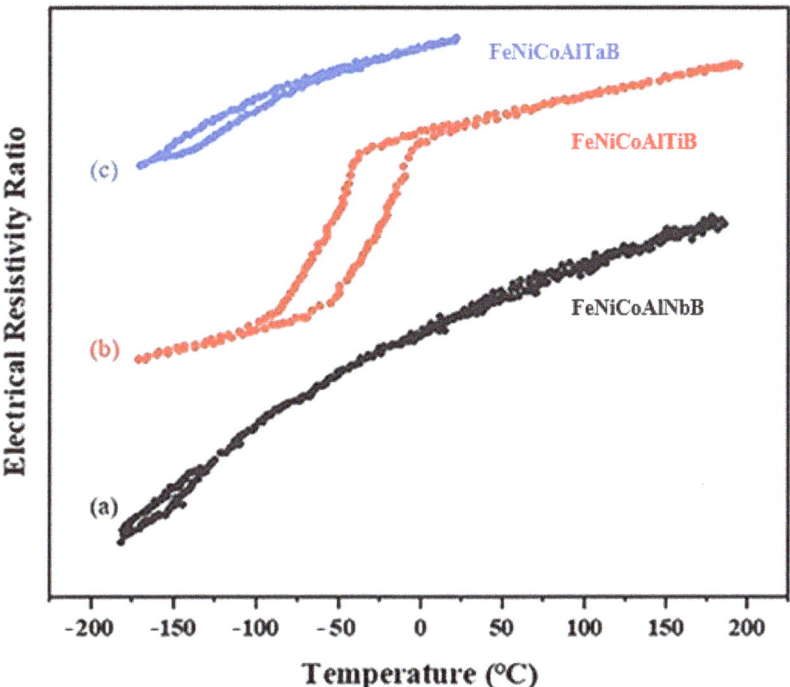

Figure 16. Resistivity curves: (a) FeNiCoAlNbB; (b) FeNiCoAlTiB; (c) FeNiCoAlTaB.

After the addition of alloying elements, the properties of the alloys will change. As mentioned in the former, the addition of alloying elements (Ni, C, Ti, Nb, Ta, B, Cu) can affect the formation of precipitates and grain sizes, and further affect superelasticity. Although the addition of Cr inhibits the abnormal growth of grains, Fe-Mn-Al-Ni-Cr alloy can show excellent superelasticity in an extremely wide temperature window through proper heat treatment [119]. The relation between the critical stress and temperature of Fe-Mn-Al-Ni-Cr alloy is almost zero, indicating that the temperature has no effect on the critical stress [119]. This kind of alloy with a wide range of applications and small critical stress changes is expected to become a candidate for aerospace parts.

7. Conclusion and Outlook

Fe-based superelastic alloys are favored by virtue of their good superelasticity, large superelastic temperature range, low temperature dependence, low price and easy processing, possessing significant development prospects. The thermoelastic martensitic transformation of Fe-based superelastic alloy is affected by factors such as structure and grain. Therefore, factors such as precipitate size, grain size, grain orientation, and grain boundary characteristic may all affect the superelasticity. Obtaining good superelasticity requires that both the precipitates and grains have appropriate sizes; more martensite variants can be activated during deformation; a good grain boundary state, etc. Grains with appropriate size can reduce the constraints of grain boundaries during phase transformation. Precipitates with appropriate size can coherently strengthen the parent phase. Activating more martensite variants can make martensitic transformation and its recovery to easily occur, reducing the formation of dislocations at the phase interface, and thereby reduce the pinning effect of dislocations on martensite. A good grain boundary characteristic can ensure that the material does not break before exhibiting superelasticity. Although the research of Fe-based superelastic alloy has made some progress, there are still some problems that have not been solved. The properties are difficult to meet the requirements of practical applications. The relevant basic theoretical researches are still needed and mainly as follows:

(1) The superelasticity of Fe-based superelastic alloy depends on the thermoelastic martensitic transformation, which can be induced by two factors (temperature and stress). Although there are some studies on the thermodynamics and kinetics of temperature-induced Fe-based thermoelastic martensitic transformation, stress-induced martensitic transformation has not been studied in depth. However, the thermodynamic and kinetic mechanism of thermoelastic martensitic transformation induced by stress is rarely reported.

(2) The low-energy grain boundary can inhibit the precipitation of brittle phases at the grain boundary, and ensure that the alloy obtains good superelasticity. At present, the only way to suppress the precipitation of brittle phases at grain boundaries is the addition of B element combined with cold-rolling with large deformation to obtain strong texture. However, some brittle phases will still precipitate and weaken the grain boundaries. In addition, cold-rolling with large deformation amount is generally difficult to achieve. Therefore, it is urgent to find a simple and practical method to obtain low-energy grain boundaries.

(3) It is necessary to further investigate the influence of composition and thermomechanical treatment on the superelasticity, since microstructure is determined by the composition and thermomechanical treatment.

(4) The current precipitation strengthening is to make the martensitic transformation into thermoelastic by precipitating single coherent ordered phases. It is theoretically feasible to study and design two or multiple coherent ordered precipitates to synergistically strengthen the alloy to obtain good superelasticity.

Author Contributions: Formal analysis, Z.L. and K.D.; funding acquisition, Y.Z. and Z.Z.; investigation, Y.Z. and Z.Z.; supervision, Y.Z. and Z.Z.; writing original draft, Z.L. and K.D. All authors have read and agreed to the published version of the manuscript.

Funding: This work was supported by the National Key Research and Development Project, grant numbers 2018YFE0115800, 2020YFE0202600, Youth Talent Project of China National Nuclear Corporation, grant numbers CNNC2019YTEP-HEU01, CNNC2021YTEP-HEU01, the NSFC Funding grant numbers 52001083, 52171111, U2141207, and Heilongjiang Touyan Innovation Team Program.

Institutional Review Board Statement: Not applicable.

Informed Consent Statement: Not applicable.

Data Availability Statement: Not applicable.

Conflicts of Interest: The authors declare no conflict of interest.

References

1. Jin, X.J.; Jin, M.J.; Geng, Y.H. Recent development of martensitic transformation in ferrous shape memory alloys. *Mater. Chin.* **2011**, *30*, 32–41.
2. Jani, J.M.; Leary, M.; Subic, A.; Gibson, M.A. A review of shape memory alloy research, applications and opportunities. *Mater. Des.* **2014**, *56*, 1078–1113. [CrossRef]
3. Ikeda, T. The use of shape memory alloys (SMAs) in aerospace engineering. *Shap. Mem. Superelasticity* **2011**, *11*, 125–140.
4. Habu, T. Applications of superelastic alloys in the telecommunications industry. *Shap. Mem. Superelasticity* **2011**, *13*, 163–168.
5. Arciniegas, M.; Manero, J.M.; Espinar, E.; Llamas, J.M.; Barrera, J.M.; Gil, F.J. New Ni-free superelastic alloy for orthodontic applications. *Mater. Sci. Eng. C* **2013**, *33*, 3325–3328. [CrossRef]
6. Huang, Z.W.; Wang, Z.G.; Zhu, S.J.; Yuan, F.H.; Wang, F.G. Thermomechanical fatigue behavior and life prediction of a cast nickel-based superalloy. *Mater. Sci. Eng. A* **2006**, *432*, 308–316. [CrossRef]
7. Roh, H.; Reinhorn, A.M. Hysteretic behavior of precast segmental bridge piers with superelastic shape memory alloy bars. *Eng. Struct.* **2010**, *32*, 3394–3403. [CrossRef]
8. Luo, Y.; Kodaira, S.; Zhang, Y.; Takagi, T. The application of superelastic SMAs in less invasive haemostatic forceps. *Smart Mater. Struct.* **2007**, *16*, 1061. [CrossRef]
9. Predki, W.; Klönne, M. Superelastic NiTi-alloys under torsional loading. *J. Phys. IV* **2003**, *112*, 807–810. [CrossRef]
10. Shabalovskaya, S.A. On the nature of the biocompatibility and on medical applications of NiTi shape memory and superelastic alloys. *Biomed. Mater. Eng.* **1996**, *6*, 267–289. [CrossRef]
11. Shaw, J.A.; Grummon, D.S.; Foltz, J. Superelastic NiTi honeycombs: Fabrication and experiments. *Smart Mater. Struct.* **2007**, *16*, S170–S178. [CrossRef]
12. Yamauchi, K.; Ohkata, I.; Tsuchiya, K.; Miyazaki, S. *Shape Memory and Superelastic Alloys*; Woodhead Publishing: Cambridge, UK, 2011.
13. Gómez-Cortés, J.F.; Fuster, V.; Pérez-Cerrato, M.; Lorenzo, P.; Ruiz-Larrea, I.; Breczewski, T.; No, M.L.; San Juan, J.M. Superelastic damping at nanoscale in ternary and quaternary Cu-based shape memory alloys. *J. Alloys Compd.* **2021**, *883*, 160865. [CrossRef]
14. Chumlyakov, Y.; Kireeva, I.; Panchenko, E.; Timofeeva, E.; Kretinina, I.; Kuts, O.; Karaman, I.; Maier, H. Shape memory effect and superelasticity in single crystals of high-strength ferromagnetic alloys. *Adv. Mater. Res.* **2014**, *1013*, 15–22. [CrossRef]
15. Popa, M.; Mihalache, E.; Cojocaru, V.D.; Gurau, C.; Gurau, G.; Cimpoesu, N.; Pricop, B.; Comaneci, R.; Vollmer, M.; Kroo, P.; et al. Effects of thermomechanical processing on the microstructure and mechanical properties of Fe-based alloys. *J. Mater. Eng. Perform.* **2020**, *29*, 7–8. [CrossRef]
16. Chowdhury, P.; Canadinc, D.; Sehitoglu, H. On deformation behavior of Fe-Mn based structural alloys. *Mater. Sci. Eng. R* **2017**, *122*, 1–28. [CrossRef]
17. Breinan, E.M.; Ansell, G.S. The influence of austenite strength upon the austenite-martensite transformation in alloy steels. *Metall. Trans.* **1970**, *1*, 1513–1520. [CrossRef]
18. Thomas, G.; Vercaemer, C. Enhanced strengthening of a spinodal Fe-Ni-Cu alloy by martensitic transformation. *Metall. Mater. Trans. B* **1972**, *3*, 2501–2506. [CrossRef]
19. Yong, L. Strengthening of virgin martensite through cryogenic deformation. *Metall. Mater. Trans. A* **2002**, *33*, 3576–3578.
20. Kishino, T.; Nagaki, S.; Inoue, T. On transformation kinetics, heat conduction and elastic-plastic stresses during quenching of steel. *J. Soc. Mater. Sci. Jpn.* **1979**, *28*, 861–867. [CrossRef]
21. Montroll, E.W. A note on the theory of diffusion controlled reactions with application to the quenching of fluorescence. *J. Chem. Phys.* **1946**, *14*, 202–211. [CrossRef]
22. Najbar, J. The fluorescence quenching rate constant for the distance-dependent quenching processes in the presence of diffusion. *Chem. Phys.* **1988**, *120*, 367–373. [CrossRef]
23. Kurdjumov, G.V.; Khachaturyan, A.G. Nature of axial ratio anomalies of the martensite lattice and mechanism of diffusionless transformation. *Scr. Metall.* **1975**, *23*, 1077–1088.
24. Pereloma, E.V.; Miller, M.K.; Timokhina, I.B. On the decomposition of martensite during bake hardening of thermomechanically processed transformation-induced plasticity steels. *Metall. Mater. Trans. A* **2008**, *39*, 3210–3216. [CrossRef]
25. Stormvinter, A.; Hedstrom, P.; Borgenstam, A. A transmission electron microscopy study of plate martensite formation in high-carbon low alloy steels. *J. Mater. Sci. Technol.* **2013**, *29*, 373–379. [CrossRef]
26. Luo, Z.J.; Shen, J.C.; Su, H.; Ding, Y.H.; Yang, C.F.; Zhu, X. Effect of substructure on toughness of lath martensite/bainite mixed structure in low-carbon steels. *J. Iron Steel Res.* **2010**, *17*, 40–48. [CrossRef]
27. Zhao, H.Z.; Lee, S.J.; Lee, Y.K.; Liu, X.H.; Wang, G.D. Effects of applied stresses on martensite transformation in AISI4340 steel. *J. Iron Steel Res.* **2007**, *14*, 63–67. [CrossRef]
28. Zhang, S.; Hidenori, T.; Yu-Ichi, K. In-situ observation of martensite transformation and retained austenite in supermartensitic stainless steel. *Trans. JWRI* **2010**, *39*, 115–117. [CrossRef]
29. Galligan, J.M.; Garosshen, T. On the nucleation of the martensite transformation. *Nature* **1978**, *274*, 674. [CrossRef]
30. Wang, G.Z. Effects of notch geometry on stress–strain distribution, martensite transformation and fracture behavior in shape memory alloy NiTi. *Mater. Sci. Eng. A* **2006**, *434*, 269–279. [CrossRef]
31. Xu, Z.Y. Classification of the martensitic transformations. *Acta Metall. Sin.* **1997**, *33*, 45–53.

32. Bando, Y. Characteristics of phase transformation in metallic fine particles (Martensitic transformation of Fe-Ni alloys, and ordering of CuAu and Cu$_3$Au alloys). *J. Phys. IV* **2006**, *5*, 135–141.
33. Shih, C.H.; Averbach, B.L.; Cohen, M. Some characteristics of the isothermal martensitic transformation. *J. Met.* **1955**, *7*, 183–187. [CrossRef]
34. Gu, N.; Wang, B.; Li, H.; Wen, C.; Song, X. The characteristics of lattice deformation in ferrous martensitic transformation. *J. Phys. IV* **2003**, *112*, 315–318. [CrossRef]
35. Yan, S.M.; Pu, J.; Chi, B.; Li, J. Prediction of crystallographic characteristics of martensitic transformation in a Ni–Mn–Ga based Heusler alloy. *Intermetallics* **2011**, *19*, 1630–1633. [CrossRef]
36. Lin, C.X.; Wang, G.X.; Wang, J.G. Martensitic transformation during cyclic deformation and strain fatigue characteristics of Fe-Mn-Si alloy. *Trans. Mater. Heat Treat.* **2007**, *28*, 62–65.
37. Sato, A.; Chishima, E.; Soma, K.; Mori, T. Shape memory effect in γ↔ε transformation in Fe-30Mn-1Si alloy single crystals. *Acta Metall.* **1982**, *30*, 1177–1183. [CrossRef]
38. Kajiwara, S.; Ogawa, K. Mechanism of improvement of shape memory effect by training in Fe-Mn-Si-based alloys. *Mater. Sci. Forum* **2000**, *327*, 211–214. [CrossRef]
39. Sato, A.; Chishima, E.; Yamaji, Y.; Mori, T. Orientation and composition dependencies of shape memory effect IN Fe-Mn-Si alloys. *Acta Metall.* **1984**, *32*, 539–547. [CrossRef]
40. L'vov, V.A.; Rudenko, A.A.; Chernenko, V.A.; Cesari, E.; Pons, J.; Kanomata, T. Stress-induced martensitic transformation and superelasticity of alloys: Experiment and theory. *Mater. Trans.* **2005**, *46*, 790–797.
41. Tanaka, Y.; Himuro, Y.; Kainuma, R.; Sutou, Y.; Omori, T.; Ishida, K. Ferrous polycrystalline shape-memory alloy showing huge superelasticity. *Science* **2010**, *327*, 1488–1490. [CrossRef]
42. Xu, Z.Y. *Martensite Transformation and Martensite*; Science Press: Beijing, China, 1999.
43. Xu, Z.Y.; Jiang, H.B. *Shape Memory Materials*; Shanghai Jiao Tong University Press: Shanghai, China, 2000.
44. Omori, T.; Kainuma, R. Martensitic transformation and superelasticity in Fe–Mn–Al-based shape memory alloys. *Shap. Mem. Superelasticity* **2017**, *3*, 322–334. [CrossRef]
45. Khalil, W.; Saint-Sulpice, L.; Chirani, S.A.; Bouby, C.; Mikolajczak, A.; Ben Zineb, T. Experimental analysis of Fe-based shape memory alloy behavior under thermomechanical cyclic loading. *Mech. Mater.* **2013**, *63*, 1–11. [CrossRef]
46. Krooß, P.; Holzweissig, M.J.; Niendorf, T.; Somsen, C.; Schaper, M.; Chumlyakov, Y.I.; Maier, H.J. Thermal cycling behavior of an aged FeNiCoAlTa single-crystal shape memory alloy. *Scr. Mater.* **2014**, *81*, 28–31. [CrossRef]
47. Zhao, L.C.; Cai, W.; Zheng, Y. *Shape Memory Effect and Superelasticity in Alloys*; National Defense Industry Press: Beijing, China, 2002.
48. Du, X.W.; Sun, G.; Sun, S.S. Piecewise linear constitutive relation for pseudo-elasticity of shape memory alloy (SMA). *J. Shanghai Jiaotong Univ.* **2005**, *393*, 332–337. [CrossRef]
49. Zhang, X.H.; Ping, F.; He, Y.J.; Yu, T.X.; Sun, Q.P. Experimental study on rate dependence of macroscopic domain and stress hysteresis in NiTi shape memory alloy strips. *Int. J. Mech. Sci.* **2010**, *52*, 1660–1670. [CrossRef]
50. Dong, L.; Sun, Q.P. Stress hysteresis and domain evolution in thermoelastic tension strips. *Acta Mech. Solida Sin.* **2009**, *22*, 400–406. [CrossRef]
51. Otsuka, K.; Wayman, C.M. *Shape Memory Materials*; Cambridge University Press: Cambridge, UK, 1998.
52. Ortín, J.; Planes, A. Thermodynamic analysis of thermal measurements in thermoelastic martensitic transformations. *Acta Metall.* **1988**, *36*, 1873–1889. [CrossRef]
53. Cui, S.S.; Wan, J.F.; Zuo, X.W.; Chen, N.L.; Zhang, J.H.; Rong, Y.H. Three-dimensional, non-isothermal phase-field modeling of thermally and stress-induced martensitic transformations in shape memory alloys. *Int. J. Solids Struct.* **2017**, *109*, 1–11. [CrossRef]
54. Boyer, L.L.; Kaxiras, E.; Mehl, M.J. Energy barrier for 'magic-strain' transformations in crystals with FCC lattices. *MRS Online Proc. Libr.* **1990**, *205*, 447–452. [CrossRef]
55. Mehl, M.J.; Boyer, L.L. Calculation of energy barriers for physically allowed lattice-invariant strains in aluminum and iridium. *Phys. Rev. B Condens. Matter Mater. Phys.* **1991**, *43*, 9498. [CrossRef]
56. Feng, Y.C.; Liu, M.; Shi, Y.P.; Ma, H.; Li, D.Z.; Li, Y.Y.; Lu, L. High-throughput modeling of atomic diffusion migration energy barrier of FCC metals. *Prog. Nat. Sci. Mater. Int.* **2019**, *29*, 101–108. [CrossRef]
57. Koval, Y.N.; Monastyrsky, G.E. Reversible martensite transformation and shape memory effect in Fe-Ni-Nb alloys. *Scr. Metall. Mater.* **1993**, *28*, 41–46. [CrossRef]
58. Maki, T.; Kobayashi, K.; Minato, M.; Tamura, I. Thermoelastic martensite in an ausaged Fe-Ni-Ti-Co alloy. *Scr. Metall.* **1984**, *18*, 1105–1109. [CrossRef]
59. Omori, T.; Ando, K.; Okano, M.; Xu, X.; Tanaka, Y.; Ohnuma, I.; Kainuma, R.; Ishida, K. Superelastic effect in polycrystalline ferrous alloys. *Sci.* **2011**, *333*, 68–71. [CrossRef] [PubMed]
60. Tanaka, Y.; Kainuma, R.; Omori, T.; Ishida, K. Alloy design for Fe-Ni-Co-Al-based superelastic alloys. *Mater. Today Proc.* **2015**, *2*, S485–S492. [CrossRef]
61. Omori, T.; Nagasako, M.; Okano, M.; Endo, K.; Kainuma, R. Microstructure and martensitic transformation in the Fe-Mn-Al-Ni shape memory alloy with B2-type coherent fine particles. *Appl. Phys. Lett.* **2012**, *101*, 1966. [CrossRef]
62. Kokorin, V.V.; Kozlova, L.E.; Titenko, A.N.; Perekos, A.; Levchuk, Y.S. Characteristics of thermoelastic martensitic transformation in ferromagnetic Fe-Co-Ni-Ti alloys alloyed with Cu. *Phys. Met. Metall.* **2008**, *105*, 564–567. [CrossRef]

63. Lee, D.; Omori, T.; Kainuma, R. Microstructure and mechanical properties in B-doped Fe-31.9Ni-9.6Co-4.7Ti alloys. *Shap. Mem. Superelasticity* **2016**, *2*, 228–234. [CrossRef]
64. Ando, K.; Omori, T.; Ohnuma, I.; Kainuma, R.; Ishida, K. Ferromagnetic to weak-magnetic transition accompanied by bcc to fcc transformation in Fe–Mn–Al alloy. *Appl. Phys. Lett.* **2009**, *95*, 595. [CrossRef]
65. Kajiwara, S.; Liu, D.; Kikuchi, T.; Shinya, N. Remarkable improvement of shape memory effect in Fe-Mn-Si based shape memory alloys by producing NbC precipitates. *Scr. Mater.* **2001**, *44*, 2809–2814. [CrossRef]
66. Eyméoud, P.; Huang, L.; Maugis, P. Impact of Ni alloying on Fe-C martensite ageing: An atomistic investigation. *Scr. Mater.* **2021**, *205*, 114182. [CrossRef]
67. Dumay, A.; Chateau, J.-P.; Allain, S.; Migot, S.; Bouaziz, O. Influence of addition elements on the stacking-fault energy and mechanical properties of an austenitic Fe–Mn–C steel. *Mater. Sci. Eng. A* **2008**, *483*, 184–187. [CrossRef]
68. Omori, T.; Abe, S.; Tanaka, Y.; Lee, D.Y.; Ishida, K.; Kainuma, R. Thermoelastic martensitic transformation and superelasticity in Fe–Ni–Co–Al–Nb–B polycrystalline alloy. *Scr. Mater.* **2013**, *69*, 812–815. [CrossRef]
69. Lee, D.; Omori, T.; Kainuma, R. Ductility enhancement and superelasticity in Fe–Ni–Co–Al–Ti–B polycrystalline alloy. *J. Alloys Compd.* **2014**, *617*, 120–123. [CrossRef]
70. Jin, M.; Geng, Y.; Zuo, S.; Jin, X. Precipitation and its effects on martensitic transformation in Fe-Ni-Co-Ti alloys. *Mater. Today Proc.* **2015**, *2*, S837–S840. [CrossRef]
71. La Roca, P.; Baruj, A.; Sobrero, C.; Malarría, J.; Sade, M. Nanoprecipitation effects on phase stability of Fe-Mn-Al-Ni alloys. *J. Alloys Compd.* **2017**, *708*, 422–427. [CrossRef]
72. Walnsch, A.; Kriegel, M.J.; Motylenko, M.; Korpalab, G.; Prahl, U.; Leineweber, A. Thermodynamics of martensite formation in Fe–Mn–Al–Ni shape memory alloys. *Scr. Mater.* **2021**, *192*, 26–31. [CrossRef]
73. Tong, H.C.; Wayman, C.M. Characteristic temperatures and other properties of thermoelastic martensites. *Acta Metall.* **1974**, *22*, 887–896. [CrossRef]
74. Lee, J.K.; Barnett, D.M.; Aaronson, H.I. The elastic strain energy of coherent ellipsoidal precipitates in anisotropic crystalline solids. *Metall. Trans. A* **1977**, *8*, 963–970. [CrossRef]
75. Chen, H.; Xiao, F.; Liang, X.; Li, Z.X.; Li, Z.; Jin, X.J.; Min, N.; Fukuda, T. Improvement of the stability of superelasticity and elastocaloric effect of a Ni-rich Ti-Ni alloy by precipitation and grain refinement. *Scr. Mater.* **2018**, *162*, 230–234. [CrossRef]
76. Vallejos, J.M.; Giordana, M.F.; Sobrero, C.E.; Malarria, J.A. Phase stability of three Fe–Mn–Al–Ni superelastic alloys with different Al:Ni ratios. *Shap. Mem. Superelasticity* **2021**, *7*, 362–372. [CrossRef]
77. Walnsch, A.; Kriegel, M.J.; Fabrichnaya, O.; Leineweber, A. Thermodynamic assessment and experimental investigation of the systems Al–Fe–Mn and Al–Fe–Mn–Ni. *Calphad* **2019**, *66*, 101621. [CrossRef]
78. Kaputkina, L.M.; Svyazhin, A.G.; Kaputkin, D.E.; Bazhenov, V.E.; Bronz, A.V.; Smarygina, I.V. Effect of Mn, Al, Ni, and C content on the equilibrium phase composition of alloys based on the Fe–Mn–Al–Ni–C system. *Metallurgist* **2016**, *59*, 1075–1080. [CrossRef]
79. Nduka, J.K.; Amuka, J.; Onwuka, J.C.; Udowelle, N.A.; Orisakwe, O.E. Human health risk assessment of lead, manganese and copper from scrapped car paint dust from automobile workshops in Nigeria. *Environ. Sci. Pollut. Res. Int.* **2016**, *23*, 20341–20349. [CrossRef]
80. Sivasubramanian, K.; Rao, M.N. Significance of alloying element levels in realizing the specified tensile properties in 18 wt % nickel maraging steel. *Mater. Sci. Appl.* **2011**, *2*, 1116–1120.
81. Tseng, L.W.; Ma, J.; Hornbuckle, B.C.; Karaman, I.; Thompson, G.B.; Luo, Z.P.; Chumlyakov, Y.I. The effect of precipitates on the superelastic response of [100] oriented FeMnAlNi single crystals under compression. *Acta Mater.* **2015**, *97*, 234–244. [CrossRef]
82. Evirgen, A.; Ma, J.; Karaman, I.; Luo, Z.P.; Chumlyakov, Y.I. Effect of aging on the superelastic response of a single crystalline FeNiCoAlTa shape memory alloy. *Scr. Mater.* **2012**, *67*, 475–478. [CrossRef]
83. Sehitoglu, H.; Efstathiou, C.; Maier, H.J.; Chumlyakov, Y.I. Hysteresis and deformation mechanisms of transforming FeNiCoTi. *Mech. Mater.* **2006**, *38*, 538–550. [CrossRef]
84. Titenko, A.N.; Demchenko, L.D.; Babanli, M.B.; Sharai, I.V.; Titenko, Y.A. Effect of thermomechanical treatment on deformational behavior of ferromagnetic Fe–Ni–Co–Ti alloy under uniaxial tension. *Appl. Nanosci.* **2019**, *9*, 937–943. [CrossRef]
85. Geng, Y.H.; Jin, M.J.; Ren, W.J.; Zhang, W.M.; Jin, X.J. Effects of aging treatment on martensitic transformation of Fe–Ni–Co–Al–Ta–B alloys. *J. Alloys Compd.* **2013**, *577*, S631–S635. [CrossRef]
86. Jin, M. *Phase Transformations and Structure Evolution in Fe–Ni–Co–Ti and Au–Cu–Al Alloys*; Shanghai Jiao Tong University: Shanghai, China, 2010.
87. Tanaka, Y.; Himuro, Y.; Omori, T.; Sutou, Y.; Kainuma, R.; Ishida, K. Martensitic transformation and shape memory effect in ausaged Fe–Ni–Si–Co alloys. *Mater. Sci. Eng. A* **2006**, *438*, 1030–1035. [CrossRef]
88. Ozcan, H.; Ma, J.; Karaman, I.; Chumlyakov, Y.I.; Santamarta, R.; Brown, J.; Noebe, R.D. Microstructural design considerations in Fe-Mn-Al-Ni shape memory alloy wires: Effects of natural aging. *Scr. Mater.* **2018**, *142*, 153–157. [CrossRef]
89. Vollmer, M.; Kroo, P.; Karaman, I.; Niendorf, T. On the effect of titanium on quenching sensitivity and pseudoelastic response in Fe-Mn-Al-Ni-base shape memory alloy. *Scr. Mater.* **2017**, *126*, 20–23. [CrossRef]
90. Kokorin, V.V.; Samsonov, Y.I.; Chernenko, V.A.; Shevchenko, O.M. Superelasticity in Fe-Ni-Co-Ti alloys. *Phys. Metall.* **1989**, *67*, 202–204.
91. Sims, C.T.; Stoloff, N.S.; Hagel, W.C. *Superalloys II: High-Temperature Materials for Aerospace and Industrial Power*; Wiley-Interscience: Hoboken, NJ, USA, 1987.

92. Vollmer, M.; Segel, C.; Krooß, P.; Gunther, J.; Tseng, L.W.; Karaman, I.; Weidner, A.; Biermann, H.; Niendorf, T. On the effect of gamma phase formation on the pseudoelastic performance of polycrystalline Fe–Mn–Al–Ni shape memory alloys. *Scr. Mater.* **2015**, *108*, 23–26. [CrossRef]
93. Neuhaus, D.H.; Münzer, A. Industrial silicon wafer solar cells. *Adv. OptoElectron.* **2007**, *2007*, 1–15. [CrossRef]
94. Fromm, B.S.; Adams, B.L.; Ahmadi, S.; Knezevic, M. Grain size and orientation distributions: Application to yielding of α-titanium. *Acta Mater.* **2009**, *57*, 2339–2348. [CrossRef]
95. Zhang, S.Q.; Dong, T.; Geng, Z.; Lin, H.; Hua, Z.; Jun, H.; Yi, L.; Liu, M.X.; Hu, Y.H.; Wei, Z. The influence of grain size on the magnetic properties of Fe_3O_4 nanocrystals synthesized by solvothermal method. *J. Sol-Gel Sci. Technol.* **2018**, *98*, 422–429. [CrossRef]
96. Wu, J.X.; Jiang, B.H.; Hsu, T.Y. Influence of grain size and ordering degree of the parent phase on M_s in a CuZnAl alloy containing boron. *Acta Metall.* **1988**, *36*, 1521–1526.
97. Dvorak, I.; Hawbolt, E.B. Transformational elasticity in a polycrystalline Cu-Zn-Sn alloy. *Metall. Trans. A* **1975**, *6*, 95. [CrossRef]
98. Sutou, Y.; Omori, T.; Yamauchi, K.; Ono, N.; Kainuma, R.; Ishida, K. Effect of grain size and texture on pseudoelasticity in Cu–Al–Mn-based shape memory wire. *Acta Mater.* **2005**, *53*, 4121–4133. [CrossRef]
99. Sutou, Y.; Omori, T.; Kainuma, R.; Ishida, K. Grain size dependence of pseudoelasticity in polycrystalline Cu–Al–Mn-based shape memory sheets. *Acta Mater.* **2013**, *61*, 3842–3850. [CrossRef]
100. Saburi, T. *Ti-Ni Shape Memory Alloy*; Shape Memory Materials: Ibaraki, Japan, 1998.
101. Tseng, L.W.; Ma, J.; Wang, S.J.; Karaman, I.; Kaya, M.; Luo, Z.P.; Chumlyakov, Y.I. Superelastic response of a single crystalline FeMnAlNi shape memory alloy under tension and compression. *Acta Mater.* **2015**, *89*, 374–383. [CrossRef]
102. Tseng, L.W.; Ma, J.; Wang, S.J.; Karaman, I.; Chumlyakov, Y.I. Effects of crystallographic orientation on the superelastic response of FeMnAlNi single crystals. *Scr. Mater.* **2016**, *116*, 147–151. [CrossRef]
103. Omori, T.; Okano, M.; Kainuma, R. Effect of grain size on superelasticity in Fe-Mn-Al-Ni shape memory alloy wire. *APL Mater.* **2013**, *1*, 032103. [CrossRef]
104. Taylor, G.I. Analysis of plastic strain in a cubic crystal. *Metals*. **1938**, *62*, 218–224.
105. Sachs, G. Plasticity problems in metals. *Trans. Faraday Soc.* **1928**, *24*, 84–92. [CrossRef]
106. Tseng, L.W.; Ma, J.; Vollmer, M.; Krooss, P.; Niendorf, T.; Karaman, I. Effect of grain size on the superelastic response of a FeMnAlNi polycrystalline shape memory alloy. *Scr. Mater.* **2016**, *125*, 68–72. [CrossRef]
107. Yang, C.; Baker, I. Effect of soluble particles on microstructural evolution during directional recrystallization. *Acta Mater.* **2020**, *188*, 288–301. [CrossRef]
108. Omori, T.; Iwaizako, H.; Kainuma, R. Abnormal grain growth induced by cyclic heat treatment in Fe-Mn-Al-Ni superelastic alloy. *Mater. Des.* **2016**, *101*, 263–269. [CrossRef]
109. Vollmer, M.; Arold, T.; Kriegel, M.J.; Klemm, V.; Degener, S.; Freudenberger, J.; Niendorf, T. Promoting abnormal grain growth in Fe-based shape memory alloys through compositional adjustments. *Nat. Commun.* **2019**, *10*, 2337. [CrossRef] [PubMed]
110. Vallejos, J.M.; Malarría, J.A. Growing Fe-Mn-Al-Ni single crystals by combining directional annealing and thermal cycling. *J. Mater. Process. Technol.* **2020**, *275*, 116317. [CrossRef]
111. Gall, K.; Sehitoglu, H.; Chumlyakov, Y.I. Tension-compression asymmetry of the stress-strain response in aged single crystal and polycrystalline NiTi. *Acta Mater.* **1999**, *47*, 1203–1217. [CrossRef]
112. Chu, C.; James, R.D. Analysis of microstructures in Cu-14.0%Al-3.9%Ni by energy minimization. *J. Phys. IV* **1995**, *5*, 143–149. [CrossRef]
113. Kuruvilla, S.P.; Menon, C.S. Lattice thermal expansion of the shape memory alloys Cu-Al-Ni, Cu-Al-Zn, Cu-Al-Be and Cu-Al-Pd. *Adv. Mater. Res.* **2008**, *52*, 135–142. [CrossRef]
114. Tseng, L.W.; Ma, J.; Chumlyakov, Y.I.; Karaman, I. Orientation dependence of superelasticity in FeMnAlNi single crystals under compression. *Scr. Mater.* **2019**, *166*, 48–52. [CrossRef]
115. Fu, H.D.; Zhao, H.M.; Zhang, Y.X.; Xie, J.X. Enhancement of superelasticity in Fe-Ni-Co-based shape memory alloys by microstructure and texture control. *Procedia Eng.* **2017**, *207*, 1505–1510. [CrossRef]
116. Kokorin, V.V. *Martensite Transformations in Heterogeneous Solid Solutions*; Naukova Dumka: Kiev, Ukraine, 1987.
117. Kalinin, V.M.; Kornyakov, V.A. Young's modulus of alloyed iron–nickel invars. *Fiz. Met. Metalloved.* **1981**, *51*, 1110–1113.
118. Titenko, A.N.; Demchenko, L.D. Superelastic deformation in polycrystalline Fe-Ni-Co-Ti-Cu alloys. *J. Mater. Eng. Perform.* **2012**, *21*, 2525–2529. [CrossRef]
119. Xia, J.; Noguchi, Y.; Xu, X.; Odaira, T.; Kimura, Y.; Nagasako, M.; Omori, T.; Kainuma, R. Iron-based superelastic alloys with near-constant critical stress temperature dependence. *Science* **2020**, *396*, 855–858. [CrossRef]

Communication

Microstructural Effects on Thermal-Mechanical Alleviation of Cold Dwell Fatigue in Titanium Alloys

Songlin Shen [1], Mei Zhan [1], Pengfei Gao [1], Wenshuo Hao [1], Fionn P. E. Dunne [2] and Zebang Zheng [1,*]

[1] State Key Laboratory of Solidification Processing, Shaanxi Key Laboratory of High-Performance Precision Forming Technology and Equipment, School of Materials Science and Engineering, Northwestern Polytechnical University, Xi'an 710072, China; shensonglin@mail.nwpu.edu.cn (S.S.); meizhan@nwpu.edu.cn (M.Z.); gaopengfei@nwpu.edu.cn (P.G.); wenshuoh@mail.nwpu.edu.cn (W.H.)

[2] Department of Materials, Imperial College London, London SW7 2AZ, UK; fionn.dunne@imperial.ac.uk

* Correspondence: zebang.zheng@nwpu.edu.cn

Abstract: Cold dwell fatigue is a well-known problem in the titanium components of aircraft engines. The high temperature and low dwell stress of in-service conditions have been reported to give rise to dwell fatigue resistance through a thermal-mechanical alleviation process. Here, dwell fatigue tests at room temperature and the component operating temperature were performed on IMI834 titanium alloy to assess the microstructural effects on thermal-mechanical alleviation of cold dwell fatigue while eliminating the effect of chemical composition. The ratcheting strain rates under different loading conditions were quantitatively investigated to aid the understanding of thermal-mechanical alleviation.

Keywords: low cycle fatigue; microstructure; titanium alloys; thermal-mechanical alleviation

1. Introduction

Titanium alloys are widely used in manufacturing high-stressed components of aero-engines due to their high strength-to-weight ratio and good corrosion resistance [1–5]. These titanium components have been found to suffer from cold dwell fatigue since the 1970s [6]. Serious lifetime reduction is observed when loading cycles contain a dwell period where the stress is held at a high magnitude. The increasing peak stress at the soft-hard grain boundaries induced by ratcheting strain accumulation is argued to contribute to the facet crack nucleation under dwell fatigue loading and cause early failure of key gas turbines components [7–12]. The ratcheting behavior under cyclic loadings of titanium alloys is related to their strong strain rate sensitivity (SRS) at near-ambient temperatures [13].

Experimental [14–17] and analytical [18–22] observations have shown that the dwell and strain rate sensitivity is associated with the microstructure of titanium alloys. The microstructural heterogeneity in dual-phase titanium alloys arises from the anisotropy of α and β phase properties. The rate-dependent properties of the two phases have been found to be remarkably different at near-room temperatures [23]. As a result, alloys with different morphology and texture show different rate sensitivity and, hence, dwell sensitivity. Shen et al. [24–26] have shown that titanium alloys with different microstructures have remarkably different fatigue crack resistance. Experimental work on dwell fatigue sensitivities of Ti-6Al-2Sn-4Zr-xMo (Ti-624x) alloys by Qiu et al. [27] found that the Mo content apparently had a large influence on the dwell fatigue life debit. However, both chemical composition and microstructure were different in the Ti-624x series alloys considered, although the alloys were designed to have the same nominal type of microstructure, but the average grain sizes and phase volume fractions were different. In addition, the atomistic simulations suggest that Mo content does not give rise to the observed differences in SRS of Ti-6242 and Ti-6246 alloys [28]. Hence, for the purpose of investigating microstructural effects on cold dwell fatigue, it is important and necessary to eliminate the composition effects of alloys.

On the other hand, the temperature has been found to significantly affect dwell-sensitivity by altering the strain rate sensitivity [29,30]. At near-room temperature, near-α Ti alloys (such as Ti-6242 and Ti-6Al) are known to demonstrate strong rate and dwell sensitivities. With the temperature increasing, the load shedding is found to peak at about 120 °C and progressively diminishes when the temperature increases to ≥230 °C. Other types of Ti alloys also show a similar trend of dwell sensitivity with increasing temperatures, but the most significant temperature may be shifted (e.g., the most dwell sensitive temperature for Ti-6246 alloy has been suggested to be about 300 °C) [30]. The diminution of SRS at in-service temperatures has been reported to give rise to resistance to dwell fatigue [31]. In addition, the hoop stress (primary loading) under in-service conditions has been found to be much lower than the macroscopic yield strengths of the alloy. The combinations of high operating temperatures and low operating stresses are argued to resist the facet nucleation by inhibiting plastic strain ratcheting due to cyclic loading, which is termed as thermal-mechanical alleviation [31].

The thermal-mechanical alleviation in IMI834 alloy with a bimodal microstructure has been systematically investigated using a dual-phase crystal plasticity model with α-β morphology explicitly represented [31]. The reduction in dwell sensitivity has been clearly demonstrated, but the microstructural effects on thermal-mechanical alleviation have not yet been fully addressed. Thus, the motivation of the present work is to explore the thermal-mechanical alleviation in four commonly observed microstructures of titanium alloys and to evaluate the microstructural effects on dwell sensitivity.

2. Materials and Methods

The near-α titanium alloy IMI834 was supplied as an ingot consisting of mainly equiaxed α grains (with an average grain size ≈40 µm) and some lamellar structures with thin β laths located near grain boundaries. Specimens of approximately 20 mm × 20 mm × 100 mm cuboids were cut from the ingot for heat treatment. Four different heat treatment routes were used to achieve the commonly observed microstructures: an equiaxed structure, a Widmanstätten structure, a bi-modal structure, and a trimodal structure. The hot working processes and the corresponding phase contents and microstructures are presented in Figure 1. The heat treatment was conducted in the air atmosphere. The equiaxed microstructure was obtained by thermal processing at near-β temperatures (~10–20 °C below β-transus temperature, $T_\beta \approx 1055$ °C for IMI834) for 1 h and then furnace cooled (FC) to room temperature. The resulting average grain size was ~50 µm. The Widmanstätten microstructure was achieved by heating up to ~15–30 °C above T_β for 1 h, followed by furnace cooling. The bimodal microstructure was obtained by a two-step heat treatment: specimens were first heated to near-β temperatures and then were annealed at ~800 °C. This microstructure consists of equiaxed α grains (α_p) and transformed β matrix (β_t). For the trimodal microstructure, there is an extra step before annealing: the temperature was kept at ~30–50 °C below T_β for 1 h, which allows lamellar α grains (α_l) to grow from the transformed β matrix. The cooling conditions for both bimodal and trimodal were air cooling (AC). Scanning electron microscopy (SEM) examinations were carried out to characterize the morphology of each microstructure. The phase volume fraction was then calculated using ImageJ software [32]. Electron Backscattering Diffraction (EBSD) characterizations were conducted on all the obtained microstructures using a scanning step of 0.3 mm and an acceleration voltage of 20 kV. EBSD results on the four microstructures demonstrated a similar strong texture with the basal poles tilted ±30° from the normal direction (ND).

A set of stress relaxation tests were first carried out at room temperature (e.g., 25 °C) and at in-service temperature (e.g., ~350 °C [31]) to investigate the effect of temperature and microstructures on the SRS of the alloy. Tensile specimens with a gauge length of 40 mm and a diameter of 6 mm were machined with the loading direction parallel to ND, as shown in Figure 2. Specimens with different microstructures were uniaxially stretched up to 2 mm within 5 s, which results in a constant strain rate of $\dot{\varepsilon} = 1 \times 10^{-2}$ s^{-1}, and then held at the maximum strain for 60 s. Substantial stress relaxation is expected during the hold if the specimen experiences strong rate-dependent plasticity. In order to obtain reproducible results, the stress relaxations of each microstructure were repeated three times.

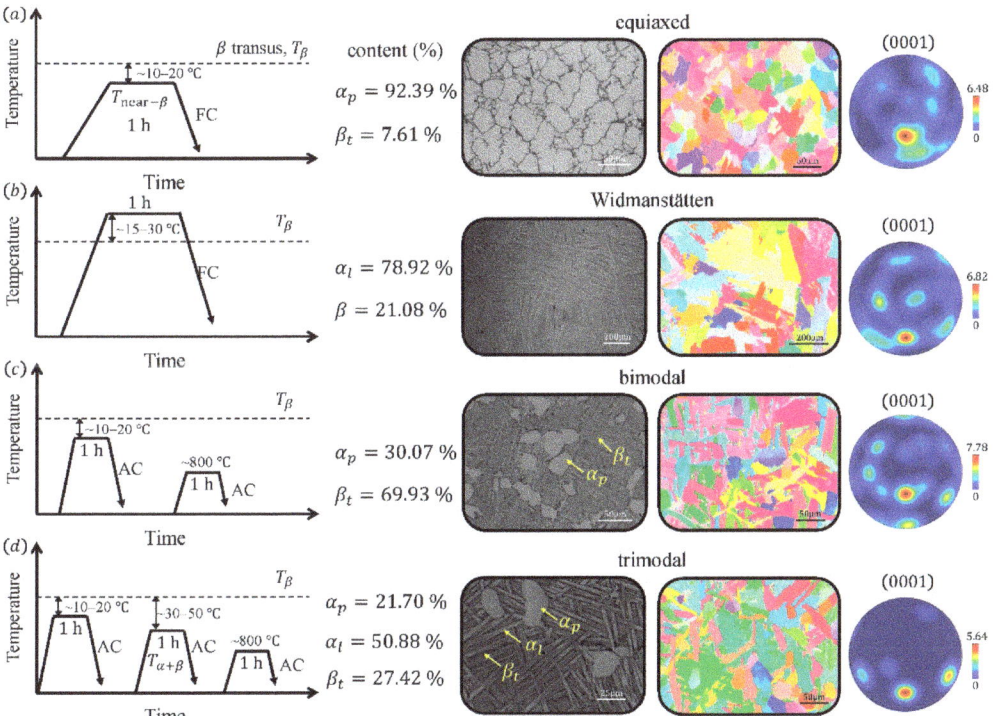

Figure 1. Heat treatment schedules and the obtained morphologies and textures for (**a**) equiaxed microstructure; (**b**) Widmanstätten microstructure; (**c**) bimodal microstructure; and (**d**) trimodal microstructure.

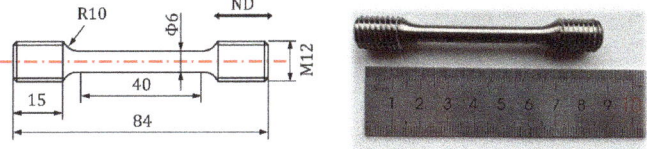

Figure 2. Specimen geometries for the stress relaxation tests.

3. Results and Discussion

In order to isolate the effect of microstructure from the effect of chemical composition on dwell sensitivity, energy dispersive spectroscopy (EDS) was performed to compare the element distribution of different microstructures. By analyzing five main alloying elements, Al, Sn, Zr, Nb, and Mo, although there is segregation between the α and β phases at the microstructural scale, the overall composition was not changed after different heat treatment processes.

Figure 3 shows the engineering stress-time curves obtained at 25 °C and 350 °C, respectively, for the four microstructures. From the first 5-s uniaxial tension part, the 0.2 pct proof stress $\sigma_{0.2}$ can be extracted. The strain rate sensitivity exponent m can be determined from the stress-drop during the strain hold period given by [33]

$$m = \frac{d \ln(\sigma)}{d \ln(-\dot{\sigma})}. \tag{1}$$

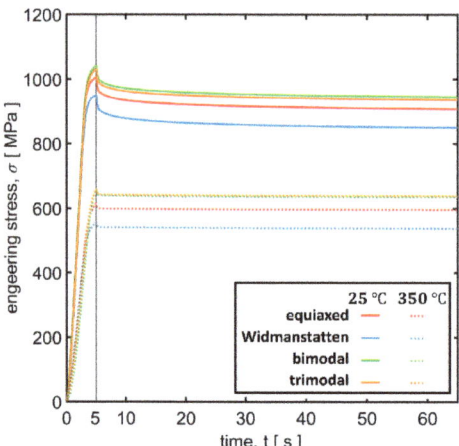

Figure 3. Engineering stress-time curves for four different microstructures at 25 °C and 350 °C.

The first 1 s of the stress relaxation data, where the stress-drop is the most significant, was used to calculate the m value for all the specimens. Resultant proof stresses and SRS exponents are summarized in Figure 4a,b, respectively. The error bars in Figure 4a represent the scatters of the repeated tests. Although a higher scatter in the proof stress can be observed at the room temperature compared to 350 °C, the general trend of the strength between microstructures is unchanged. In addition, due to the relatively good repeatability during the stress relaxation for all the repeated tests, there is no noticeable difference (<0.0002) in the measured strain rate sensitivity exponent. Overall, for all the considered microstructures, both proof stresses and the m values are lower at the high temperature as expected. The bimodal and trimodal structures show a higher strength under both temperatures, while the Widmanstätten microstructure has the lowest strength. Material strength is one of the aspects that affect the in-service performance, but, more importantly, the strain-rate-dependent plasticity is fundamentally related to the dwell fatigue life. The equiaxed and Widmanstätten structures are the most rate-sensitive, while the bimodal and trimodal are relatively less sensitive to strain rate. However, at 350 °C, all the SRS exponents are reduced to about 0.003, which means the strain rate sensitivity of IMI834 alloy is negligible at this temperature. The stress relaxation curves in Figure 3 also suggest that there is almost no stress drop at the high temperature. Other evidence in the literature has demonstrated that the rate sensitivity diminishes progressively at temperatures above 230 °C for some Ti alloys [29,30]. The diminishment of SRS at the high temperature was attributed to the energy barrier of the pinned dislocation from the obstacle under the considered loading rates being easily overcome of. Since the low strain rate regime where the SRS is predominated by the thermal activation process [34], a high deformation temperature can lead to a high frequency of dislocation escape from the pinned obstacle. As a result, the average dislocation velocity is much higher than the value to accommodate the applied deformation rate; hence, the variation in strain rates does not alter the flow stress of the material at high temperatures.

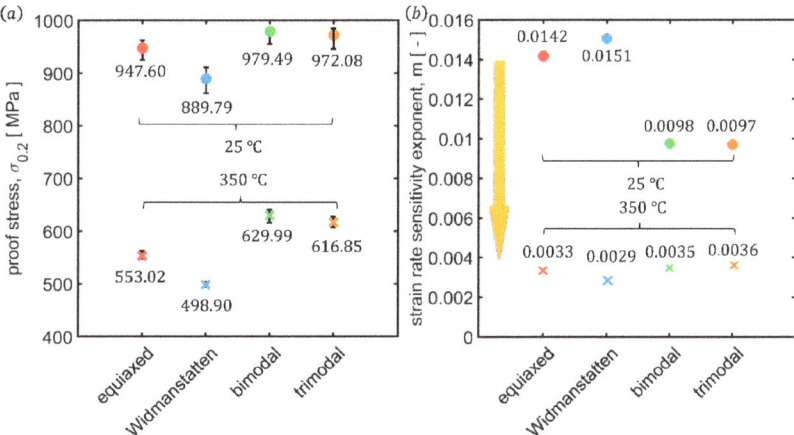

Figure 4. The extracted (**a**) proof stresses and (**b**) strain rate sensitivity exponents from the stress relaxation curves in Figure 3.

In order to investigate the microstructural effect on dwell fatigue, low cycle fatigue (LCF) and low cycle dwell fatigue (LCDF) tests were performed at room (25 °C) temperature on specimens with a 15 × 6 mm diameter cylindrical gauge section, as shown in the inset of Figure 5. The peak stress for both loading modes was identical and chosen to be 95% of the proof stress $\sigma_{0.2}$, as determined in Figure 4a, i.e., 900 MPa, 845 MPa, 931 MPa, and 923 MPa for the equiaxed, Widmanstätten, bimodal, and trimodal microstructure, respectively. The rise and fall time for both LCF and LCDF were chosen to be identical as 1 s. The dwell period in the LCDF tests was 60 s, and the stress ratio R in both loading modes was 0.

The accumulated plastic strain at the end of each loading cycle was recorded, as shown in Figure 5. The differing microstructures under LCF showed different plastic strain accumulations at 25 °C, as shown in Figure 5a. The plastic strains-to-failure for the Widmanstätten and bimodal structures was about 3%, while those for the equiaxed and trimodal structures were about 2%. On the other hand, the plastic strain accumulation under LCDF at 25 °C in Figure 5b displays both higher ratcheting rates and strains-to-failure. The results clearly demonstrate that all the considered microstructures experience a strong dwell sensitivity at room temperature. In order to investigate this phenomenon quantitatively, the LCF and LCDF lives and the dwell debit (N_{LCF}/N_{LCDF}) are summarized in Figure 5c. The maximum dwell debit is 28.66 in the Widmanstätten structure, followed by a value of 24.76 in the equiaxed structure. Generally, the fatigue responses in the bimodal and trimodal structure are similar to each other, but that for the trimodal is slightly less sensitive to dwell fatigue. Lei et al. [35] have also reported that the fatigue life of the trimodal structure of TA-15 alloy is slightly longer than that of the bimodal structure. Stroh [36] proposed a dislocation pile-up model to explain fatigue crack nucleation. The progressive build-up of dislocations within a favorably oriented grain under cyclic loading is argued to be responsible for the crack nucleation at the boundary with the neighboring "hard" grain. The mean-free path of dislocations in the trimodal microstructure is effectively reduced by the thick lamellar α_l grains. Hence, the peak stress achieved at the grain boundaries is lower, which makes the fatigue life longer. Discrete dislocation plasticity modeling [37] has demonstrated that thick laths within a lamellar structure can reduce the size of the pile-ups and change the SRS of the material, which then leads to different dwell fatigue susceptibility. The Widmanstätten structure in the present study was designed to have thick α laths and very thin β laths (≤2 μm). The average size of dislocation pile-ups is larger in the microstructure with a larger α grain size and thinner β laths. Thus, the resistance to dislocation motions provided the grain and phase boundaries is lower. The local resolved shear stress acting on the leading dislocation of each pile-up group is higher, which promotes the thermally-activated escaping from the pinned obstacle. As a result, the average velocity of dislocations is higher, and the strain rate sensitivity and dwell-sensitivity of the Widmanstätten and equiaxed structures are stronger.

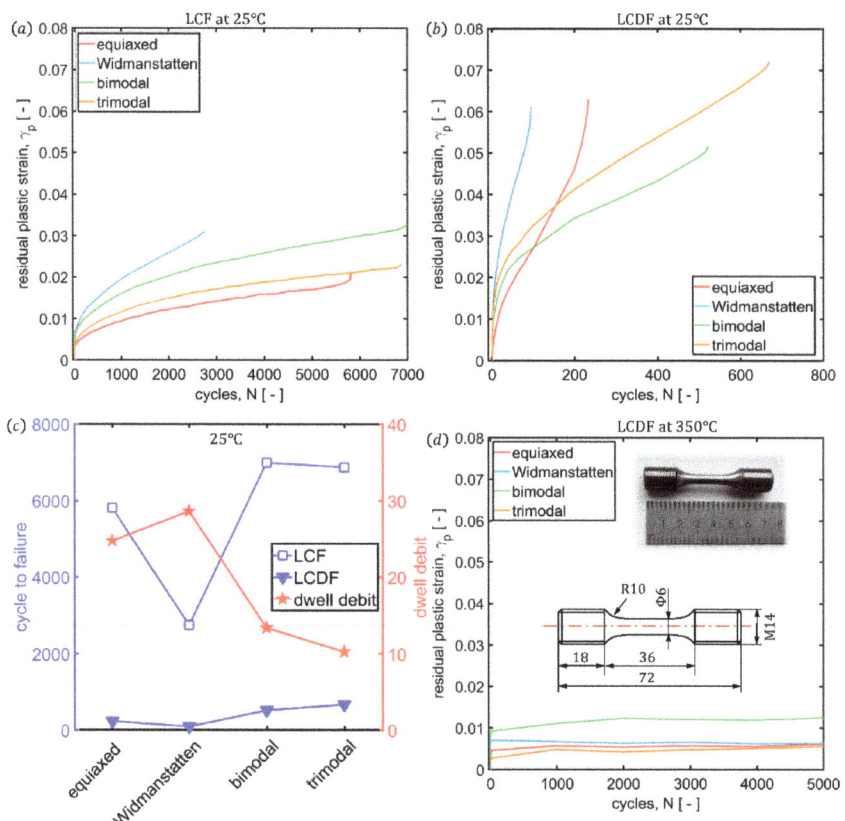

Figure 5. Plastic strain accumulation at 25 °C under (**a**) LCF and (**b**) LCDF. (**c**) The dwell life debit at 25 °C for the differing microstructures. (**d**) Residual plastic strain during LCDF at 350 °C.

The analysis of the fatigue responses has so far been carried out at room temperature and at relatively high stress. In order to understand the behavior of the different microstructures under in-service loading conditions, the LCDF tests were performed at a high temperature and low maximum stress. The thermal-mechanical loading histories of IMI834 rig components indicate that the first dwell period of a long haul flight consists of a stress hold at magnitude very much below the macroscopic yield stress and at a temperature of about 340–370 °C [31]. Hence, the temperature for the LCDF tests was chosen to be 350 °C, and the maximum stress chosen was 85% of the proof stress $\sigma_{0.2}$ at the corresponding temperature. The evolutions of the plastic strains are shown in Figure 5d. No specimen at 350 °C was found to fail under LCDF loadings; hence, the tests were terminated after 5000 cycles because of the very low strain accumulations. The dwell fatigue responses were totally different compared to the low-temperature scenario. The accumulated plastic strain was found barely to increase after the first loading cycle, and the maximum residual plastic strain after 5000 loading cycles was found to remain below 1.5% for all the specimens. The resistance to dwell fatigue through the thermal-mechanical alleviation process was clearly demonstrated, for the first time, in four different types of microstructure. The observed reduction in the dwell sensitivity under in-service loading conditions can be ascribed to two aspects: on the thermal side, the strain rate sensitivity was substantially reduced at the high temperature, while, on the mechanical side, the lower hold-stress made the amount of plastic deformation much lower. The former is argued to be the main reason for the observed thermal-mechanical alleviation in dwell fatigue. In addition, the maximum in-service stress can even be lower than 0.85 $\sigma_{0.2}$, but, due to the difference in the stress-multiaxiality and the temperature histories between the laboratory fatigue tests and the in-service rig spin tests, lower stress may not be able to trigger plasticity in the laboratory LCDF tests. Although low applied stress can effectively reduce the amount of plastic deformation and prolong the dwell fatigue life of the material, fundamentally, it does not change the sensitivity to rate or stress-hold, and the residual plasticity strain is still accumulated with the cycles. Hence, the reduction in the maximum applied stress cannot be ignored in the thermal-mechanical alleviation observed here but only plays a secondary role.

Comparing the accumulated plastic strain evolution curves under differing loading conditions in Figure 5, there are three main ratcheting behaviors observed: (1) a relatively low ratcheting strain rate under low-temperature LCF; (2) a high ratcheting strain rate under low-temperature LCDF; and (3) a very low ratcheting strain rate under high-temperature LCDF. These three types of ratcheting behaviors in fatigue have been recently reviewed [38]. The ratcheting strain rates for different microstructures under different loading conditions are extracted from the steady stage of the plastic strain accumulation curves and are summarized in Figure 6. Broadly, the ratcheting strain rates show similar trends to the fatigue life of the specimen: a higher ratcheting rate leads to faster strain accumulation, which eventually causes plastic instability and shorter fatigue life. In particular, the equiaxed structure has an LCDF ratcheting rate about 110 times higher than that under LCF. Nevertheless, the ratcheting rates under in-service (elevated temperature) conditions for all the considered microstructures are reduced to a negligible level where no obvious additional ratcheting strain accumulation was noticed after the initial few cycles. Since the energy barrier to thermally activated dislocation escape is significantly reduced with higher temperature, a stable dislocation structure is established at the early stage of the loading. In addition, the applied stress is lower compare to the room temperature LCDF, and there is no driving force for new dislocations to be nucleated, so the alloy is no longer sensitive to strain rate or dwell loading, regardless of the microstructure.

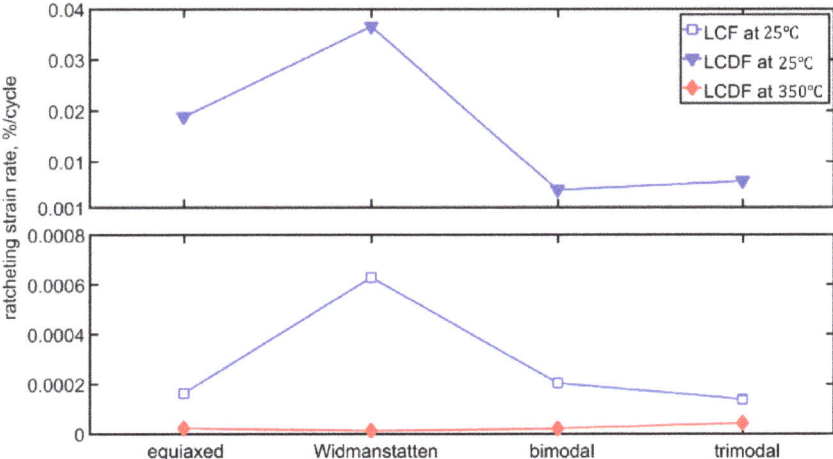

Figure 6. The variation of ratcheting strain rate with microstructure under different loading conditions.

4. Conclusions

In summary, stress relaxation tests were carried out on IMI834 alloys with four commonly observed microstructures at room temperature and 350 °C. The strain rate sensitivity at room temperature is highly microstructure-dependent but is reduced significantly at elevated temperature and becomes nearly independent of microstructure in this alloy. The low cycle fatigue and low cycle dwell fatigue results at room temperature suggest that all the microstructures experience a strong dwell sensitivity. The LCDF tests under in-service loading conditions demonstrate the thermal-mechanical alleviation in all the considered microstructures. The dwell fatigue lives under in-service loading conditions are much longer than that at room temperature, indicating temperature and the applied stress play a key role in cold dwell fatigue.

Author Contributions: Conceptualization, Z.Z.; methodology, M.Z. and P.G.; formal analysis and investigation, S.S.; writing—original draft preparation, S.S. and Z.Z.; writing—review and editing, F.P.E.D.; visualization, W.H.; funding acquisition, M.Z. and Z.Z. All authors have read and agreed to the published version of the manuscript.

Funding: This research was funded by the National Natural Science Foundation Of China, grant number 91860130 and U1910213.

Institutional Review Board Statement: Not applicable.

Informed Consent Statement: Not applicable.

Data Availability Statement: Not applicable.

Acknowledgments: F.P.E.D. wishes to acknowledge gratefully the provision of funding for his Royal Academy of Engineering/Rolls-Royce research chair. S.S. would like to thank Z. Lei for useful discussions.

Conflicts of Interest: The authors declare no conflict of interest.

References

1. Lütjering, G.; Williams, J.C. *Titanium, Engineering Materials and Process*; Springer: Manchester, UK, 2003.
2. Donachie, M.J. *Titanium: A Technical Guide*; ASM International: Almere, The Netherlands, 2000; ISBN 161503062X.
3. Gao, P.; Yu, C.; Fu, M.; Xing, L.; Zhan, M.; Guo, J. Formability enhancement in hot spinning of titanium alloy thin-walled tube via prediction and control of ductile fracture. *Chin. J. Aeronaut.* **2022**, *35*, 320–331. [CrossRef]
4. Gao, P.F.; Yan, X.G.; Li, F.G.; Zhan, M.; Ma, F.; Fu, M.W. Deformation mode and wall thickness variation in conventional spinning of metal sheets. *Int. J. Mach. Tools Manuf.* **2022**, *173*, 103846. [CrossRef]
5. Zheng, Z.; Zhao, P.; Zhan, M.; Shen, S.; Wang, Y.; Fu, M.W. The roles of rise and fall time in load shedding and strain partitioning under the dwell fatigue of titanium alloys with different microstructures. *Int. J. Plast.* **2022**, *149*, 103161. [CrossRef]
6. Bache, M. A review of dwell sensitive fatigue in titanium alloys: The role of microstructure, texture and operating conditions. *Int. J. Fatigue* **2003**, *25*, 1079–1087. [CrossRef]
7. Sinha, V.; Mills, M.J.; Williams, J.C.; Spowart, J.E. Observations on the faceted initiation site in the dwell-fatigue tested ti-6242 alloy: Crystallographic orientation and size effects. *Metall. Mater. Trans. A* **2006**, *37*, 1507–1518. [CrossRef]
8. Dunne, F.P.E.; Rugg, D. On the mechanisms of fatigue facet nucleation in titanium alloys. *Fatigue Fract. Eng. Mater. Struct.* **2008**, *31*, 949–958. [CrossRef]
9. Ozturk, D.; Pilchak, A.L.; Ghosh, S. Experimentally validated dwell and cyclic fatigue crack nucleation model for α–titanium alloys. *Scr. Mater.* **2017**, *127*, 15–18. [CrossRef]
10. Zheng, Z.; Balint, D.S.; Dunne, F.P.E. Discrete dislocation and crystal plasticity analyses of load shedding in polycrystalline titanium alloys. *Int. J. Plast.* **2016**, *87*, 15–31. [CrossRef]
11. Ozturk, D.; Shahba, A.; Ghosh, S. Crystal plasticity FE study of the effect of thermo-mechanical loading on fatigue crack nucleation in titanium alloys. *Fatigue Fract. Eng. Mater. Struct.* **2016**, *39*, 752–769. [CrossRef]
12. Cuddihy, M.A.; Stapleton, A.; Williams, S.; Dunne, F.P.E. On cold dwell facet fatigue in titanium alloy aero-engine components. *Int. J. Fatigue* **2017**, *97*, 177–189. [CrossRef]
13. Conrad, H. Effect of interstitial solutes on the strength and ductility of titanium. *Prog. Mater. Sci.* **1981**, *26*, 123–403. [CrossRef]
14. Sinha, V.; Mills, M.J.; Williams, J.C. Crystallography of fracture facets in a near-alpha titanium alloy. *Metall. Mater. Trans. A* **2006**, *37*, 2015–2026. [CrossRef]
15. Pilchak, A.L.; Williams, J.C. Observations of Facet Formation in Near-α Titanium and Comments on the Role of Hydrogen. *Metall. Mater. Trans. A* **2011**, *42*, 1000–1027. [CrossRef]

16. Woodfield, A.P.; Gorman, M.D.; Sutliff, J.A.; Corderman, R.R. *Effect of Microstructure on Dwell Fatigue Behavior of Ti-6242*; GE Aircraft Engines: Cincinnati, OH, USA, 1999.
17. Bandyopadhyay, R.; Mello, A.W.; Kapoor, K.; Reinhold, M.P.; Broderick, T.F.; Sangid, M.D. On the crack initiation and heterogeneous deformation of Ti-6Al-4V during high cycle fatigue at high R ratios. *J. Mech. Phys. Solids* **2019**, *129*, 61–82. [CrossRef]
18. Waheed, S.; Zheng, Z.; Balint, D.S.; Dunne, F.P.E. Microstructural effects on strain rate and dwell sensitivity in dual-phase titanium alloys. *Acta Mater.* **2019**, *162*, 136–148. [CrossRef]
19. Zhang, Z.; Dunne, F.P.E. Microstructural heterogeneity in rate-dependent plasticity of multiphase titanium alloys. *J. Mech. Phys. Solids* **2017**, *103*, 199–220. [CrossRef]
20. Dunne, F.P.E.; Rugg, D.; Walker, A. Lengthscale-dependent, elastically anisotropic, physically-based hcp crystal plasticity: Application to cold-dwell fatigue in Ti alloys. *Int. J. Plast.* **2007**, *23*, 1061–1083. [CrossRef]
21. Bache, M.; Cope, M.; Davies, H.; Evans, W.; Harrison, G. Dwell sensitive fatigue in a near alpha titanium alloy at ambient temperature. *Int. J. Fatigue* **1997**, *19*, 83–88. [CrossRef]
22. Ghosh, S.; Anahid, M. Homogenized constitutive and fatigue nucleation models from crystal plasticity FE simulations of Ti alloys, Part 1: Macroscopic anisotropic yield function. *Int. J. Plast.* **2013**, *47*, 182–201. [CrossRef]
23. Zhang, Z.; Jun, T.-S.; Britton, T.B.; Dunne, F.P.E. Determination of Ti-6242 α and β slip properties using micro-pillar test and computational crystal plasticity. *J. Mech. Phys. Solids* **2016**, *95*, 393–410. [CrossRef]
24. Shen, W.; Soboyejo, W.O.; Soboyejo, A.B.O. Microstructural effects on fatigue and dwell-fatigue crack growth in α/β Ti-6Al-2Sn-4Zr-2Mo-0.1Si. *Metall. Mater. Trans. A* **2004**, *35*, 163–187. [CrossRef]
25. McBagonluri, F.; Akpan, E.; Mercer, C.; Shen, W.; Soboyejo, W.O. An investigation of the effects of microstructure on dwell fatigue crack growth in Ti-6242. *Mater. Sci. Eng. A* **2005**, *405*, 111–134. [CrossRef]
26. Shen, W.; Soboyejo, W.; Soboyejo, A.B. An investigation on fatigue and dwell-fatigue crack growth in Ti-6Al-2Sn-4Zr-2Mo-0.1Si. *Mech. Mater.* **2004**, *36*, 117–140. [CrossRef]
27. Qiu, J.; Ma, Y.; Lei, J.; Liu, Y.; Huang, A.; Rugg, D.; Yang, R. A Comparative Study on Dwell Fatigue of Ti-6Al-2Sn-4Zr-xMo (x = 2 to 6) Alloys on a Microstructure-Normalized Basis. *Metall. Mater. Trans. A* **2014**, *45*, 6075–6087. [CrossRef]
28. Ready, A.J.; Haynes, P.D.; Grabowski, B.; Rugg, D.; Sutton, A.P. The role of molybdenum in suppressing cold dwell fatigue in titanium alloys. *Proc. R. Soc. A Math. Phys. Eng. Sci.* **2017**, *473*, 20170189. [CrossRef]
29. Zhang, Z.; Cuddihy, M.A.; Dunne, F.P.E. On rate-dependent polycrystal deformation: The temperature sensitivity of cold dwell fatigue. *Proc. R. Soc. A Math. Phys. Eng. Sci.* **2015**, *471*, 20150214. [CrossRef]
30. Zheng, Z.; Balint, D.S.; Dunne, F.P.E. Mechanistic basis of temperature-dependent dwell fatigue in titanium alloys. *J. Mech. Phys. Solids* **2017**, *107*, 185–203. [CrossRef]
31. Zheng, Z.; Stapleton, A.; Fox, K.; Dunne, F.P.E. Understanding thermal alleviation in cold dwell fatigue in titanium alloys. *Int. J. Plast.* **2018**, *111*, 234–252. [CrossRef]
32. Schneider, C.A.; Rasband, W.S.; Eliceiri, K.W. NIH Image to ImageJ: 25 years of image analysis. *Nat. Methods* **2012**, *9*, 671–675. [CrossRef]
33. Hedworth, J.; Stowell, M.J. The measurement of strain-rate sensitivity in superplastic alloys. *J. Mater. Sci.* **1971**, *6*, 1061–1069. [CrossRef]
34. Zheng, Z.; Balint, D.S.; Dunne, F.P.E. Rate sensitivity in discrete dislocation plasticity in hexagonal close-packed crystals. *Acta Mater.* **2016**, *107*, 17–26. [CrossRef]
35. Lei, Z.N.; Gao, P.F.; Li, H.W.; Cai, Y.; Li, Y.X.; Zhan, M. Comparative analyses of the tensile and damage tolerance properties of tri-modal microstructure to widmanstätten and bimodal microstructures of TA15 titanium alloy. *J. Alloys Compd.* **2019**, *788*, 831–841. [CrossRef]
36. Stroh, A.N. The formation of cracks as a result of plastic flow. *Proc. R. Soc. Lond. Ser. A Math. Phys. Sci.* **1954**, *223*, 404–414.
37. Zheng, Z.; Waheed, S.; Balint, D.S.; Dunne, F.P.E. Slip transfer across phase boundaries in dual phase titanium alloys and the effect on strain rate sensitivity. *Int. J. Plast.* **2018**, *104*, 23–38. [CrossRef]
38. Paul, S.K. A critical review of experimental aspects in ratcheting fatigue: Microstructure to specimen to component. *J. Mater. Res. Technol.* **2019**, *8*, 4894–4914. [CrossRef]

Article

Effect of Current on Corrosion Resistance of Duplex Stainless Steel Layer Obtained by Plasma Arc Cladding

Juan Pu [1,*], Peng Xie [1], Weimin Long [2], Mingfang Wu [1], Yongwang Sheng [3] and Jie Sheng [3]

1. School of Materials Science and Engineering, Jiangsu University of Science and Technology, Zhenjiang 212003, China; 15160709894@163.com (P.X.); wu_mingfang@163.com (M.W.)
2. Zhengzhou Research Institute of Mechanical Engineering, Zhengzhou 450001, China; brazelong@163.com
3. Zhejiang Yongwang Welding Material Manufacturing Co., Ltd., Jinhua 321000, China; papervip@163.com (Y.S.); s_jkd9078@163.com (J.S.)
* Correspondence: pu_juan84@163.com; Tel.: +86-15952815816

Abstract: In order to repair or strengthen stainless steel structural parts, the experiment was conducted by using plasma arc cladding technology to prepare 2205 duplex stainless steel (DSS) layers on the surface of Q345 steel. Their macro morphology and microstructure were observed by an optical microscope and the phase composition of microstructure was analyzed by an X-ray diffractometer instrument (XRD). The electrochemical behavior of 2205 DSS cladding layer under different current in 3.5% NaCl etching solution was studied by the potentiodynamic polarization, the electrochemical impedance spectroscopy (EIS) and X-ray photoelectron spectrometer (XPS). The results showed that when the current was 100 A, the forming of cladding layer was continuous, complete and fine with the dilution ratio of 11.43%. The mass ratio of austenite to ferrite in the microstructure increased with the increase of current and it was up to the optimum of 1.207 with the current of 100 A. Under such conditions, the self-corrosion potential of the cladding layer was up to the maximum while its corrosion current density reached the minimum, thus the corrosion resistance of the cladding layer reached the optimum. It was attributed to the existence of a large amount of Cr^{3+} and Mo^{6+} in the passive film of cladding layer, which can stabilize the passive film and promote the formation of Cr_2O_3 in the passive film.

Keywords: current; plasma arc cladding technology; 2205 duplex stainless steel; potentiodynamic polarization curve; passive film composition

1. Introduction

2205 duplex stainless steel (DSS) has good plastic toughness and weldability of austenitic steel and high strength and good corrosion resistance of ferritic steel [1]. Therefore, it is widely used in the cooling water heat exchangers, large oil storage tanks, the third-generation nuclear power fuel pool and seawater desalination systems in China [2–4]. During service, stainless steel structural parts (i.e. stainless steel-clad plate, using Q345 or 16Mn as the base metal and the stainless steel as the cladding material) are often damaged in the form of wear and corrosion [5,6], so it is important for parts to be repaired and strengthened. Surface remanufacturing technology is a means of using a heat source to clad(surface) alloy powder or wire on the surface of a workpiece to modify the properties of materials, such as wear resistance, corrosion resistance and oxidation resistance of coating (surfacing layer) [7–9]. It is beneficial to prolong the service life of large stainless steel structural parts, save precious and rare metal materials, finally reduce costs and improve benefits, which accords with the strategic policy of developing recycle economy and realizing sustainable development in China.

At present, the researchers mainly use the technology of laser cladding, CMT (Cold Metal Transfer Cladding) and TIG (Tungsten Inert Gas Arc Cladding) to prepare 2205 DSS cladding layer. Jing Ming et al. [10,11] successfully prepared 2205 DSS/TiC layer on 16Mn substrate with the help of laser cladding technology. The addition of TiC particles improved the hardness and wear resistance of duplex stainless-steel coating, but the layer started to crack when the mass fraction of TiC reached 15%. Liu Shao et al. [12] studied the optimal process parameter of 2205 DSS prepared by CMT cladding with double wires. The optimal parameter was MAG (Metal Active Gas Arc Cladding) current 220~290 A of front wire and CMT + P current 160~240 A of rear wire. The cladding layer was composed of 40~60% austenite, which met the service requirements. Liu Yu et al. [13] prepared 2205 DSS surfacing layer by TIG using ER2209 as surfacing materials and studied the effect of sensitization treatment time on pitting corrosion resistance of surfacing layer. He believes that with the increase of sensitization treatment time, the pitting corrosion resistance of surfacing layer decreases. However, when the sensitization time exceeds 15 min, the ferrite phase α occurs eutectoid reaction and $\sigma + \gamma_2$ forms. The formation of σ decreases the pitting corrosion resistance of the surfacing layer. The literature [14–16] showed that the corrosion resistance will be deteriorated if the austenite phase and ferrite phase was imbalance in the microstructure of 2205 DSS cladding layer. In summary, the above research showed that optimizing the process parameters of surfacing technology to possess the optimal microstructure was the critical factor to obtain best comprehensive performance for DSS cladding layer. Additionally, how to improve the corrosion resistance of DSS cladding layer in various environments (such as atmospheric, salt solution, acid and so on) was also the key point to prolong its service life.

Plasma arc cladding technology as a high-energy beam surface cladding technology and a metal surface treatment technology, is one of the latest technologies developed after surfacing technology and laser cladding technology. It uses a high-temperature plasma arc to melt the alloy powder or wire and base metal. Then, with the transfer of plasma arc, the molten metal solidifies rapidly to form a metallurgical bonding layer. This technology has the advantages of low dilution rate, dense microstructure, small heat-affected zone and good combination between surfacing layer and substrate. Compared with the laser cladding layer, the plasma arc cladding layer is not easy to produce defects such as pores, microstructure segregation and cracks [17–19]. So far, there are few reports on the preparation of 2205 duplex stainless steel layer by plasma arc cladding technology. In this paper, 2205 DSS layer will be prepared on the surface of Q345 steel by plasma arc cladding technology. The effects of different current on the macro morphology, austenite/ferrite mass ratio and corrosion resistance of the cladding layer will be studied. The corrosion resistance mechanism of the cladding layer will be revealed. The optimal process parameter of preparing 2205 DSS layer by plasma arc cladding will be acquired, which will guide the application of 2205 DSS cladding layer in engineering practice.

2. Materials and Methods

2.1. Materials

ER2209 stainless steel wire with a diameter of 1.2 mm was selected as the cladding wire, which was prepared by Jiangsu Jiuzhou New Material Technology Co., Ltd. A Q345 low alloy steel plate purchased on the market with the size of 160 mm × 70 mm × 10 mm was used as the base metal. Their chemical composition was listed in Table 1. 2205 DSS cladding layer was prepared on the surface of Q345 steel using 99.99% argon as the shielding gas.

Table 1. Chemical compositions of wire and base metal.

Materials	C	Si	Mn	P	S	Cr	Ni	Mo	Fe
ER2209	0.02	0.53	1.90	0.02	0.004	21.73	9.12	2.67	Balance
Q345	≤0.20	≤0.50	≤1.70	≤0.035	≤0.035	≤0.30	≤0.50	≤0.10	Balance

2.2. Preparation of 2205 DSS Cladding Layer

2205 DSS cladding layer was prepared on the surface of Q235 steel by DML-V03AD plasma arc cladding equipment. The specific process parameters were wire feeding speed of 29 mm/s, welding speed of 6 mm/s, ion gas flow of 1 L/min, gas flow of 20 L/min and Nozzle height of 10 mm. The current was designed as 80 A, 90 A, 100 A, 110 A and 120 A to obtain the optimal parameters. The thickness of cladding layer was about 6 mm. The sample of structure analysis was intercepted by wire cutting along the cross-section direction of the cladding layer. The sample of corrosion resistance analysis was cut along the direction perpendicular to the cross-section of the cladding layer.

2.3. Macro Morphology and Microstructure Analysis

The macro morphology of 2205 DSS cladding layer was observed by an Optical Microscope (OM, Oberkochen, Germany). It measured the melting width W, melting depth H and height h of the cladding layer. The dilution ratio was calculated according to the formula $\psi = H/(H + h)$. The schematic diagram was shown in Figure 1. The metallographic specimens with the size of 25 mm × 10 mm were taken along the cladding layer. After grinding and polishing, the sample was corroded for 10 s in Behara reagent. Behara reagent was composed of 88 ml H_2O + 12 ml HCl + 1.2 g $K_2S_2O_5$. The microstructure of cladding layer was observed by a ZEISS optical microscope (OM, Oberkochen, Germany) and a JSM-6480 scanning electron microscope (SEM, JEOL, Tokyo, Japan). The elements migration through the interface of the cladding layer and Q345 steel was tested by Energy Dispersive Spectrometer (EDS). The proportion calculation of austenite phase in the microstructure was obtained by Image Pro Analysis Software. The phase composition of cladding layer was analyzed by an X-ray diffractometer instrument (XRD-6000, Shimadzu, Kyoto, Japan) with Cu-K_α radiation and scanning angles (2θ) between 10° and 90°.

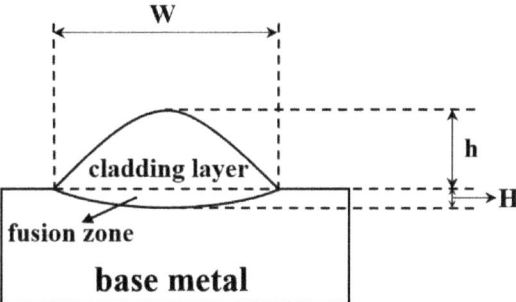

Figure 1. The schematic diagram of the dilution ratio calculation.

2.4. Electrochemical Measurements

The electrochemical measurements were carried out with the EGM283 Electrochemical Workstation with a three-electrode cell system, while the cladding layer acted as a working electrode, a saturated calomel electrode (SCE) as a reference electrode and a platinum plate as an auxiliary electrode. The specimens with dimensions of 10 mm × 10 mm × 5 mm for electrochemical tests, and the exposed measurement area was 10 × 10 mm retained by the epoxy resin. Before the experiment, the samples were polished with silicon carbide (SiC) emery papers down to 2000#, then ultrasonically cleaned in acetone and rinsed in distilled water. 3.5% NaCl solution was taken as the etching solution.

The potentiodynamic polarization curves were recorded with a scanning rate of 1 mV/s, starting from −0.5 V to 1.5 V. The electrochemical impedance spectroscopy (EIS) data were obtained with the frequency range from 100 kHz to 10 mHz and a sinusoidal potential perturbation of 5 mV at the open-circuit potentials.

The experimental results were interpreted based on an equivalent electrical circuit. The capacitance measurements on the passive films were performed with a fixed frequency of 1 kHz, and the potential range from 1.0 V to 1.0 V (vs. SCE). All measurements were carried out at an ambient temperature of approximately 25.

2.5. XPS Analysis

In order to explore the corrosion mechanism of cladding layer under different currents, the chemical composition of passive film on the surface of the sample was tested. The samples were placed in the constant passivation potential of 0.3 V and polarized for 2 h to form a stable passive film on the surface of cladding layer. The chemical composition of the cladding layer passive films was investigated by X-ray photoelectron spectrometer (XPS) with a monochromatic Al Ka radiation source and a hemispherical electron analyzer operated at the pass energy of 25 eV. The element composition and content were analyzed by comparing with the standard spectra of elements from XPS company's Perkin-Elmer data handbook and International Inc. XPS website. The fitting curve was performed with the commercial software of Avantage.

3. Results and discussion

3.1. Effect of Current on Macro Morphology and Microstructure of Cladding Layers

Dilution ratio and macro morphology are two important indexes to evaluate the formation quality of cladding layers [20]. Figure 2 shows the macro morphology of 2205 DSS cladding layer on the condition of different current. Figure 3 shows the corresponding dilution rate change curve of 2205 DSS cladding layer. When the current was 80 A, melting amount of wire was not enough to produce a full and continuous cladding layer, and some pits appeared on the surface of cladding layer. As the current was 100 A, a fine cladding layer formed. Its corresponding dilution ratio of cladding layer was 11.43%. When the current was increased to 110 A, the dilution ratio of cladding layer increased to 30.56%. The dilution rate of the cladding layer determines the utilization rate of wire or powder and the quality of the cladding layer. On the premise of ensuring the perfect formation of the cladding layer and a good combination between the cladding layer and the substrate, the lower the dilution rate is, the higher the utilization rate of wire or powder and the higher the performance of the cladding layer is. If the dilution ratio is too small, the bonding performance of transition zone between the cladding layer and the substrate is poor. If the dilution ratio is too high, the alloy elements of base metal will deeply dilute the alloy composition of cladding layer, which will damage the performance of cladding layer [21–23].

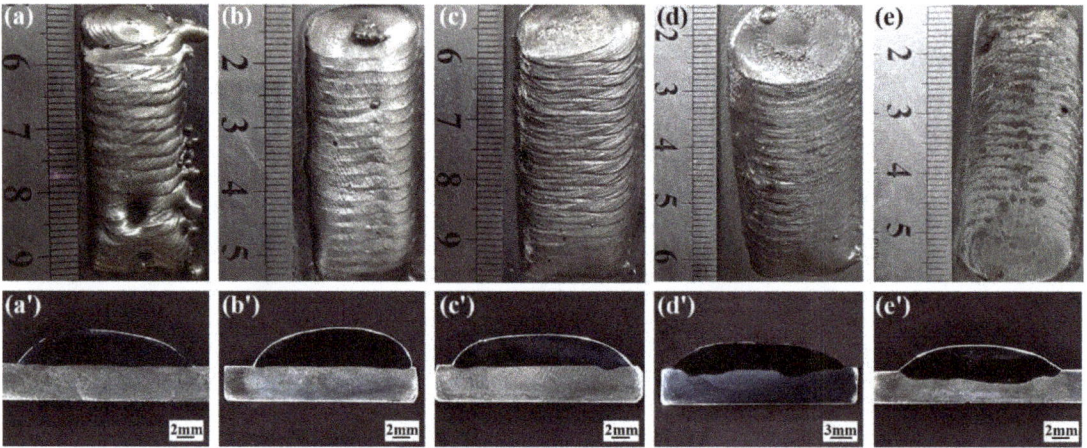

Figure 2. Macro morphology of 2205 DSS cladding layer under different current. (**a**,**a′**) 80 A; (**b**,**b′**) 90 A; (**c**,**c′**) 100 A; (**d**,**d′**) 110 A; (**e**,**e′**) 120 A.

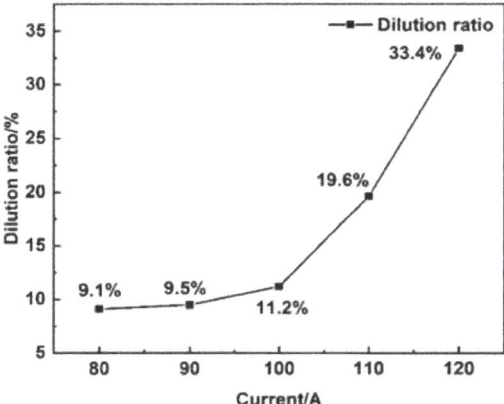

Figure 3. Variation curve on dilution ratio of 2205 DSS cladding layer under different current.

Figure 4 shows the SEM results of specimens under different current. The line scanning results show that the elements migration occur at the interface of the 2205 DSS cladding layer and the substrate, which suggests an excellent metallurgical combination between them. It can be seen from Figure 4a–d that with the increase of the current, the grain size increased.

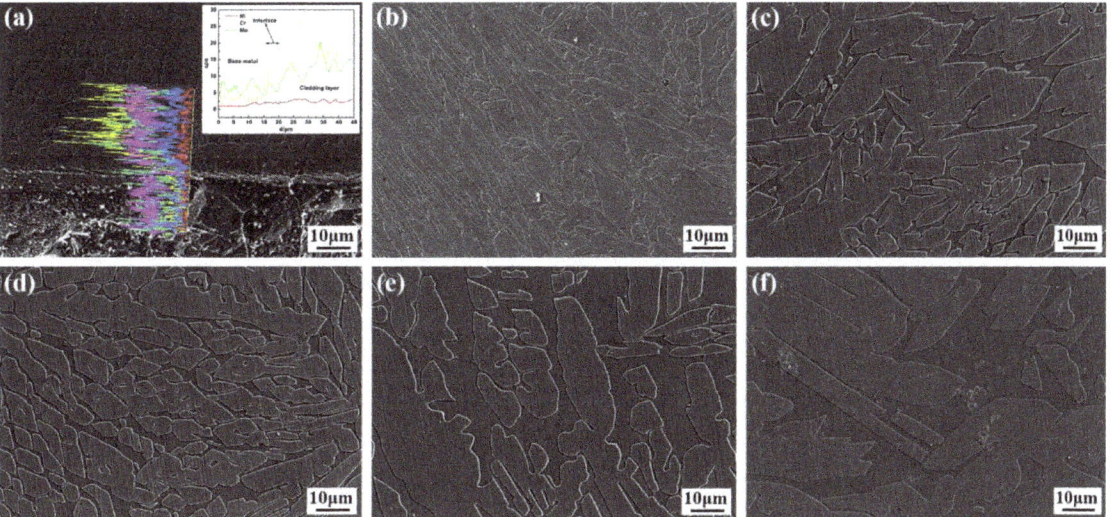

Figure 4. The SEM results of specimens under different current. (**a**) the Line Scanning result (80 A); (**b**) 80 A; (**c**) 90 A; (**d**) 100 A; (**e**) 110 A; (**f**) 120 A.

In order to analyze the microstructural composition in details, the cross-sectional microstructure of the specimen and the top microstructure of the cladding layer on the condition of different current and austenite/ferrite mass ratio in mirostructure are shown in Figure 5. The austenite phase was shown as white while the gray region was the ferrite phase.

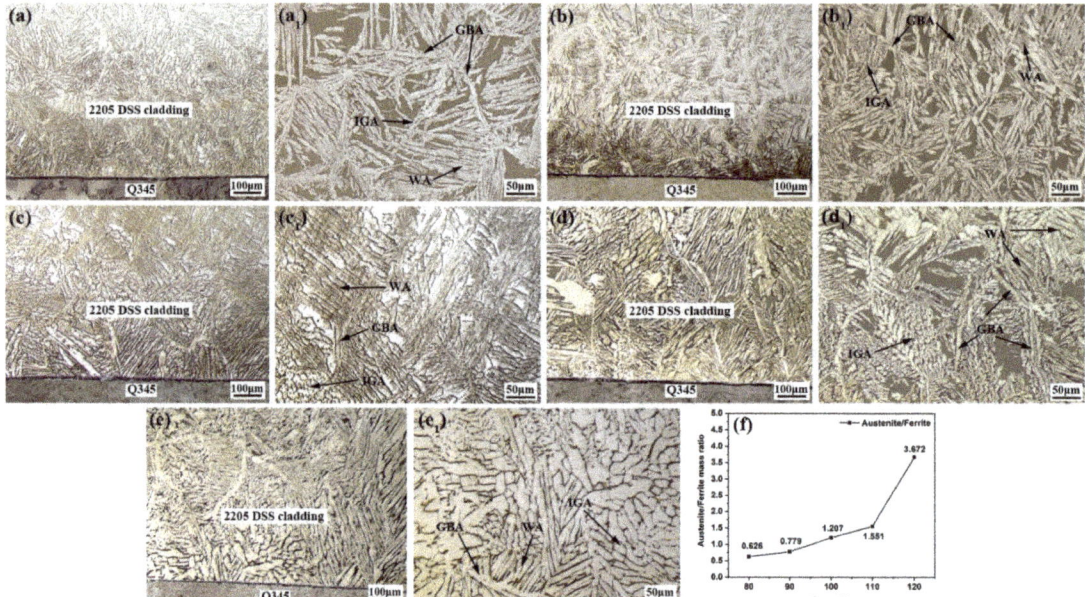

Figure 5. Microstructure of specimens under different current and Austenite/ Ferrite mass ratio. (**a**,**a₁**) 80 A; (**b**,**b₁**) 90 A; (**c**,**c₁**) 100 A; (**d**,**d₁**) 110 A; (**e**,**e₁**) 120 A; (**f**) Austenite/ Ferrite mass ratio in mirostructure.

When the current was 80 A (see Figure 5a,a₁), the austenite phase was composed of Grain-Boundary Austenite (GBA), Widmanstatten Austenite (WA) and Intragranular Austenite (IGA). The austenite phase and ferrite phase were relatively fine, with the average grain size of 63.50 μm and 21.80 μm, respectively. The austenite/ferrite mass ratio in microstructure was 0.626 (see Figure 5f). With the increase of current, the austenite/ferrite mass ratio continued to grow and the grain size of austenite and ferrite increased. As the current was increased to 100 A (see Figure 5c,c₁), it can be observed that a large amount of IGA formed in the microstructure. The austenite/ferrite mass ratio was 1.207 (see Figure 5f). The average grain size of austenite was 79.25 μm while that of ferrite was 44.81 μm. When the current was 110 A and 120 A (see Figure 5d,d₁,e,e₁), the austenite/ferrite mass ratio were 1.551 and 3.672, respectively (see Figure 5f). The average grain size of austenite were 98.52 μm and 119.70 μm while that of ferrite were 45.81 μm and 51.60 μm.

In general, the structure solidification of 2205 DSS cladding layer is ferrite mode [14,15]. At the beginning of solidification, ferrite is entirely formed. While the temperature falls below the solid solution line, the ferrite starts to transfer to austenite. Since austenitizing stable elements (such as C, Mn, Ni and so on) are easily enriched in the grain boundary and sub-grain boundary of ferrite, GBA preferentially grows along the grain boundary of ferrite. After GBA completely covering the grain boundary of ferrite, WA begins to grow along the direction perpendicular to the grain boundary of ferrite. Meanwhile, IGA nucleates and grows up in the grains of ferrite. As can be seen from the above figures, with the increase of current, the heat input increases, it means the transformation time from ferrite into austenite increases, so the amount of austenite in microstructure increases. Moreover, the content of austenitizing stable elements (such as Mn and Ni) in ER2209 wire is increased, which promote the formation of austenite. The microstructure is gradually coarsened with the increase of the current.

Figure 6 shows the XRD analysis results of 2205 DSS cladding layer with the current of 90–110 A. It also confirmed that the microstructure was mainly composed of the ferrite phase and the austenite phase.

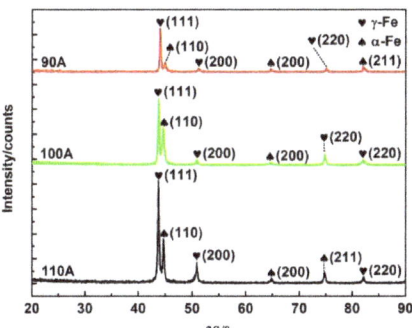

Figure 6. XRD results of 2205 DSS cladding layer under different current.

3.2. Analysis on Potentiodynamic Polarization Curve of 2205 DSS Cladding Layer under Different Current

In order to analyze the effect of current on the corrosion resistance of 2205 DSS cladding layer, an electrochemical corrosion test was carried out. The potentiodynamic polarization curves of cladding layer under different current are shown in Figure 7. On the condition of different current, the passive film formed in 3.5% NaCl solution was very stable. The passive potential of cladding layer was in the potential region from −0.03 to 0.40 VSCE (0.41 VSCE, 0.46 VSCE). At the initial stage of corrosion, the dissolution rate of passive film is close to its regeneration rate, so the passive film is in the equilibrium state of dissolution and regeneration [13]. With the increase of potential, the corrosion current increases and the dissolution rate of passive film increases. When the dissolution rate is greater than the regeneration rate, the passive film breaks down and the cladding layer is corroded.

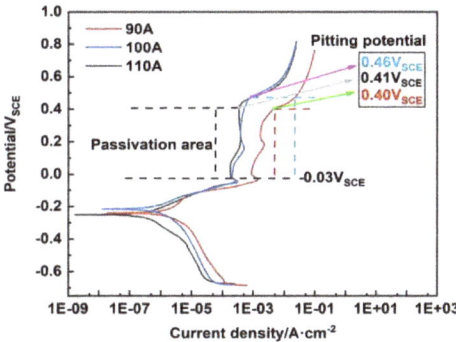

Figure 7. Potentiodynamic polarization curves of samples under different current.

The polarization curve was fitted by CView analysis software to obtain the self-corrosion potential and corrosion current density. The fitting results are shown in Table 2. The relationship between current and self-corrosion potential was 100 A (−0.213 V) > 90 A (−0.241 V) > 110 A (−0.251 V), and the relationship between current and corrosion current density was 100 A (1.95×10^{-7}) < 110 A (3.69×10^{-7}) < 90 A (3.09×10^{-6}). The self-corrosion potential represents the difficulty of corrosion of the cladding layer. The greater the self-corrosion potential, the less likely corrosion will occur. The corrosion current density represents the actual corrosion rate of cladding layer. The smaller the corrosion current density is, the smaller the corrosion rate is [24]. Therefore, when the current was 100 A, the self-corrosion potential of cladding layer was the maximum while the corrosion current density was the minimum, indicating that the corrosion resistance of cladding layer was optimal.

Table 2. Fitting parameters obtained by using CView analysis software to deal with the potentiodynamic polarization curves.

Current	Potential/V_{SCE}	Current Density/$A \cdot cm^{-2}$
90	−0.241	3.09×10^{-6}
100	−0.213	1.95×10^{-7}
110	−0.251	3.69×10^{-7}

3.3. Chemical Stability of Passive Film Formed on the Cladding Layer under Different Current

In order to investigate the relative stability of passive films formed on the surface of 2205 DSS cladding layer under different current, the samples were in 3.5wt.% NaCl etching solution for passivated 1h at 0.2 VSCE. Electrochemical impedance spectroscopy (EIS) measurements were carried out after passive film generation. Figure 8 displays the EIS results of three species passive film formed under the current of 90 A, 100 A and 110 A. It can be observed that the three species passive films have similar impedance characteristics from the Nyquist plots (Figure 8a), and there is a capacitive arc with different radius in the test frequency range. These curves present the influence of different current on the impedance behavior of passive film. As previously reported in the literature [25–28], the polarization resistance of passive film is relevant to the diameter of the semicircular arc in impedance measurement. The increase of the semicircular arc means the enhancement of the passive film stability. The results demonstrated that the radius of capacitive arc of passive film with the current of 100 A reaches up to the maximum, which means that the corrosion resistance of passive film is optimal.

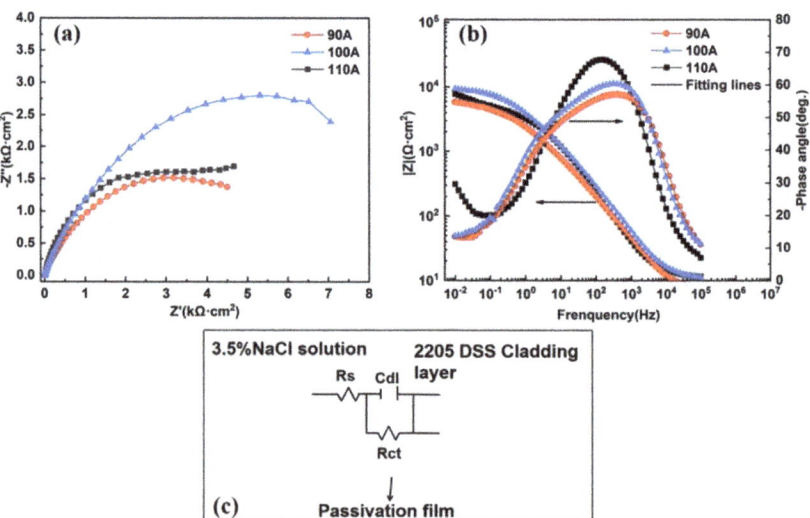

Figure 8. EIS date of passive film on the surface of 2205 DSS cladding layer under different current. (a) Nyquist plots; (b) Bode plots; (c) Equivalent circuit.

Figure 8b shows the Bode plots of passive films under different current. Figure 8c presents the equivalent circuit which is used to fit the impedance data of passive film. In this model [13], **R_s** represents the solution resistance, **R_{ct}** is the charge transfer resistance of passive film and C_{dl} is the corresponding double-layer capacitance. Table 3 lists the electrochemical fitting parameters based on the equivalent circuit displayed in Figure 8c. The dispersive exponent **n** means the deviation from the ideal capacitance, which always lies from 0.5 to 1. As shown in Table 3, with the increase of current, the R_{ct} value of passive film increased firstly and then decreased. It reached the maximum of 9078 $\Omega \cdot cm^{-2}$ with the current of 100 A, which suggested that the compactness and stability of passive film was the optimum.

Table 3. Fitting parameters of EIS results obtained from a proposed equivalent model.

Current	Rs ($\Omega \cdot cm^{-2}$)	Cdl ($F \cdot cm^{-2}$)	Rct ($\Omega \cdot cm^{-2}$)	n
90	10.3	3.50×10^{-5}	5443	0.84
100	7.8	4.77×10^{-5}	9078	0.92
110	4.7	8.12×10^{-5}	5203	0.81

3.4. XPS Results and Corrosion Mechanism of Cladding Layer

The excellent corrosion resistance of 2205 duplex stainless steel is due to the spontaneous formation of a dense passive film on the surface in a corrosive environment. In order to explore the corrosion resistance mechanism of 2205 duplex stainless steel cladding layer under different current, XPS analysis was undertaken to provide more information on the chemical composition of the passive films, finally the optimal process parameters were obtained.

The results of XPS spectra obtained for the samples passivated at 0.3 VSCE in 3.5% NaCl solution sputtered for 2 h are shown in Figure 9. The results indicated that the constitution of the three species passive film were similar even if the current was different. The observed spectra indicated the presence of Cr, Fe, Mo, Ni, O and C peaks. According to the peak strength, Cr, Fe, Mo and Ni dominate the components of the passive film. The peak intensity of the C element signal was small, so the C element may be an impurity element introduced in the preparation process.

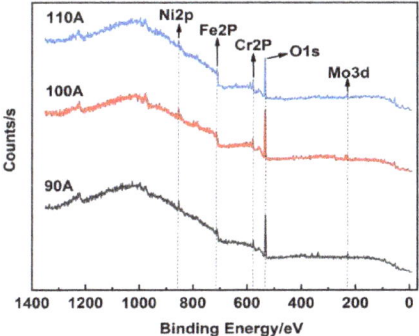

Figure 9. XPS spectra of passive film on the surface of 2205 DSS cladding layer after passivation for 2 h at 0.3 VSCE in 3.5% NaCl solutions.

Figures 10–14 present the Cr2p, Fe2p, Mo2p, Ni2p and O2p XPS spectra of three different passive films, respectively. It can be seen from Figure 10 that the signal of Cr2p performed primary three peaks, which represented the metal Cr (574.3 eV), Cr_2O_3 (576.4 eV) and $Cr(OH)_3$ (577.3 eV). The composition proportion of Cr, Cr_2O_3 and $Cr(OH)_3$ in the passive film of cladding layer was different on the condition of different currents. The content of Cr_2O_3 in the passive film increased firstly and then decreased with the increase of current. The content of Cr_2O_3 reached up to the maximum with the current of 100 A.

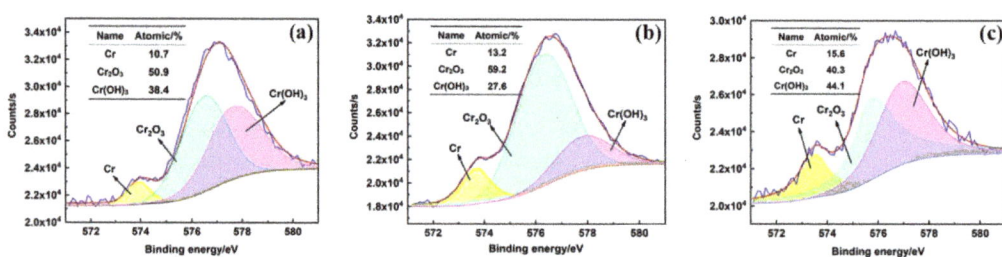

Figure 10. Cr2p peak fitting of 2205 DSS cladding layer under different current. (**a**) 90 A; (**b**) 100 A; (**c**)110 A.

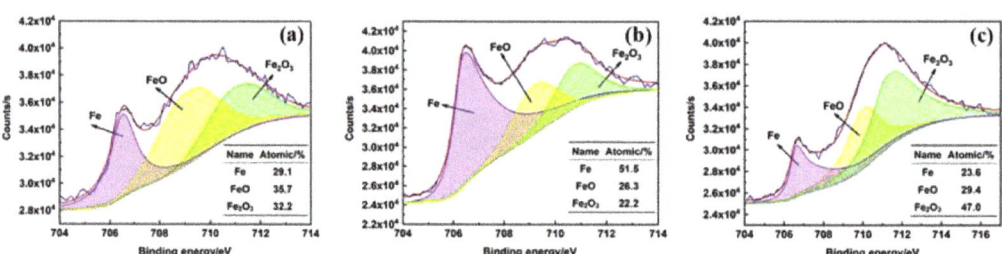

Figure 11. Fe2p peak fitting of 2205 DSS cladding layer under different current. (**a**) 90 A; (**b**) 100 A; (**c**) 110 A.

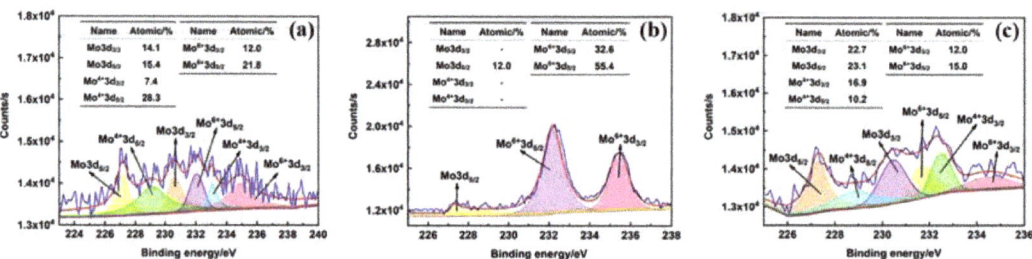

Figure 12. Mo2p peak fitting of 2205 DSS cladding layer under different current. (**a**) 90 A; (**b**) 100 A; (**c**) 110 A.

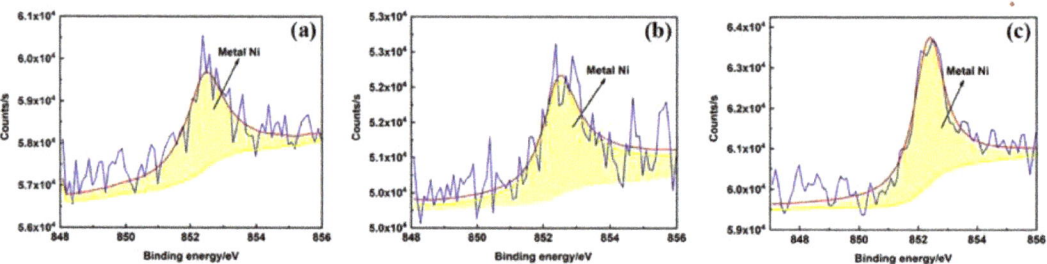

Figure 13. Ni2p peak fitting of 2205 DSS cladding layer under different current. (**a**) 90 A; (**b**) 100 A; (**c**) 110 A.

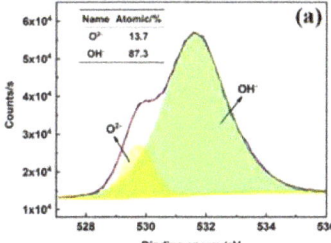 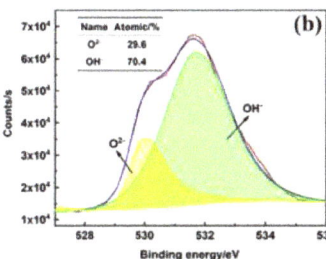 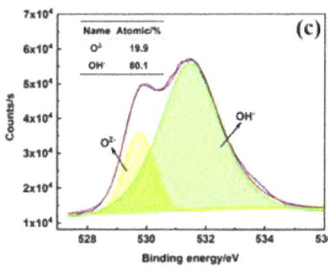

Figure 14. O2p peak fitting of 2205 DSS cladding layer under different current. (**a**) 90 A; (**b**) 100 A; (**c**) 110 A.

Figure 11 presents the iron profile performs three primary peaks: Fe (706.7 eV), FeO (709.3 eV) and Fe_2O_3 (711.3 eV), which indicates Fe^{2+} and Fe^{3+} are the main types of iron oxides in the passive film. The proportion of metal Fe was up to the maximum of 51.5% when the current was 100 A.

Figure 12 shows that the molybdenum profile peaks of metals Mo, Mo^{4+} and Mo^{6+} could be detected in the passive film. Obviously, the metal Mo can be clearly observed in the passive film, due to Mo being not easily oxidized such as Fe and Cr [29]. Mo can enhance the compactness of passive film so as to improve the corrosion resistance of cladding layer. Meanwhile, some literature has proved that the Mo element in passive film of 2205 DSS can prevent the adsorption of Cl^- to the surface of passive film and reduce the migration rate of Cl^- through the passive film [30]. As the current was 100 A, Mo^{6+} was the primary constituent of passive film.

Figure 13 shows that the nickel profile performs one characteristic peak of metal Ni (852.8 eV). The Ni element, which is an austenite forming element, can increase the self-corrosion potential to enhance the corrosion resistance of steel [31].

Figure 14 shows the spectra of the passive film formed in the O2p region. Oxygen species, O^{2-} and OH^- in passive film, play a role of connecting metal ions. O2p spectra are split into O^{2-} (530.2 eV) and OH^- (531.8 eV). It can be seen that OH^- and O^{2-} are the primary constituents of passive film, OH^- corresponds to the formation of $M_x(OH)_y$ (M-Metal), and O^{2-} corresponds to the forming of M_xO_y.

When the cladding layer sample was immersed in 3.5% NaCl solution, a corrosion mirocell was formed on the surface of the sample. The depolarization reaction of oxygen took place at the cathode of micro cell, as shown in formula (1). In contrast, the dissolution reaction of the metal matrix (Fe, Cr) occurred at the anode of micro cell, as shown in formula (2~9) [32,33]. Based on EET theory, the smaller the covalent electrons number of the strongest bond, the easier it is to form a new phase [32]. Many previous literatures has proved that the covalent electron number n_A of the strongest bond of Cr_2O_3 is 0.95284, while n_A of Fe_2O_3 is 1.12898. Obviously, the n_A of Cr_2O_3 is smaller. Therefore, OH^- in the corrosion solution is adsorbed to the surface of cladding layer, it preferentially reacts with Cr to form Cr_2O_3, and then it interacts with Fe to form Fe_2O_3 and FeO.

Cathode reaction:

$$O_2 + H_2O + 4e \rightarrow 4OH^- \tag{1}$$

Anode reaction for Cr:

$$Cr + H_2O \rightarrow Cr(OH)_{ads} + H^+ + 2e^- \tag{2}$$

$$Cr(OH)_{ads} + OH^- \rightarrow Cr(OH)^{2+} + 2e^- \tag{3}$$

$$Cr(OH)^{2+} + 2OH^- \rightarrow Cr(OH)_3 \tag{4}$$

$$2Cr(OH)_3 \rightarrow Cr_2O_3 + 3H_2O \tag{5}$$

Anode reaction for Fe:

$$Fe + H_2O \rightarrow Fe(OH)_{ads} + H^+ + e^- \tag{6}$$

$$Fe(OH)_{ads} + OH^- \rightarrow Fe(OH)_2 + e^- \tag{7}$$

$$Fe(OH)_2 + OH^- \rightarrow Fe(OH)_3 + e^- \tag{8}$$

$$2Fe(OH)_3 + 2OH^- \rightarrow Fe_2O_3 + FeO + 4H_2O + 2e^- \tag{9}$$

After the corrosion reaction, many active sites formed on the surface of the passive film, which resulted in the instability of oxide film [34]. However, Mo^{6+} formed the hydroxide of Mo at these active sites, which helped to create a stable passive film. Some literature shows that the presence of Mo^{6+} promotes the formation of Cr_2O_3 [35,36]. By comparison, different currents have no effect on the element composition of passive film. When the current was 100 A, chromium enrichment occurred in the oxide film protective layer, especially the proportion of Cr_2O_3 increased significantly. Meanwhile, the proportion of Mo^{6+} in the passive film was up to the maximum. Cr_2O_3 plays an essential role in the corrosion resistance of passive film [37]. Therefore, the optimal process parameters of 2205 DSS cladding layer obtained by plasma arc cladding technology are wire feeding speed of 29 mm/s, welding speed of 6 mm/s, ion gas flow of 1 L/min, gas flow of 20 L/min, Nozzle height of 10 mm and the current of 100 A.

4. Conclusions

The 2205 DSS cladding layer was prepared on the surface of a low carbon steel by plasma arc cladding technology. The effects of different current on the macro morphology, austenite/ferrite mass ratio and corrosion resistance of the cladding layer were studied. The main conclusions were drawn as followings:

(1) The current increased from 80 A to 120 A, the dilution ratio of cladding layer increased, and its macro morphology varied. When the current was 100 A, the forming of cladding layer was continuous, complete and fine with the dilution ratio of 11.43%.

(2) Regardless of the current, the microstructure of cladding layer was composed of austenite and ferrite. The mass ratio of austenite to ferrite in the microstructure increased with the increase of current. It was up to the optimum of 1.207 with the current of 100 A.

(3) When the current was 100 A, the self-corrosion potential of cladding layer was the maximum while the corrosion current density was the minimum, and the corrosion resistance of cladding layer was the optimum. The main reason was the existence of Mo^{6+} in the passive film, which stabilized the passive film and promoted the formation of Cr_2O_3.

Author Contributions: Conceptualization: J.P., W.L. and M.W.; methodology: P.X. and J.P.; formal analysis: J.P. and P.X.; resources: Y.S. and J.S.; data curation: P.X. and J.S.; writing—original draft preparation: J.P. and P.X.; writing—review and editing: J.P. and W.L.; writing—manuscript finalization: J.P.; W.L. and M.W. supervision: J.P. and W.L.; project administration: W.L. and Y.S.; funding acquisition: Y.S. and J.S. All authors have read and agreed to the published version of the manuscript.

Funding: This research was supported by State Key Laboratory of Advanced Brazing Filler Metals and Technology, the Natural Science Foundation of Jiangsu Province (Grant No. BK20191458), the Technology and Innovation Program for Undergraduates of Jiangsu University of Science and Technology, the Research and Practice innovation plan for Postgraduates in Jiangsu Province(No. SJCX 20-1457), and the open research foundation of the State Key Laboratory of Advanced Welding and Joining (No. AWJ-21M15).

Institutional Review Board Statement: Not applicable.

Informed Consent Statement: Not applicable.

Data Availability Statement: Not applicable.

Conflicts of Interest: The authors declare no conflict of interest. The funder had no role in the design of the study; in the collection, analyses, or interpretation of data; in the writing of the manuscript; or in the decision to publish the results.

References

1. Fu, H.W.; Dönges, B.; Krupp, U.; Pietsch, U.; Fritzen, C.-P.; Yun, X.B.; Christ, H.-J. Evolution of the residual stresses of types I, II, and III of duplex stainless steel during cyclic loading in high and very high cycle fatigue regimes. *Int. J. Fatigue* **2020**, *142*, 105972. [CrossRef]
2. Yuya, H.; Kobayashi, R.; Otomo, K.; Yabuuchi, K.; Komura, A. Microstructure and mechanical properties of HAZ of RPVS clad with duplex stainless steel. *J. Nucl. Mater.* **2021**, *545*, 152756. [CrossRef]
3. Zhang, Z.Q.; Jing, H.Y.; Xu, L.Y.; Zhang, T.G.; Xu, Y.T. Research progress on microstructure and properties of welded joint of ferrite/austenite duplex stainless steel. *Trans. Mater. Heat Treat.* **2020**, *41*, 13–27.
4. Rajesh-Kannan, A.; Siva-Shanmugam, N.; Rajkumar, V.; Vishnukumar, M. Insight into the microstructural features and corrosion properties of wire arc additive manufactured super duplex stainless steel (ER2594). *Mater. Lett.* **2020**, *270*, 127680. [CrossRef]
5. Dong, C.F.; Luo, H.; Xiao, K.; Sun, T.; Liu, Q.; Li, X.G. Effect of temperature and Cl^- concentration on pitting of 2205 duplex stainless steel. *J. Wuhan Univ. Technol. Mater. Sci. Ed.* **2011**, *26*, 641–647. [CrossRef]
6. Khattak, M.A.; Zaman, S.; Kazi, S.; Ahmed, H.; Habib, H.M.; Ali, H.M.; Tamin, M.N. Failure investigation of welded 430 stainless steel plates for conveyor belts. *Eng. Fail. Anal.* **2020**, *116*, 104754. [CrossRef]
7. Pu, J.; Li, Z.P.; Hu, Q.X.; Wang, Y.X. Effect of heat treatment on microstructure and wear resistance of high manganese steel surfacing layer. *Int. J. Mod. Phys. B* **2019**, *33*, 1940035. [CrossRef]
8. Wu, L.; Pu, J.; Wu, M.F.; Long, W.M.; Zhong, S.J.; Hu, Q.X.; Lan, Y. Effects of different tungsten carbide contents on microstructure and properties of Ni-based tungsten carbide cladding layer by plasma arc cladding technology. *Mater. Rep.* **2021**, *35*, 16111–16114, 16119.
9. Jin, M.; He, D.Y.; Wang, Z.J.; Zhou, Z.; Wang, G.H.; Li, X.X. Microstructure and properties of laser cladded 2205 dual-phase stainless steel/TiC composite coatings. *Laser. Optoelectron. P* **2018**, *055*, 285–290.
10. Pu, J.; Rao, J.W.; Shen, Y.F.; Liu, C.; Zhang, L.; Wang, Y.X. Effect of Ti element on the microstructure and properties of high chromium surfacing layer. *Int. J. Mod. Phys. B* **2020**, *34*, 2040028. [CrossRef]
11. Chen, L.; Richter, B.; Zhang, X.Z.; Ren, X.D.; Frank, E.; Pfefferkorn, B. Modification of surface characteristics and electrochemical corrosion behavior of laser powder bed fused stainless-steel 316L after laser polishing. *Addit. Manuf.* **2020**, *32*, 101013. [CrossRef]
12. Liu, S.; Fang, W.P.; Yi, Y.Y.; Deng, Y.L. Study on microstructure and properties of flux cored twin-wire CMT welded joint of 2205 duplex stainless steel. *H. Work. Technol.* **2020**, *49*, 16–19, 24.
13. Liu, Y.; Bao, Y.F.; Song, Q.N.; Jiang, Y.F. Influence of sensitization on pitting corrosion in surfacing layer of 2209 duplex stainless steel. *Trans. China Weld. Inst.* **2020**, *41*, 33–38, 98–99.
14. Brunner-Schwer, C.; Petrat, T.; Graf, B.; Rethmeier, M. Highspeed-plasma-laser-cladding of thin wear resistance coatings: A process approach as a hybrid metal deposition-technology. *Vacuum* **2019**, *166*, 123–126. [CrossRef]
15. Ding, H.; Dai, J.W.; Dai, T.; Sun, Y.W.; Lu, T.; Jia, X.J.; Huang, D.S. Effect of preheating/post-isothermal treatment temperature on microstructures and properties of cladding on U75V rail prepared by plasma cladding method. *Surf. Coat. Technol.* **2020**, *399*, 126122. [CrossRef]
16. Huang, J.K.; Liu, S.; Yu, S.R.; An, L.; Yu, X.Q.; Fan, D.; Yang, F.Q. Cladding Inconel 625 on cast iron via bypass coupling micro-plasma arc welding. *J. Manuf. Process.* **2020**, *56*, 106–115. [CrossRef]
17. Cui, S.; Pang, S.; Pang, D.; Zhang, Z. Influence of Welding Speeds on the Morphology, Mechanical Properties, and Microstructure of 2205 DSS Welded Joint by K-TIG Welding. *Materials* **2021**, *14*, 3426. [CrossRef] [PubMed]
18. Rodriguez, B.R.; Miranda, A.; Gonzalez, D. Maintenance of the Austenite/Ferrite Ratio Balance in GTAW DSS Joints Through Process Parameters Optimization. *Materials* **2020**, *13*, 780. [CrossRef]
19. Hany, S.; Asiful, H.S.; Jabair, A.M.; Tauriq, U. Ameliorative Corrosion Resistance and Microstructure Characterization of 2205 Duplex Stainless Steel by Regulating the Parameters of Pulsed Nd:YAG Laser Beam Welding. *Metals* **2021**, *11*, 1206.
20. Syed, W.U.H.; Pinkerton, A.J.; Li, L. A comparative study of wire feeding and powder feeding in direct diode laser deposition for rapid prototyping. *Appl. Surf. Sci.* **2005**, *247*, 268–276. [CrossRef]
21. Lin, C.M. Parameter optimization of laser cladding process and resulting microstructure for the repair of tenon on steam turbine blade. *Vacuum* **2015**, *115*, 117–123. [CrossRef]
22. Wang, H.; Woo, W.; Kim, D.-K.; Em, V.; Lee, S.Y. Effect of chemical dilution and the number of weld layers on residual stresses in a multi-pass low-transformation-temperature weld. *Mater. Des.* **2018**, *160*, 384–394. [CrossRef]
23. Hemmati, I.; Ocelik, V.; De-Hosson, J.T.M. Dilution effects in laser cladding of Ni-Cr-B-Si-C hardfacing alloys. *Mater. Lett.* **2012**, *84*, 69–72. [CrossRef]
24. Kang, D.H.; Lee, H.W. Study of the correlation between pitting corrosion and the component ratio of the dual phase in duplex stainless steel welds. *Corros. Sci.* **2013**, *74*, 396–407. [CrossRef]
25. Abreu, C.M.; Cristóbal, M.J.; Losada, R.; Nóvoa, X.R.; Pena, G.; Pérez, M.C. The effect of Ni in the electrochemical properties of oxide layers grown on stainless steels. *Electrochim. Acta* **2006**, *51*, 2991–3000. [CrossRef]

26. Guiñón-Pina, V.; Igual-Muñoz, A.; García-Antón, J. Influence of pH on the electrochemical behaviour of a duplex stainless steel in highly concentrated LiBr solutions. *Corros. Sci.* **2011**, *53*, 575–581. [CrossRef]
27. Jin, T.Y.; Cheng, Y.F. In situ characterization by localized electrochemical impedance spectroscopy of the electrochemical activity of microscopic inclusions in an X100 steel. *Corros. Sci.* **2011**, *53*, 850–853. [CrossRef]
28. Boissy, C.; Alemany-Dumont, C.; Normand, B. EIS evaluation of steady-state characteristic of 316L stainless steel passive film grown in acidic solution. *Electrochem. Commun.* **2013**, *26*, 10–12. [CrossRef]
29. Montemor, M.F.; Simões, A.M.P.; Ferreira, M.G.S.; Belo, M.D.C. The role of Mo in the chemical composition and semiconductive behaviour of oxide films formed on stainless steels. *Corros. Sci.* **1999**, *41*, 17–34. [CrossRef]
30. Zhang, B.; Wei, X.X.; Wu, B.; Wang, J.; Shao, X.H.; Yang, L.X.; Zheng, S.J.; Zhou, Y.T.; Jin, Q.Q.; Oguzie, E.E.; et al. Chloride attack on the passive film of duplex alloy. *Corros. Sci.* **2019**, *154*, 123–128. [CrossRef]
31. Potgieter, J.H.; Olubambi, P.A.; Cornish, L.; Machio, C.N.; Sherif, E.-S.M. Influence of nickel additions on the corrosion behaviour of low nitrogen 22% Cr series duplex stainless steels. *Corros. Sci.* **2008**, *50*, 2572–2579. [CrossRef]
32. Wang, Y.; Li, C.F.; Lin, Y.H. Electronic theoretical study of the influence of Cr on corrosion resistance of Fe-Cr alloy. *Acta Metall. Sin.* **2017**, *53*, 622–630.
33. Li, F.; Wu, Y.S.; Yang, D.J.; Zhang, W.Q. The passive film and pitting propagation models of Fe-Cr alloy. *J. Beijing Univ. Iron Steel Technol.* **1986**, *1*, 125–134.
34. Cheng, X.Q.; Wang, Y.; Li, X.G.; Dong, C.F. Interaction between austen-ferrite phases on passive performance of 2205 duplex stainless steel. *J. Mater. Sci. Technol.* **2018**, *34*, 2140–2148. [CrossRef]
35. Betova, I.; Bojinov, M.; Karastoyanov, V.; Kinnunen, P.; Saario, T. Estimation of kinetic and transport parameters by quantitative evaluation of EIS and XPS data. *Electrochim. Acta* **2010**, *55*, 6163–6173. [CrossRef]
36. Pardo, A.; Merino, M.C.; Coy, A.E.; Viejo, F.; Arrabal, R.; Matykina, E. Pitting corrosion behaviour of austenitic stainless steels–combining effects of Mn and Mo additions. *Corros. Sci.* **2008**, *50*, 1796–1806. [CrossRef]
37. Abreu, C.M.; Cristóbal, M.J.; Losada, R.; Nóvoa, X.R.; Pena, G.; Pérez, M.C. High frequency impedance spectroscopy study of passive films formed on AISI 316 stainless steel in alkaline medium. *J. Electroanal. Chem.* **2004**, *572*, 335–345. [CrossRef]

Article

Solid Energetic Material Based on Aluminum Micropowder Modified by Microwave Radiation

Andrei Mostovshchikov [1,2], Fedor Gubarev [3,*], Pavel Chumerin [1], Vladimir Arkhipov [4], Valery Kuznetsov [4] and Yana Dubkova [4]

[1] School of Nuclear Science & Engineering, National Research Tomsk Polytechnic University, 634050 Tomsk, Russia; avmost@tpu.ru (A.M.); chumerinpy@tpu.ru (P.C.)
[2] Department of Electronic Technique, Tomsk State University of Control Systems and Radioelectronics, 634050 Tomsk, Russia
[3] Research School of Chemistry & Applied Biomedical Sciences, National Research Tomsk Polytechnic University, 634050 Tomsk, Russia
[4] Scientific Research Institute of Applied Mathematics and Mechanics, National Research Tomsk State University, 634050 Tomsk, Russia; leva@niipmm.tsu.ru (V.A.); dv_1@mail.ru (V.K.); y.a.dubkova@niipmm.tsu.ru (Y.D.)
* Correspondence: gubarevfa@tpu.ru

Abstract: The paper discusses the application of pulsed microwave radiation for the modification of crystalline components of a high-energy material (HEsM). The model aluminized mixture with increased heat of combustion was studied. The mixture contained 15 wt.% aluminum micron powder, which was modified by microwave irradiation. It was found that the HEM thermogram has an exo-effect with the maximum at 364.3 °C. The use of a modified powder in the HEM composition increased the energy release during combustion by 11% from 5.6 kJ/g to 6.2 kJ/g. The reason for this effect is the increase in the reactivity of aluminum powder after microwave irradiation. In this research, we confirmed that the powders do not lose the stored energy, even as part of the HEM produced on their basis. A laser projection imaging system with brightness amplification was used to estimate the speed of combustion front propagation over the material surface. Measurement of the burning rate revealed a slight difference in the burning rates of HEMs based on irradiated and non-irradiated aluminum micropowders. This property can be demanded in practice, allowing a greater release of energy while maintaining the volume of energetic material.

Keywords: aluminum micron powder; microwave radiation; energetic material; aluminized high-energy material; high-speed imaging; laser monitor

Citation: Mostovshchikov, A.; Gubarev, F.; Chumerin, P.; Arkhipov, V.; Kuznetsov, V.; Dubkova, Y. Solid Energetic Material Based on Aluminum Micropowder Modified by Microwave Radiation. *Crystals* **2022**, *12*, 446. https://doi.org/10.3390/cryst12040446

Academic Editor: Yang Zhang

Received: 27 February 2022
Accepted: 21 March 2022
Published: 23 March 2022

Publisher's Note: MDPI stays neutral with regard to jurisdictional claims in published maps and institutional affiliations.

Copyright: © 2022 by the authors. Licensee MDPI, Basel, Switzerland. This article is an open access article distributed under the terms and conditions of the Creative Commons Attribution (CC BY) license (https://creativecommons.org/licenses/by/4.0/).

1. Introduction

Nowadays, many studies are devoted to investigating aluminum powders and high-energy materials on their basis [1–10]. One of the main research tasks is to increase the specific heat effect of the oxidation of aluminum powders. The solution to this problem is possible with the use of high-energy effects: gamma radiation, neutron flux, electric explosion as a method for producing aluminum nanopowders, etc. [10–12]. An increase in the specific thermal effect of oxidation of metal powders (in particular, aluminum) is explained by the generation and accumulation of defects in the crystal structure of the particles, as well as a change in the configuration of atoms at the grain boundaries and the surface of the particles and is called 'stored energy' [11,12]. In particular, in aluminum nanopowders, through such influences, it is possible to increase the specific thermal effect of oxidation by 2.5 kJ/g [13]. In our previous study, we established that aluminum powder significantly increases the thermal oxidation effect (~2.4 kJ/g) after irradiation with the electron beam [14] and exposure to microwave (MW) radiation [15]. The use of MW radiation for these purposes is more expedient and environmentally friendly due to the relative safety of working with MW radiation, compared to the use of gamma radiation, high-energy electron beams, or neutrons. Other researchers have demonstrated that continuous MW radiation makes it possible to control the combustion process of high-energy materials [16]. At the same time, high-power short-pulse MW radiation makes it possible to generate defects in the crystal structure and on the surface of metal particles but not to ignite them [15]. Such an accumulation of particle structure defects can be useful for their application in high-energy compositions [12,13]. Therefore, it is necessary to determine whether it is possible to improve the energy properties of high-energy materials containing irradiated micron aluminum powders.

The aim of this work is to prove that MW irradiation is effective in increasing the exothermal combustion effect of aluminized energetic material based on aluminum micron powder. At the same time, it is important to understand whether the burning rate of aluminized high-energy material changes when components irradiated with MW radiation are used in its composition. Therefore, the aim of the work is also to estimate the burning rate by visualizing the propagation of the combustion wave front on the surface of the energetic material.

2. Experimental Technique and Material Characteristics

In the experiments, we used the aluminum micron powder produced by spraying liquid aluminum in an inertial medium. The microscopic image of the aluminum micron powder obtained by the JEOL JEM-2100F transmission electron microscope is presented in Figure 1. The particle size distribution was near lognormal with a maximum of 3.5 µm. The particle size distribution was measured using a Shimadzu nanoparticle size distribution analyzer (SALD-7101). The powder contained less than 90 wt.% of active aluminum. The exothermal oxidation effect of the micron powder in air was 5.43 kJ/g; the starting temperature of oxidation was 354.8 °C.

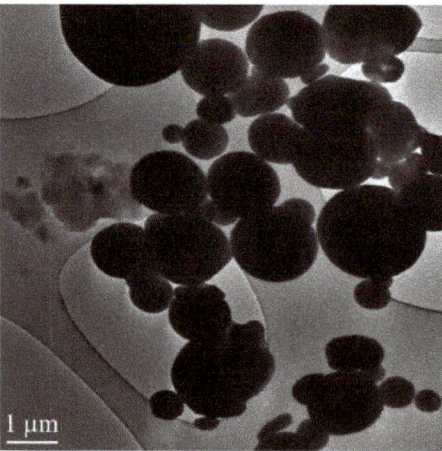

Figure 1. Microscopic image of the aluminum micron powder.

The elementary composition of the micron powder was determined by neutron activation analysis (NAA) [17]. The activation was carried out with a thermal neutrons flux of 5×10^{13} neutrons/cm^2·s in a 'dry' vertical channel of an IRT-T research nuclear reactor (Tomsk Polytechnic University, Tomsk, Russia). The analysis determined that the powder contained 14 different elements (Table 1). According to Table 1, the total impurity content in the micron powder was lower than 1.6 wt.%. Therefore, the oxidation of impurities did not make a significant contribution to the specific thermal effect of oxidation.

Table 1. Results of neutron activation analysis of aluminum micron powders.

	Element	Content, (μg/g)	Standard Enthalpy of Metal Oxidation, $-H^0$, [kJ/mol]
1	Fe	1127.1934	822.2
2	Ga	265.6707	1089.0
3	Na	114.4161	513.2
4	Zn	32.4767	350.6
5	Cr	11.6218	1140.6
6	Ce	1.9393	125.8
7	Co	1.1021	239.3
8	Mo	0.7081	745.2
9	Sb	0.6379	1007.5
10	Hf	0.2700	1175.5
11	U	0.2449	1224.0
12	W	0.1821	842.9
13	Sc	0.0947	1908.6
14	Sm	0.0493	1822.6

The aluminum micron powder was irradiated by pulsed MW radiation with a frequency of 2.85 GHz and an average power density of 8 kW/cm^2. The pulse duration was 25 ns and the pulse rate was 400 Hz. The choice of these parameters was due to the requirement that exposure to MW radiation should not lead to sintering or ignition of the powder. Thus, to initiate powder activation processes, a pulsed exposure mode was required, the parameters of which were selected experimentally. The S-band with a frequency exceeding the frequency of household microwave ovens (2.45 GHz), at which the maximum effect on water molecules occurs, was chosen as the frequency range of exposure. The pulse duration and repetition rate were selected experimentally. They were chosen as minimally possible, at which a corona discharge occurs between the particles (Video S1 in Supplementary Materials), but the particles do not sinter and significantly heat up. Irradiation was carried out in air. The powder temperature during irradiation was monitored by an IR camera and did not exceed 40 °C. Thus, it was insignificant and did not lead to the melting and sintering of the powders. We used the facility for metal powder irradiation in air by MW radiation presented in our previous work [15]. The differential thermal analysis (DTA) [18] of the samples was carried out using an SDT Q600 thermogravimetric analyzer (TA Instruments, New Castle, DE, USA). The experimental study of the burning velocity and combustion heat was conducted to analyze the MW irradiation influence on the burning characteristics of aluminized blended HEM. A similar model composition was used earlier in [8]. The model HEM was obtained by mixing aluminum micron powder with ammonium perchlorate and polymer binder. The mixtures were compacted and dried up. Samples of aluminized blended HEM contained 15.8% inert burning binder (trademark SKDM-80), 69.2% ammonium perchlorate, and 15% non-irradiated aluminum powder (mixture A) and irradiated (mixture B). This percentage composition is optimal for model studies [1]. The cylindrical samples were produced by continuous compaction with a compaction pressure ~2×10^5 Pa. The hardening was carried out for 24 h at a temperature of 20 °C. The samples were 30–35 mm high and 10 mm in diameter. Their density was 1.70–0.01 g·cm^{-3} and their mass was 1.7–2.0 g.

The complexity of visualizing the surface of burning high-energy materials is caused by the intense glow and scattering of combustion products, which requires the use of the 'through the flame' imaging techniques [19,20]. The energetic composition studied in this work belongs to such materials. In the process of combustion, the products scattered, the samples completely lost their shape. The linear burning rate in the air under atmospheric pressure was measured by high-speed laser monitoring. The laser monitor is a laser projection microscope modified with a high-speed camera [21]. The technique of burning rate measurement for aluminum nanopowder mixtures and HEM using high-speed laser monitoring was discussed previously in [20]. Figure 2 presents the laser monitor scheme.

The image in the laser monitor was formed by a system consisting of a concave mirror and a focusing lens, providing a distance to the object of the study equal to 50 cm. The optical scheme allowed us to obtain an observation area with a diameter of ~6 mm and a spatial resolution of 25 µm. The laser monitor was built based on the laboratory-made copper bromide brightness amplifier with a gas-discharge tube aperture of 1.5 cm and a length of the active area of 50 cm. The tube generated 20 µJ pulses of amplified spontaneous emission at 510.6 nm wavelengths and ~0 µJ at 578.2 nm operating at 20 kHz pulse repetition rate [22]. The laser emission of the imaging system had no visible effect on the object of study. The images were recorded using a high-speed digital camera Phantom Miro C110 allowing 900 frms/sec imaging rate with 1280 × 1024 pixels resolution. The surface area of 4.5 × 3.6 mm² was visualized.

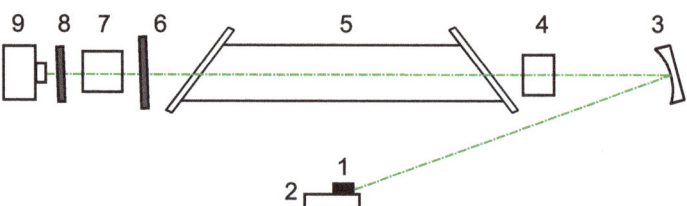

Figure 2. Scheme of the laser monitor. 1—Sample; 2—stage; 3—concave mirror; 4—lens; 5—brightness amplifier; 6 and 8—filters; 7—lens; 9—high-speed camera.

A calorimetric unit in Figure 3 was used to measure the heat of combustion. The sample was placed in a crucible, and then it was burned in the calorimeter-burning chamber, surrounded by water, under an initial oxygen pressure of 2 MPa. Then, the water temperature in the calorimeter chamber was determined. The heat of combustion of the material in the calorimeter-burning chamber was calculated based on the measurement of the water temperature in the calorimeter. The heat of combustion is given by:

$$Q = (C\Delta T \pm \Delta Q)/m \qquad (1)$$

where C—the heat capacity of water; ΔT—the change in the water temperature in the calorimetric chamber; ΔQ—the correction factor to take into account the heat exchange with the environment, which characterizes heat losses and does not exceed 1%; and m—the mass of the sample.

The effect of irradiation conditions on the content of metallic Al in the samples was studied by measuring the volume of hydrogen released during the interaction of a powder sample with an alkali solution [23]. The measuring unit consisted of a flask (250 mL), a burette (50 mL) and an equalizing funnel filled with an alkali solution and a saturated NaCl solution, respectively. When Al interacts with an alkali solution, gaseous hydrogen is released, which makes it possible to fix the volume of gas with a burette and calculate the content of metallic Al in the sample from the following chemical reaction:

$$2Al + 2NaOH + 6H_2O = 2Na[Al(OH)_4] + 3H_2 \qquad (2)$$

It has been previously suggested that the change in the thermochemical parameters of aluminum powders after irradiation is associated with the partial destruction of the particle oxide shell. This assumption is confirmed by studies of the MW radiation effect on the structure and phase transitions in aluminum oxide and hydroxide [24–26]. In this work, we carried out model studies on the effect of MW radiation on an amorphous sample of Al oxyhydroxide, which, in its phase composition, imitates the composition of the surface of aluminum powder particles. Al oxyhydroxide was obtained by precipitation from an aqueous mixture of aluminum chloride ($AlCl_3 \cdot 6H_2O$) with an ammonia solution (25%). After that, the precipitate was dried at room temperature for 24 h and then calcined at 250 °C in air for 4 h.

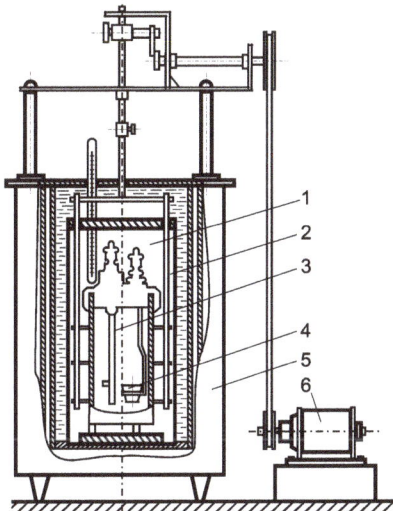

Figure 3. Scheme of the calorimetric unit. 1—Chamber; 2—agitator; 3—ignition device; 4—crucible with sample; 5—calorimeter body; 6—electric motor.

3. Results and Discussion

We applied a differential thermal analysis to the irradiated samples of the aluminum micron powder that were heated in air up to 1250 °C [14,15,17]. The advantage of the DTA method is the ability to study the state of the oxide layer on the surface of metal particles and determine the dependence of the properties of the particles on the state of the layer [27]. The specific thermal effect of their oxidation was 5.43 kJ/g before MW irradiation. Table 2 presents the DTA results for the irradiated aluminum micron powder with different MW exposure duration. According to the experimental results, the thermal oxidation effect of the aluminum micron powder depends on the time of irradiation. After irradiation for 15 s (optimal), the specific thermal effect of their oxidation increased by 56% and reached 8.48 kJ/g (sample #4). Therefore, sample #4 was chosen for further analyses and model HEM production.

Table 2. Thermochemical characteristics of the aluminum micron powders after pulsed MW irradiation.

	Microwave Exposure Durations	Exothermal Oxidation Effect, kJ/g	Oxidation Start Temperature, °C	Weight Gain Due to Oxidation, %
1	0	5.430	354.85	57.5
2	5	5.391	455.99	58.8
3	10	8.077	405.63	57.8
4	15	8.480	385.31	68.2
5	20	7.156	445.81	61.1
6	25	8.152	443.39	59.4
7	30	7.961	429.72	59.5

The aluminum powder did not change the morphology of the particles after irradiation. The particle size distribution before and after MW irradiation remained virtually unchanged (Figure 4). Aluminum micropowder has a bimodal distribution in the submicron and micron regions. In this case, under MW irradiation, the distribution maximum in the submicron region is shifted towards smaller sizes. This is probably a consequence of the deagglomeration of particles under the action of microwaves.

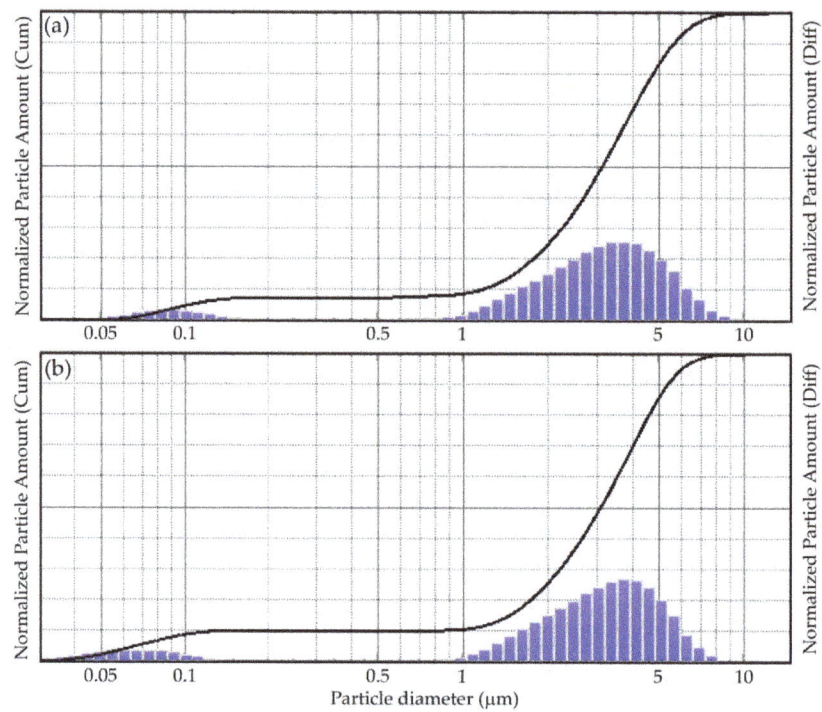

Figure 4. Particle size distributions of non-irradiated (**a**) and MW irradiated (**b**) micron powders.

Table 3 presents the results of measurements of the content of metallic aluminum by the volumetric method at different exposure times to MW radiation. The data obtained demonstrate that after exposure to MW radiation, the irradiated samples contained a metallic Al phase, which was absent in the non-irradiated sample. This effect is explained by the reduction of Al oxide in the oxide–hydroxide shell of particles with the formation of metallic Al.

Table 3. Content of metallic aluminum after MW irradiation.

	Microwave Exposure Duration, s	wt. (Al), %
1	0	90.0
2	5	91.0
3	10	91.5
4	15	93.0
5	20	91.5

Figure 5 presents the results of transmission electron microscopy (TEM) of Al oxyhydroxide samples before and after exposure to MW radiation, as well as the electron diffraction patterns of the samples obtained during microscopic examination. As a result of exposure to MW radiation in the amorphous matrix of Al oxyhydroxide, spheroidal structures up to 100 nm in size are formed, which have a higher density compared to the density of the original material. The position of the reflections on the electron diffraction pattern corresponds to the interplanar distances characteristic of the crystal lattice of metallic aluminum. Thus, there is a reviving of metallic aluminum from aluminum oxyhydroxide because of exposure to MW radiation.

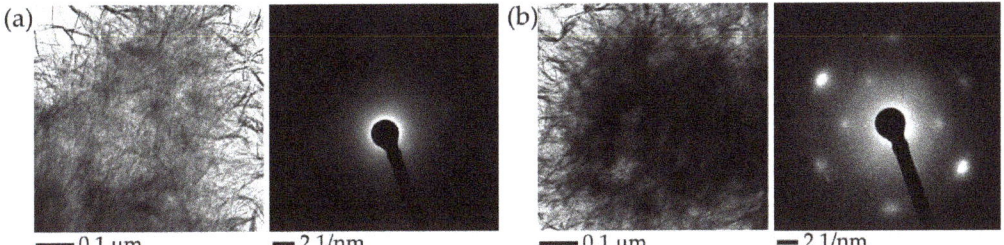

Figure 5. TEM micrographs and electron diffraction patterns of Al oxyhydroxide (**a**) before and (**b**) after MW irradiation.

The results of the study of the content of metallic Al and the results of TEM confirm the conclusion that the MW irradiation of aluminum powder increases the reactivity due to an increase in the oxide shell permeability and oxidation rate. Therefore, the partial radiation energy is stored by defects in the particle structure [11,13].

Thermograms of the model HEM were obtained with non-irradiated aluminum micron powder (Figure 6) and the same powder (Figure 7) after MW irradiation in air. According to the DTA results, a sample with irradiated aluminum powder (Figure 7) oxidized much faster than a sample with non-irradiated aluminum (Figure 6) at ~300 to 650 °C. This is evidenced by a smoother and more extended exo-effect on the thermogram in Figure 6 (indicated by 1) compared to the 'needle-like' exo-effect shape in Figure 7 (indicated by 1). The most likely cause of this 'needle-like' shape is the increase in the reactivity of the powder after MW irradiation. A faster oxidation of aluminum particles occurs due to an increase in the permeability of the oxide shell for the oxidizer [15]. In addition, the thermogram of the oxidation of the mixture with irradiated aluminum powder demonstrates the exo-effect in the temperature range from ~680 to ~1150 °C (indicated by 2). The increase in the heat of oxidation is probably due to a decrease in the agglomeration of the powder, the partial reduction of aluminum in the shell, and an increase in the content of unoxidized aluminum in the powder. This experimentally detected increase in the heat of oxidation after irradiation with short-pulse MW radiation is comparable in magnitude to the effect (~2.5 kJ/g) for aluminum nanopowder after irradiation with gamma radiation and electron beams in [13].

The burning rate was studied in this work with the samples of the HEM having the shape of a cylinder with a diameter of 8 mm and a height of 12 mm. The sample was placed horizontally and secured using a mesh form in the central part where there was an observation gap for laser monitoring, similar to work [20]. Figure 8 presents the frames of high-speed imaging obtained for non-irradiated aluminum micron powder combustion. Combustion was initiated with the gas burner; therefore, it was not possible to register the exact moment of ignition. The operator triggered the camera recording when a visible flame appeared.

The use of a laser monitor makes it possible to observe the surface of the sample during combustion. Despite the high brightness and scattering of combustion products, the laser monitor successfully records the changes in the sample surface. In particular, we can clearly monitor the propagation of the burning front and estimate its rate using this imaging technique. The visible combustion front clearly separates the homogeneous surface of the source material from the inhomogeneous surface of the burning material. The arrows and dotted lines mark the position of the burning front during combustion. For the edges' identification, we varied the brightness and contrast of the images converted to the gray scale and compared neighboring pixels. This approach was sufficient for determining the front position with an accuracy of 5–10 pixels.

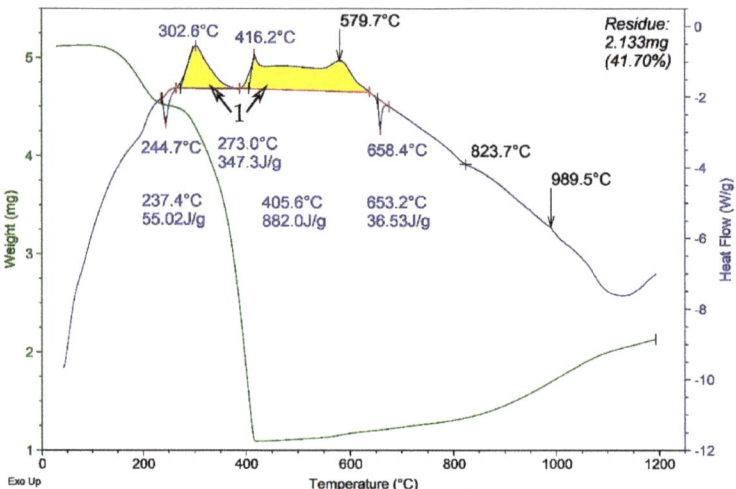

Figure 6. Thermogram of the model HEM obtained with non-irradiated aluminum micron powder (heating rate is 10 °C/min in air). 1—exo-effect.

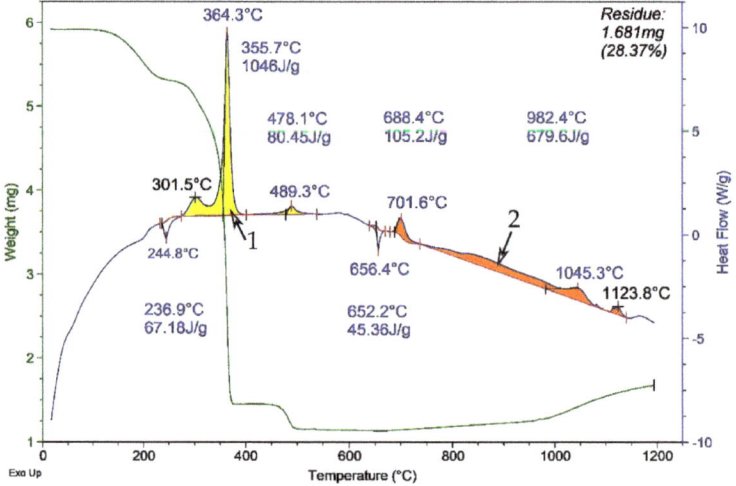

Figure 7. Thermogram of the model HEM obtained with irradiated aluminum micron powder (heating rate is 10 °C/min in air). 1, 2—exo-effects.

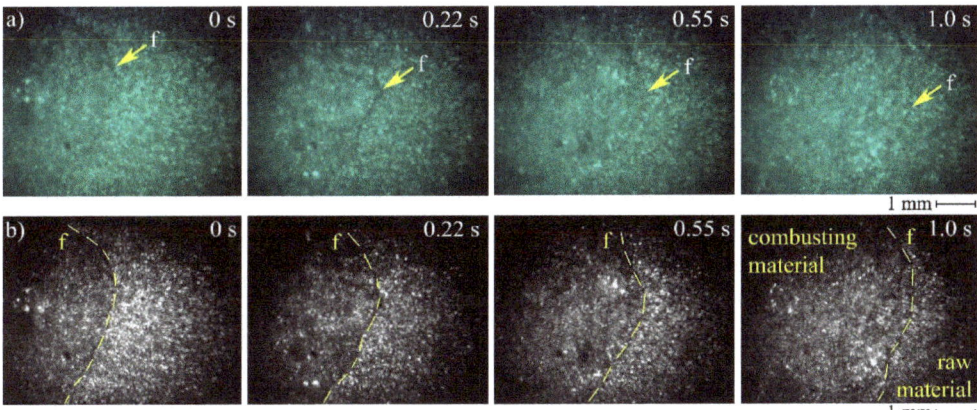

Figure 8. Frames of high-speed imaging of non-irradiated aluminum micron powder combustion: (**a**) original frames at different moments of time during combustion; (**b**) gray scale representation of the same images with brightness/contrast fitting; f—current position of the burning front.

Table 4 presents the results of the burning rate measurement using high-speed imaging with the laser monitor. Six samples with the same size were under the study. The measurement of the burning rate using a laser speed imaging system demonstrated a slight difference in the burning rate of HEMs based on irradiated and non-irradiated aluminum micropowder.

Table 4. The burning rate of the samples.

The Initial Temperature of the Sample, °C	Type of the Composition	Average Burning Rate, mm/s
20	A (non-irradiated)	1.57 ± 0.04
	B (irradiated)	1.52 ± 0.03

Table 5 presents the results of the combustion heat measurement. The analysis of measurement results demonstrates that the irradiated samples (group B) have 11% higher combustion heat than the non-irradiated samples (group A). The energy release of the HEM with irradiated powder reached 6.2 kJ/g. However, the burning rate did not depend on the type of aluminum powder (non-irradiated or irradiated).

Table 5. The combustion heat of the HEM samples.

Group	Q, kJ/g
Group A (non-irradiated)	5.6 ± 0.1
Group B (irradiated)	6.2 ± 0.1

According to the data in Tables 4 and 5, the use of irradiated aluminum micropowder in the composition of HEM allows the increase in the energy of combustion of the material, while maintaining an unchanged combustion rate. This is necessary for those applications when it is required to increase the heat release in mixtures while maintaining the same speed of the technological process or the volume of reagents (for example, for welding thermites, fuel compositions, etc.).

4. Conclusions

This work proved the results of our earlier study [15] that discovered that irradiated powders were capable of storing the radiation energy due to an increase in the surface charge of metal particles, thus increasing their reactivity. This assumption is theoretically confirmed by [28], which demonstrates that the microwave field in nanoparticles is absorbed not only by the core of the nanoparticles but also by the oxide shell of the nanoparticles. We achieved this effect in metal micron particles using short-pulse microwave radiation, in contrast to the works of other researchers who observed a similar effect using hard radiation: high-energy electrons, gamma rays, or neutrons. Exothermal oxidation effect increased from 5.43 to 8.48 kJ/g (by 1.56 times) when irradiated with a frequency of 2.85 GHz, an average power density of 8 kW/cm^2, a pulse duration of 25 ns, and a pulse rate of 400 Hz during 15 s.

In this work, based on the study of aluminum micropowder, we confirm that the stored energy during the irradiation of the initial micropowder is not lost when the micropowder becomes part of the high-energy mixture. The heat of combustion of the modified aluminized model composition in the bomb calorimeter increased by 11% (or 0.6 kJ/g) from 5.6 kJ/g for non-irradiated to 6.2 kJ/g for MW irradiated micropowder. Meanwhile, the burning velocity does not depend on the type of powder used in the HEM composition (microwave-irradiated or non-irradiated). Therefore, the application of the irradiated Al micron powder could improve the exothermal characteristics of the HEM without compromising the linear burning rate.

Supplementary Materials: The following supporting information can be downloaded at: https://www.mdpi.com/article/10.3390/cryst12040446/s1, Video S1: Video of the process of exposure to microwave radiation. The micropowder is placed in a glass vessel, which is placed in the window of the microwave waveguide. During exposure, a blue glow is observed.

Author Contributions: Conceptualization, A.M., F.G. and V.A.; methodology, A.M., F.G., V.A. and V.K.; software, F.G.; validation, V.K. and Y.D.; formal analysis, P.C.; investigation, A.M., F.G. and Y.D.; resources, A.M., P.C. and V.A.; data curation, A.M. and V.K.; writing—original draft preparation, A.M. and F.G.; writing—review and editing, A.M. and F.G.; visualization, F.G.; supervision, P.C. and V.A.; project administration, A.M.; funding acquisition, A.M. and P.C. All authors have read and agreed to the published version of the manuscript.

Funding: This research received no external funding.

Data Availability Statement: The data presented in this study are available on request from the corresponding author.

Acknowledgments: This research was supported by the Tomsk Polytechnic University development program and Tomsk State University competitiveness improvement program.

Conflicts of Interest: The authors declare no conflict of interest.

References

1. Ellern, H. *Military and Civilian Pyrotechnics*; Chemical Publisher Company Inc.: New York, NY, USA, 1968.
2. Beckstead, M.W. A summary of aluminium combustion. In Proceedings of the RTO/VKI Special Course on "Internal Aerodynamics in Solid Rocket Propulsion", Rhode-Saint-Genèse, Belgium, 27–31 May 2002.
3. Ivanov, Y.F.; Osmonoliev, M.N.; Sedoi, V.S.; Arkhipov, V.A.; Bondarchuk, S.S.; Vorozhtsov, A.B.; Korotkikh, A.G.; Kuznetsov, V.T. Productions of ultra-fine powders and their use in high energetic compositions. *Propellants Explos. Pyrotech.* 2003, 28, 319–333. [CrossRef]
4. Teipel, U. *Energetic Materials*; Wiley-VCH: Weinheim, Germany, 2004.
5. Mullen, J.C.; Brewster, M.Q. Reduced agglomeration of aluminum in wide-distribution composite propellants. *J. Propuls. Power* 2011, 27, 650–661. [CrossRef]
6. Brewster, M.Q.; Mullen, J.C. Burning-rate behavior in aluminized wide-distribution AP composite propellants. *Combust. Explos. Shock Waves* 2011, 47, 200–208. [CrossRef]
7. Wang, F.; Wu, Z.; Shangguan, X.; Sun, Y.; Feng, J.; Li, Z.; Chen, L.; Zuo, S.; Zhuo, R.; Yan, P. Preparation of mono-dispersed, high energy release, core/shell structure Al nanopowders and their application in HTPB propellant as combustion enhancers. *Sci. Rep.* 2017, 7, 5228. [CrossRef] [PubMed]
8. Korotkikh, A.G.; Glotov, O.G.; Arkhipov, V.A.; Zarko, V.E.; Kiskin, A.B. Effect of iron and boron ultrafine powders on combustion of aluminized solid propellants. *Combust. Flame* 2017, 178, 195–204. [CrossRef]

9. Nguyen, Q.; Huang, C.; Schoenitz, M.; Sullivan, K.T.; Dreizin, E.L. Nanocomposite thermite powders with improved flowability prepared by mechanical milling. *Powder Technol.* **2018**, *327*, 368–380. [CrossRef]
10. Balaji, A.B.; Ratnam, C.T.; Khalid, M.; Walvekar, R. Effect of electron beam irradiation on thermal and crystallization behavior of PP/EPDM blend. *Radiat. Phys. Chem.* **2017**, *141*, 179–189. [CrossRef]
11. Ivanov, G.V.; Tepper, F. 'Activated' aluminum as a stored energy source for propellants. *Int. J. Energet. Mater. Chem. Propuls.* **1997**, *4*, 636–645.
12. Kuo, K.K.; Acharya, R. *Applications of Turbulent and Multiphase Combustion*; John Wiley & Sons: Hoboken, NJ, USA, 2012.
13. Kuo, K.K.; Risha, G.A.; Evans, B.J.; Boyer, E. Potential usage of energetic nano-sized powders for combustion and rocket propulsion. *MRS Online Proc. Libr.* **2003**, *800*, 3–14. [CrossRef]
14. Mostovshchikov, A.V.; Ilyin, A.P.; Egorov, I.S. Effect of electron beam irradiation on the thermal properties of the aluminum nanopowder. *Radiat. Phys. Chem.* **2018**, *153*, 156–158. [CrossRef]
15. Mostovshchikov, A.V.; Il'in, A.P.; Chumerin, P.Y.; Yushkov, Y.G. Parameters of iron and aluminum nano- and micropowder activity upon oxidation in air under microwave irradiation. *Tech. Phys.* **2018**, *63*, 1223–1227. [CrossRef]
16. Barkley, S.J.; Zhu, K.; Lynch, J.E.; Michael, J.B.; Sippel, T.R. Microwave plasma enhancement of multiphase flames: On-demand control of solid propellant burning rate. *Combust. Flame* **2019**, *199*, 14–23. [CrossRef]
17. Moreira, E.G.; Vasconcellos, M.B.A.; Saiki, M. Uncertainty assessment in instrumental neutron activation analysis of biological materials. *J. Radioanal. Nucl. Chem.* **2006**, *269*, 377–382. [CrossRef]
18. Wendlandt, W.W. *Thermal Methods of Analysis*, 2nd ed.; John Wiley & Sons: New York, NY, USA, 1974.
19. Zepper, E.T.; Pantoya, M.L.; Bhattacharya, S.; Marston, J.O.; Neuber, A.A.; Heaps, R.J. Peering through the flames: Imaging techniques for reacting aluminum powders. *Appl. Opt.* **2017**, *56*, 2535–2541. [CrossRef] [PubMed]
20. Li, L.; Mostovshchikov, A.V.; Ilyin, A.P.; Antipov, P.A.; Shiyanov, D.V.; Gubarev, F.A. Gubarev, In Situ nanopowder combustion visualization using laser systems with brightness amplification. *Proc. Combust. Inst.* **2021**, *38*, 1695–1702. [CrossRef]
21. Evtushenko, G.S. *Methods and Instruments for Visual and Optical Diagnostics of Objects and Fast Processes*; Nova Science Publishers: New York, NY, USA, 2018.
22. Li, L.; Shiyanov, D.V.; Gubarev, F.A. Spatial–temporal radiation distribution in a CuBr vapor brightness amplifier in a real laser monitor scheme. *Appl. Phys. B Lasers Opt.* **2020**, *126*, 155. [CrossRef]
23. Korshunov, A.V.; Yosypchuk, B.; Heyrovský, M. Voltammetry of aqueous chloroauric acid with hanging mercury drop electrode. *Coll. Czech. Chem. Commun.* **2011**, *76*, 929–936. [CrossRef]
24. Standish, N.; Worner, H. Microwave application in the reduction of metal oxides with carbon. *J. Microw. Power Electromagn. Energy* **1990**, *25*, 177–180. [CrossRef]
25. Jung, Y.H.; Jang, S.O.; You, H.J. Hydrogen generation from the dissociation of water using microwave plasmas. *Chin. Phys. Lett.* **2013**, *30*, 065204. [CrossRef]
26. Lu, Y.-H.; Chen, H.-T. Hydrogen generation by the reaction of H_2O with Al_2O_3-based materials: A computational analysis. *Phys. Chem. Chem. Phys.* **2015**, *17*, 6834. [CrossRef]
27. Pisharath, S.; Fan, Z.; Ghee, A.H. Influence of passivation on ageing of nano-aluminum: Heat flux calorimetry and microstructural studies. *Thermochim. Acta* **2016**, *635*, 59–69. [CrossRef]
28. Biswas, P.; Mulholland, G.W.; Rehwoldt, M.C.; Kline, D.J.; Zachariah, M.R. Microwave absorption by small dielectric and semi-conductor coated metal particles. *J. Quant. Spectrosc. Radiat. Transf.* **2020**, *247*, 106938. [CrossRef]

Article

Mechanical Performance and Deformation Behavior of CoCrNi Medium-Entropy Alloy at the Atomic Scale

ZF Liu [1,†], N Tian [2,†], YG Tong [2,*], YL Hu [2], DY Deng [2], MJ Zhang [2], ZH Cai [3] and J Liu [3,*]

1. College of Aerospace Science and Engineering, National University of Defense Technology, Changsha 410114, China; liuzhuofeng@nudt.edu.cn
2. College of Automobile and Mechanical Engineering, Changsha University of Science and Technology, Changsha 410114, China; 18711175611@163.com (N.T.); ylh@163.com (Y.H.); ddy@163.com (D.D.); zmj@163.com (M.Z.)
3. National Engineering Research Center for Mechanical Product Remanufacturing, Army Academy of Armored Forces, Beijing 100072, China; czh@163.com
* Correspondence: tongyonggang_csust@163.com (Y.T.); xbdliu5899@163.com (J.L.)
† These authors contributed equally to this work.

Abstract: CoCrNi medium-entropy alloy has superior cryogenic properties with simultaneous growth of strength and plasticity at low temperatures. In order to observe the microstructure and deformation behavior of the alloy at the atomic scale, its mechanical properties and deformation mechanism at different temperatures and strain rates were investigated using molecular dynamics. It is indicated that the alloy's strength was enhanced at low temperatures and high strain rates due to the production of high dislocation density. The introduction of grain boundaries significantly decreased the dislocation density during the alloy's deformation and correspondingly reduced the crystal strength. However, the introduction of twin boundaries in polycrystalline grains obviously enhanced the strength of the polycrystal, especially at the twin boundary spacing of 3.08 nm. The strength's enhancement was attributed to the increasing dislocation density produced by the interaction between twin boundaries and dislocations during deformation.

Keywords: medium-entropy alloy; twin boundary; mechanical properties; molecular dynamics

1. Introduction

Medium-entropy alloys (MEAs) are a new class of metallic structural materials with great potential for application, which are composed of multiple principal elements in equal or near equal molar ratio distributed on the topologically ordered crystallographic lattices with a high chemical disorder [1,2]. Currently, MEAs are attracting extensive research interest, particularly the face-centered cubic (FCC) phase CrCoNi MEA, which has been found to display excellent mechanical properties, including high fracture toughness and high strength [3,4]. Compared to room temperature, the strength, ductility, and toughness of the CoCrNi alloy increased simultaneously at cryogenic temperatures, reaching strength of more than 1.3 GPa and failure strains up to 90% [4].

Previous studies have shown that the FCC-phase CoCrNi MEAs exhibit excellent mechanical properties due to the relatively low stacking fault energy (SFE) value (18 mJ/m^2) [5,6]. The lower SFE makes it easier to produce more nanotwins during deformation, resulting in better mechanical properties, especially at cryogenic temperatures [6]. At present, studies of SFE, dislocation, and nanotwins in MEAs are mainly conducted on the basis of experiments. Woo et al. [7] investigated dislocation density, twin fault probability, and SFE of CoCrNi-based MEAs deformed at different temperatures by using in-situ neutron diffraction and peaks profile analysis methods. Cao et al. [8] investigated plastic deformation mechanisms in FCC materials with low SFE by using transmission electron microscopy. However, experimental investigations can hardly track the microstructure changes during deformation in real time. Molecular dynamics (MDs) simulations are an effective tool to investigate the relationship between microstructure and properties at the atomic scale [9]. Atomic simulations have been demonstrated to provide real-time and atomic-scale monitoring of the deformation processes, such as nucleation and movement of dislocation [10–12]. Therefore, in the present study, we employed MDs to study the microstructure and mechanical properties of CoCrNi single crystals (SCs) at different temperatures and strain rates. Then, we inserted grain boundaries (GBs) and twin boundaries (TBs) to further study the deformation mechanism of polycrystals. The effects of GBs and TBs on the mechanical properties of MEAs were revealed by comparative analysis, which will help to design strong and highly ductile nanotwinned MEAs.

2. Simulation Methods

The atomic-scale SC model is shown in Figure 1a. The SC was oriented with its <100>, <010>, and <001> aligned respectively with the x-, y-, and z-axes. The average lattice parameter of the CoCrNi alloy is 3.559 Å, and the cohesive energy obtained at this point is -4.32 eV/atom. In order to study the effect of crystal orientation on mechanical properties, another model with different crystal orientations was established. This sample was oriented with its <001>, <1$\bar{1}$0>, and <110> aligned with the x-, y-, and z-axes. For the polycrystalline model, eight randomly oriented grains were created using the Voronoi construction method [13]. The obtained polycrystalline structure is shown in Figure 1b. For polycrystals with nanotwins, single crystals with certain twin boundary spacing (λ = 1.85, 2.47, 3.08, 3.70, and 4.32 nm) were constructed first, and then polycrystals with nanotwins were constructed through the Voronoi construction method, as shown in Figure 1c. The sizes of MEA samples were 106.77 × 106.77 × 106.77 Å in X, Y, and Z directions (~108,000 atoms). Three types of atoms were equimolarly randomly distributed. Periodic boundary conditions were employed in all three directions. The timestep was always 1fs in the simulation. The conjugate gradient algorithm was used to minimize the energy of all samples. The energy tolerance and force tolerance were 10^{-12} and 10^{-12}, respectively. Various temperatures (77, 300, 500, and 800 K) were calculated. For every temperature, the samples were relaxed by the Nose–Hoover isobaric-isothermal (NPT) ensemble for 100 ps. After the balancing process, the samples were deformed at a strain rate of 1×10^8 to 1×10^{10} s^{-1} along the Z direction.

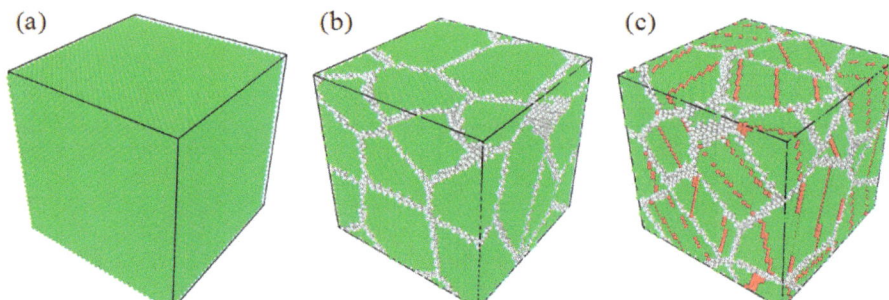

Figure 1. CoCrNi MEA model of FCC phase simulated by MD. (**a**) SC, (**b**) polycrystal, (**c**) polycrystal with TB spacing λ = 1.83 nm. FCC atoms are colored green, GBs are colored white, TBs are colored red.

All models were constructed by Atomsk software. Molecular dynamics calculations were performed using the Large-Scale Atomic/Molecular Massively Parallel Simulator (LAMMPS) [14]. The atomic interaction in CoCrNi MEA adopted the embedded atom method (EAM) potential developed by Li et al., which has been used in some MD studies involving nucleation and motion of dislocation [15]. The Ovito software was used for visualization and statistical analysis of structures [16]. Common neighbor analysis (CNA) and dislocation extraction algorithm (DXA) were used to identify lattice structures and dislocations [17–19].

3. Results and Discussion

3.1. Single Crystalline MEA

3.1.1. Effect of Temperature

The stress-strain curves of CoCrNi MEA under tension and compression at different temperatures are shown in Figure 2. With increasing the temperature from 77 to 800 K, the tensile strength along the <110> direction decreases from 13.75 to 6.60 GPa (Table 1). Different from the linear elastic deformation region that occurs in the <001> direction, the stress of the samples grows nonlinearly throughout the elastic deformation region when tensioned in the <110> direction, as shown in Figure 2a. In addition, the tensile strength of the samples along the <110> direction decreases significantly. This is presumably due to the anisotropy of materials, where the <110> direction has a higher density of atomic arrangement and is more prone to slip deformation. In this study, compression of CoCrNi at different temperatures was also calculated. Similar to the tensile strength, the compressive strength decreases with increasing temperatures, as shown in Table 1 and Figure 2b.

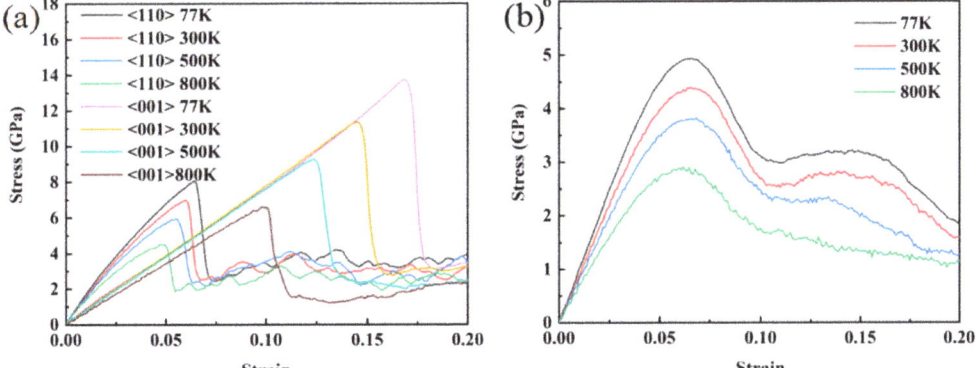

Figure 2. (a) Tensile stress-strain curves in <001> and <110> directions at different temperatures, (b) compressive stress-strain curves in <001> direction at different temperatures.

Table 1. Tensile and compressive strength of CoCrNi at different temperatures.

Temperatures	Tensile Strength (GPa)		Compressive Strength (GPa)
	<001>	<110>	<001>
77 K	13.75	8.07	4.93
300 K	11.36	7.01	4.40
500 K	9.30	5.95	3.81
800 K	6.60	4.52	2.91

In order to study the effect of temperature on the mechanical properties and deformation behavior, the stress and structural evolution during compression at different temperatures were investigated. As shown in Figure 3, FCC atoms dominate during the elastic deformation. The body-centered cubic (BCC) atoms emerge with the strain increased, leading to a decrease in the slope of the nonlinear elastic region. After reaching the maximum value, the stress starts to drop, at which time the hexagonal close-packed (HCP) atoms rapidly generate and expand from the BCC atomic aggregation. The massive formation and accumulation of dislocations, mainly Shockley dislocations (the green line in Figure 4 represents the Shockley dislocation) slow down the stress drop. Compared with the short dislocations generated at high temperatures, the long and twisted dislocations generated at low temperatures have a greater inhibitory effect on the stress drop, which results in a faster stress drop at high temperatures than at low temperatures, as shown in Figure 4.

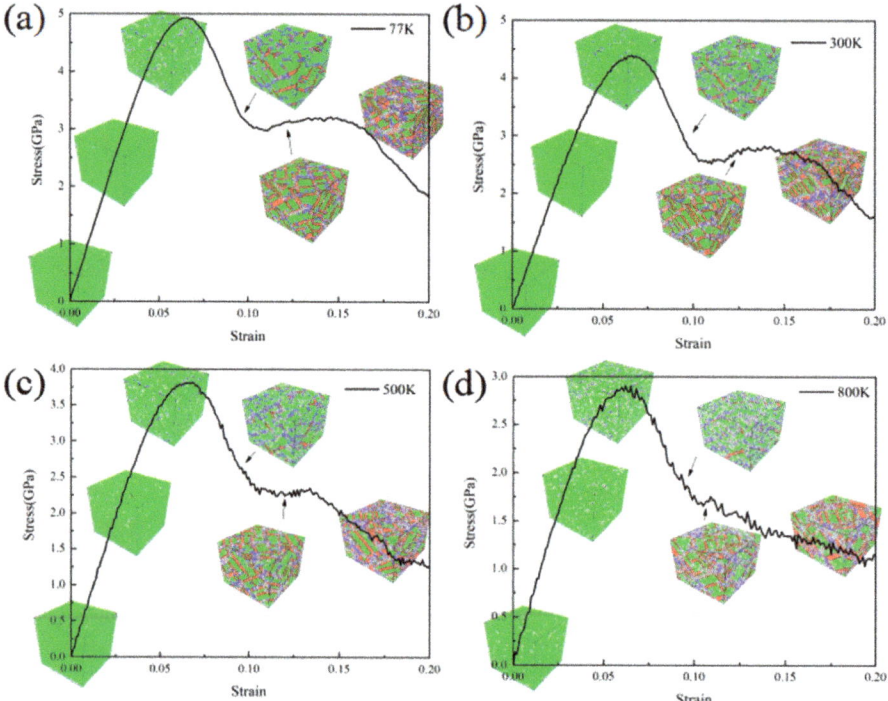

Figure 3. Stress and structural evolution of CoCrNi SCs at 77 K (**a**), 300 K (**b**), 500 K (**c**), and 800 K (**d**). Green, blue, and red represent the atoms of FCC, BCC, and HCP structures, respectively. The arrows indicate the microstructure of the alloy when the strain is 0%, 3%, 6%, 9%, 12% and 15%.

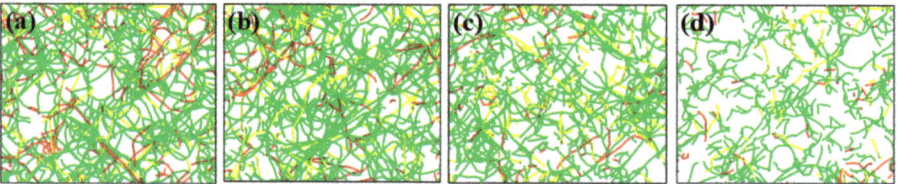

Figure 4. Dislocation distribution at 77 K (**a**), 300 K (**b**), 500 K (**c**), and 800 K (**d**) under the same strain rate.

3.1.2. Effect of Strain Rate

The mechanical properties of CoCrNi MEA at different strain rates of 1×10^8, 1×10^9, and 1×10^{10} s^{-1} are shown in Figure 5 and Table 2. As the strain rate increases, the elastic modulus is constant, while the maximum stress increases significantly. As shown in Figure 5a, the maximum tensile stress along the <110> direction is lower than that in the <001> direction because the higher density of atomic arrangement in the <110> direction makes slip deformation easier. To further analyze the mechanical and structural responses at different strain rates during the compression process, the stress and structural evolution at strain rates of 1×10^8, 1×10^9, and 1×10^{10} s^{-1} were tracked. Until the strain reaches

~5%, the atoms maintain the FCC structure and the stress increases linearly with the increase of strain. After the strain reaches ~5%, some of the atoms deviate from their equilibrium positions and rearrange themselves into BCC structure with lower coordination number, leading to a decrease in the slope of the nonlinear elastic region. At the beginning of plastic deformation at strain rates of 1×10^8 s^{-1}, the HCP structure rapidly generates and expands, resulting in a sharp decrease in stress, as shown in Figure 6a. However, after the plastic deformation begins at strain rates of 1×10^{10} s^{-1}, many small stacking faults (SFs) are formed, which intersect each other and eventually form a dense three-dimensional network of SFs, as shown in Figure 6b. The network hinders the further growth of small dislocation loop and thus, the stress drop is delayed.

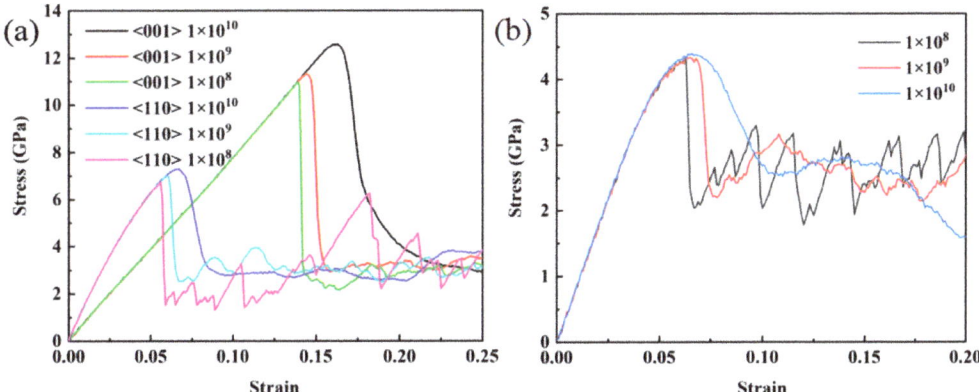

Figure 5. (**a**) Tensile stress-strain curves in direction <001> and <110> at different rates, (**b**) compressive stress-strain curves in direction <001> at different rates.

Table 2. Tensile and compressive strength of CoCrNi at different strain rates.

Strain Rates	Tensile Strength (GPa)		Compressive Strength (GPa)
	<001>	<110>	<001>
1×10^8 s^{-1}	10.96	6.76	4.27
1×10^9 s^{-1}	11.32	7.01	4.34
1×10^{10} s^{-1}	12.59	7.30	4.38

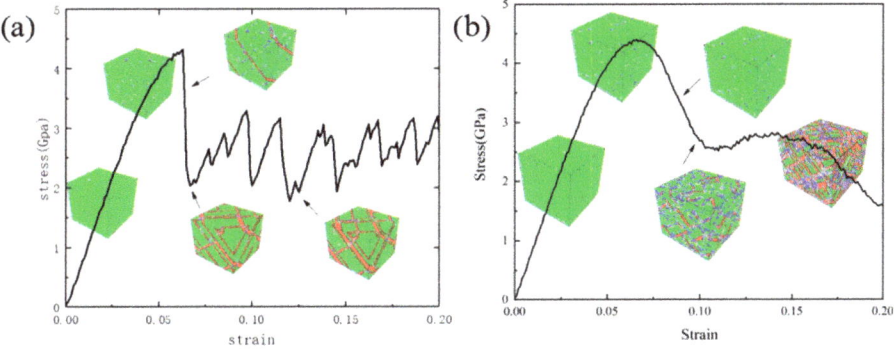

Figure 6. Stress and structural evolution of CoCrNi SCs at strain rates of 1×10^8 s^{-1} (**a**) and 1×10^{10} s^{-1} (**b**). Green, blue, and red represent the atoms of FCC, BCC, and HCP structures, respectively.

3.2. Polycrystalline MEA

Ideal SCs of a certain size are extremely rare in nature. The alloys in practical applications are usually polycrystals. Therefore, we further discussed the mechanical behavior and deformation mechanism of CoCrNi polycrystals.

Figure 7 shows the structural evolution of CoCrNi polycrystals at different strains when stretched at 77 K. During the deformation, SFs and nanotwins generate and expand. With the increase of strain, GBs slip and the proportion of atoms in the disordered state at the GBs increases. In order to consider the effect of temperature on the mechanical properties and deformation behavior of polycrystals, the stress-strain curves at different temperatures are shown in Figure 8. Similar to the effect of temperature on mechanical properties at SCs, the tensile strength and modulus of elasticity increase with decreasing temperature. By comparing the structure and dislocation distribution after tension at different temperatures, it can be found that the deformation at low temperature is dominated by dislocation movement, and the deformation at high temperature is dominated by GBs slip, as shown in Figure 9.

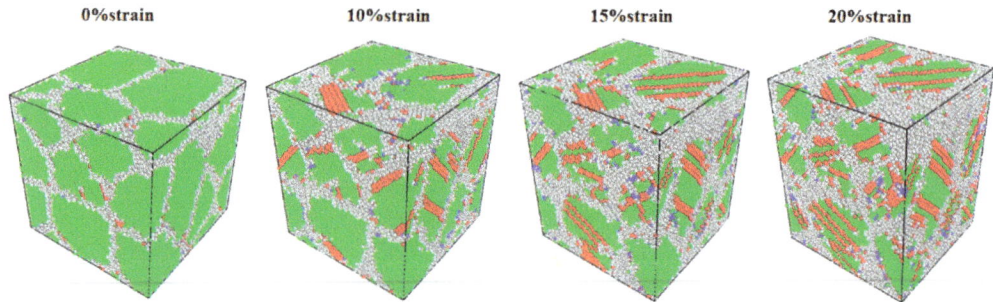

Figure 7. Structural evolution of CoCrNi polycrystals at different strains.

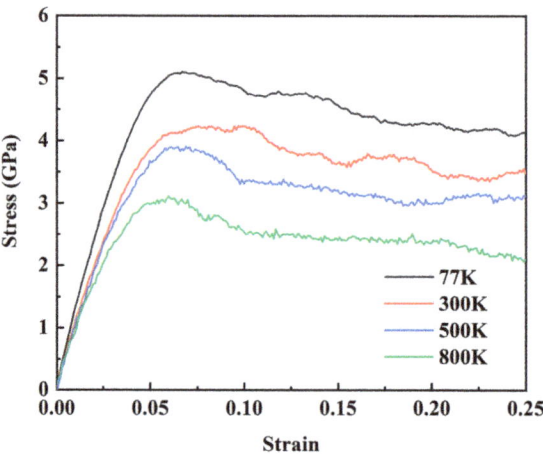

Figure 8. Stress-strain curves of CoCrNi polycrystals at different temperatures.

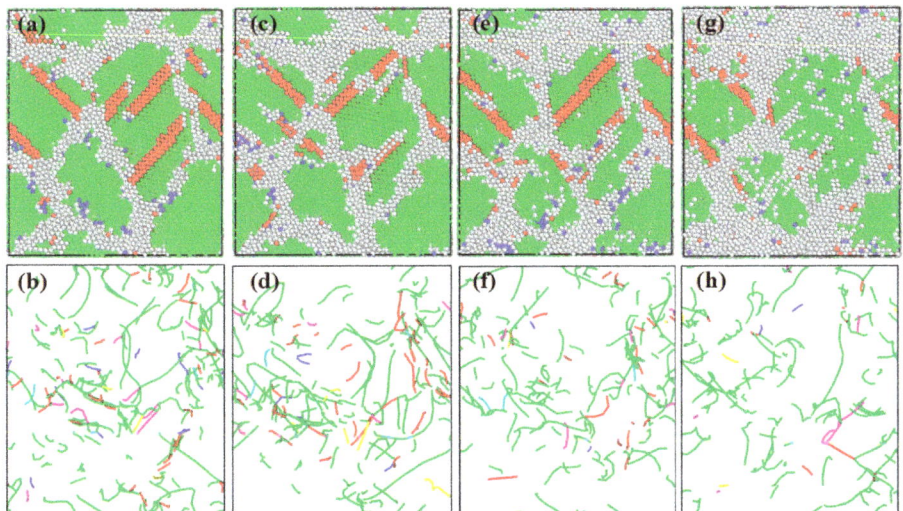

Figure 9. Structure and dislocations of CoCrNi polycrystals at 10% strain at 77 K (**a**,**b**), 300 K (**c**,**d**), 500 K (**e**,**f**), and 800 K (**g**,**h**).

3.3. Nanotwinned MEA

As can be seen from the work on SCs and polycrystals, a large number of dislocations and nanotwins are generated during the deformation process. The effect of dislocations on the mechanical properties and deformation behavior is obvious. However, the role of nanotwins is not reflected. Therefore, in this part, nanotwins with different spacing (λ) were introduced in the initial polycrystalline model to investigate the effect of TBs on mechanical performance of CoCrNi MEA.

3.3.1. Effect of Strain Rate

In this study, the effect of TB spacing on the mechanical properties was investigated using polycrystals with nanotwins as the research object. Figure 10a shows the stress-strain curves of samples with different TB spacing at 300 K. The stress of all stress-strain curves decreased after reaching maximum stress, which was commonly found in MD simulations and attributed to the nucleation of dislocations [20,21]. As shown in Figure 10b, with the increase of λ, the maximum stress increases firstly and then decreases after λ = 3.08 nm.

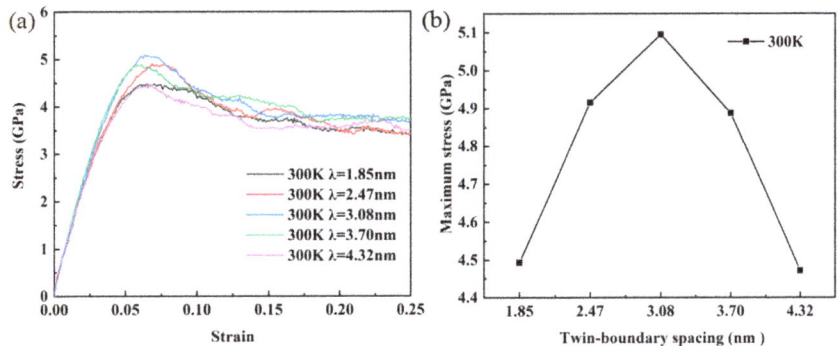

Figure 10. (**a**) Stress-strain curves of samples with different λ (from 1.85 to 4.32 nm) at 300 K, (**b**) maximum stress at different λ at 300 K.

In order to further reveal the effect of λ on mechanical properties of CoCrNi MEAs, the structures and dislocations of samples with different λ at 10% strain are shown in Figure 11d–i. For comparison, the initial structures with different λ are also displayed, as shown in Figure 11a–c. When λ = 1.85 nm, partial Shockley dislocations are perpendicular to and interacting with TBs. When λ = 3.08 nm, the dislocations are no longer perpendicular to the TBs. At this time, a large number of dislocations are entangled with each other, resulting in an increase in the density of dislocations, which improves the mechanical properties of the alloy. When λ = 4.32 nm, the interaction among TBs weakens and the dislocations reduce. Thus, the ability to delay stress drop is reduced.

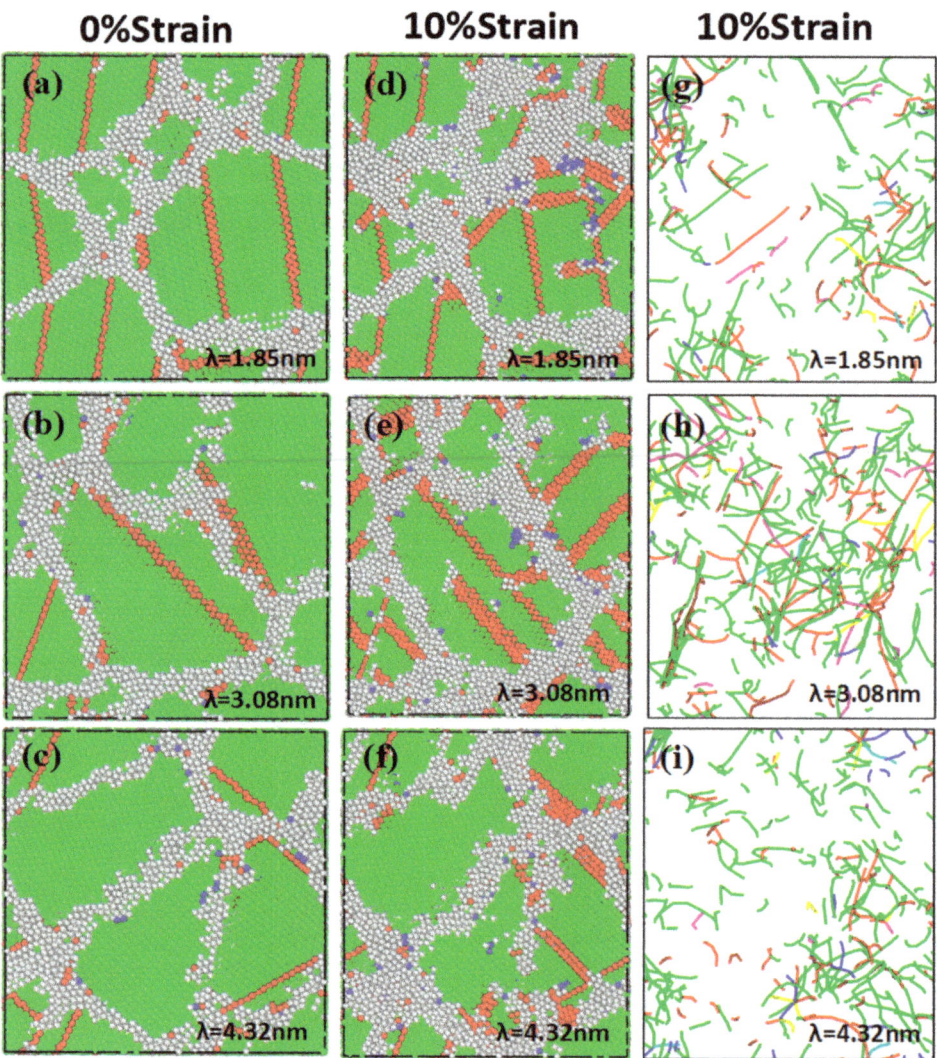

Figure 11. Initial structures at 300 K with TB spacing of (**a**) 1.85 nm, (**b**) 3.08 nm, and (**c**) 4.32 nm, structure of the samples at the strain of 10% with TB spacing of (**d**) 1.85 nm, (**e**) 3.08 nm, and (**f**) 4.32 nm, and dislocations of the samples at the strain of 10% with TB spacing of (**g**) 1.85 nm, (**h**) 3.08 nm, and (**i**) 4.32 nm.

3.3.2. Effect of Temperature

In order to investigate the effect of temperature on the mechanical properties of polycrystals with nanotwins, tensile tests at 77, 300, 500, and 800 K were conducted. The stress-strain curves of samples at different temperatures are shown in Figure 12a–d. As the temperature increases, the tensile strength decreases. Similar to the previous trend exhibited at 300 K, the trend of the maximum stress of the samples at other temperatures increases and then decreases with decreasing λ, reaching a maximum at λ = 3.08 nm (Figure 12e). To further understand the effect of temperature on the mechanical properties of polycrystals with nanotwins, the deformation mechanism and dislocation movement of the alloy at different temperatures were studied. The structure and dislocations at different temperatures are shown in Figure 13. At high temperatures, the GBs slip is obvious. The proportion of disordered atoms at the GB increases, and the dislocation density is small. At low temperatures, dislocations move more slowly, which may lead to dislocation entanglement, resulting in higher dislocation density. The high dislocation density delays the stress drop and enhances the mechanical properties of the alloy. With the decrease of temperature, the deformation mechanism dominated by the GBs slip changes to that dominated by dislocation slip.

In order to reveal the effects of GBs and TBs on the mechanical properties of CoCrNi MEAs, we compared the tensile results of SCs, polycrystals, and polycrystals with nanotwins under the same conditions. As shown in Figure 12f, SCs have the highest strength, followed by polycrystals with nanotwins and polycrystals without nanotwins. Polycrystals have different crystal orientations, complicated GBs, and many lattice defects; therefore, the polycrystals have the smaller strength compared with SCs. Then, comparing Figures 9 and 13, we found that the introduction of nanotwins increased the dislocation density of the polycrystals due to the interaction between the nanoscale TBs and dislocations, which further improves the strength of polycrystals. Liang et al. [22] investigated the deformation mechanism of the CoCrNi alloy with a high density of annealing twins by in situ transmission electron microscopy and also found that TBs not only strengthen the material by hindering the motion of dislocations but also act as a dislocation source to produce slip bands. In addition, Deng et al. [23] also tailored mechanical properties of a CoCrNi medium-entropy alloy by controlling nanotwin-HCP lamellae and annealing twins. Therefore, TBs significantly affect the mechanical properties, and introducing appropriate TB spacing will help to realize the strengthening of the alloy.

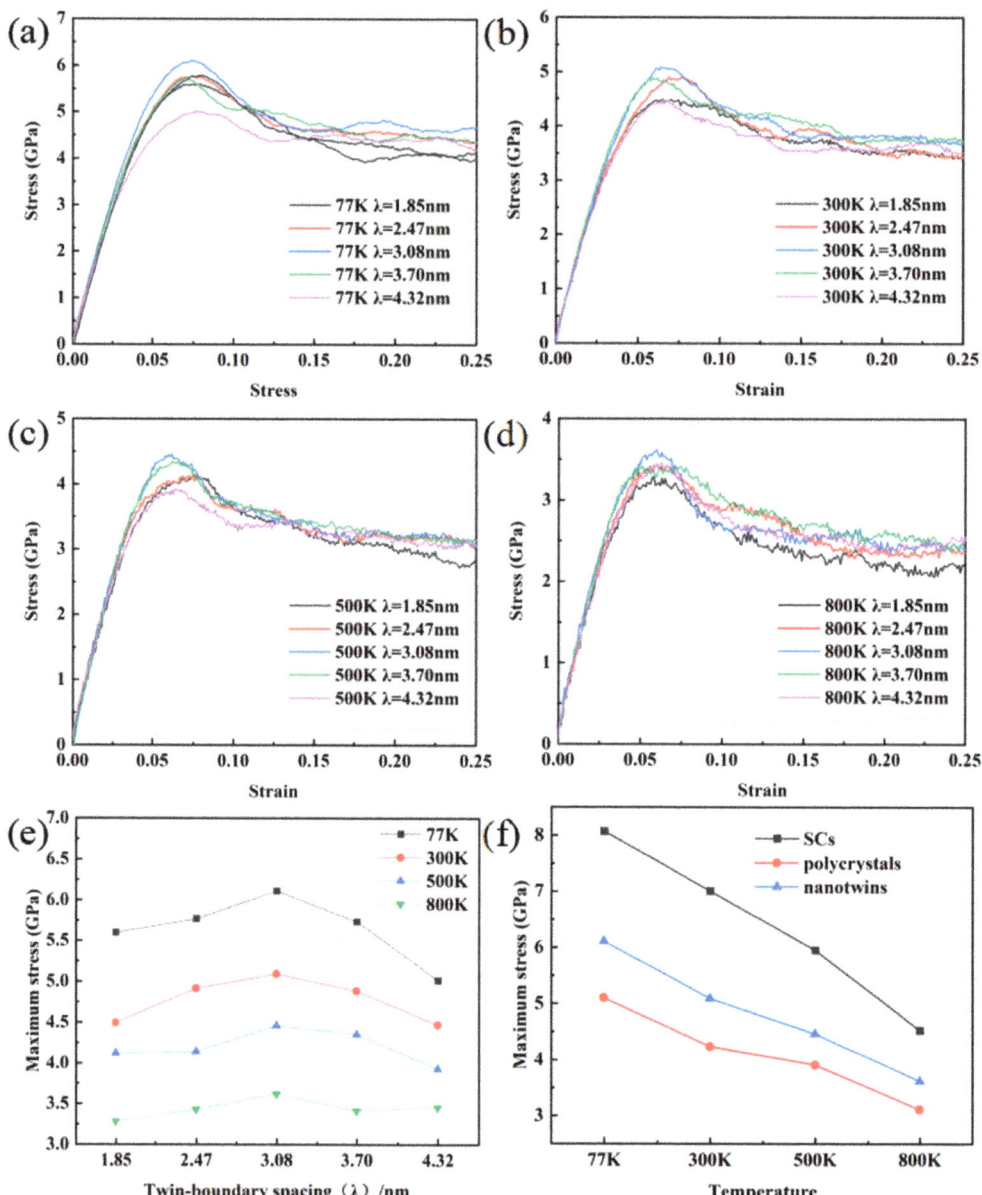

Figure 12. Stress-strain curves of samples with different TB spacing at 77 K (**a**), 300 K (**b**), 500 K (**c**) and 800 K (**d**), (**e**) maximum stress for samples with different spacing at different temperatures, (**f**) maximum stresses of SCs, polycrystals, polycrystals with nanotwins at different temperatures.

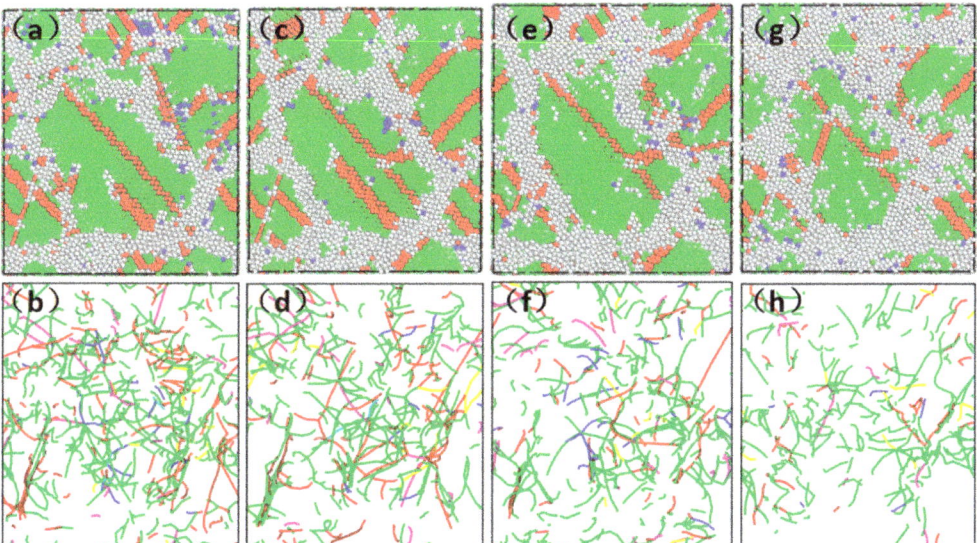

Figure 13. Structure and dislocations of CoCrNi samples with λ = 3.08 nm at 10% strain at 77 K (**a**,**b**), 300 K (**c**,**d**), 500 K (**e**,**f**), and 800 K (**g**,**h**).

4. Conclusions

In this study, the effects of temperature, strain rate, and TB spacing on the mechanical properties and deformation mechanisms of CoCrNi MEAs were investigated by MD simulations. The following conclusions were obtained through a series of comparative analyses:

(1) The strength of CoCrNi increases with the decrease of temperature and the increase of strain rate because the dislocations in the crystal are more intensive at low temperature and high strain rate;

(2) The introduction of defects such as GBs and nanotwins can greatly reduce the strength of SCs;

(3) The presence of TBs can effectively enhance the strength of the polycrystals. The strength increases with the increase of TB spacing when λ is lower than 3.08 nm, and decreases with the increase of spacing when it is higher than 3.08 nm.

Author Contributions: N.T., Z.L. and Y.T. contributed the central idea, calculations, data analysis and drafting of the paper. Y.H., D.D., M.Z., Z.C. and J.L. contributed to refining the ideas, carrying out additional analyses and finalizing this paper. All authors have read and agreed to the published version of the manuscript.

Funding: This research was funded by National Natural Science Foundation of China (No. 92166105 and 52005053), High-Tech Industry Science and Technology Innovation Leading Program of Hunan Province (No.2020GK2085), Hunan Youth Science and Technology Innovation Talent Project (2021RC3096), and Open fund of Key Laboratory of New Processing Technology for Nonferrous Metal & Materials Ministry of Education (No. 20KF-24).

Institutional Review Board Statement: Not applicable.

Informed Consent Statement: Not applicable.

Data Availability Statement: Not applicable.

Conflicts of Interest: The authors declare that they have no known competing financial interest or personal relationships that could have appeared to influence the work reported in this paper.

References

1. Cao, F.-H.; Wang, Y.-J.; Dai, L.-H. Novel atomic-scale mechanism of incipientplasticity in a chemically complex CrCoNi medium-entropy alloy associated with inhomogeneity in local chemical environment. *Acta Mater.* **2020**, *194*, 283–294. [CrossRef]
2. Jian, W.-R.; Xie, Z.; Xu, S.; Su, Y.; Yao, X.; Beyerlein, I.J. Effects of lattice distortion and chemical short-range order on the mechanisms of deformation in medium entropy alloy CoCrNi. *Acta Mater.* **2020**, *199*, 352–369. [CrossRef]
3. Gludovatz, B.; Hohenwarter, A.; Catoor, D.; Chang, E.H.; George, E.P.; Ritchie, R.O. A fracture-resistant high-entropy alloy for cryogenic applications. *Science* **2014**, *345*, 1153–1158. [CrossRef]
4. Gludovatz, B.; Hohenwarter, A.; Thurston, K.V.S.; Bei, H.; Wu, Z.; George, E.P.; Ritchie, R.O. Exceptional damage-tolerance of a medium-entropy alloy CrCoNi at cryogenic temperatures. *Nat. Commun.* **2016**, *7*, 10602. [CrossRef] [PubMed]
5. Laplanche, G.; Kostka, A.; Reinhart, C.; Hunfeld, J.; Eggeler, G.; George, E.P. Reasons for the superior mechanical properties of medium-entropy CrCoNi compared to high-entropy CrMnFeCoNi. *Acta Mater.* **2017**, *128*, 292–303. [CrossRef]
6. Liu, S.F.; Wu, Y.; Wang, H.T.; He, J.Y.; Liu, J.B.; Chen, C.X.; Liu, X.J.; Wang, H.; Lu, Z.P. Stacking fault energy of face-centered-cubic high entropy alloys. *Intermetallics* **2018**, *93*, 269–273. [CrossRef]
7. Woo, W.; Naeem, M.; Jeong, J.-S.; Lee, C.-M.; Harjo, S.; Kawasaki, T.; He, H.; Wang, X.-L. Comparison of dislocation density, twin fault probability, and stacking fault energy between CrCoNi and CrCoNiFe medium entropy alloys deformed at 293 and 140K. *Mater. Sci. Eng. A* **2020**, *781*, 139224. [CrossRef]
8. Cao, Y.Z.; Zhao, X.S.; Tu, W.D.; Yan, Y.D.; Yu, F.L. Plastic deformation mechanisms in face-centered cubic materials with low stacking fault energy. *Mater. Sci. Eng. A* **2016**, *676*, 241–245. [CrossRef]
9. Li, L.; Chen, H.; Fang, Q.; Li, J.; Liu, F.; Liu, Y.; Liaw, P.K. Effects of temperature and strain rate on plastic deformation mechanisms of nanocrystalline high-entropy alloys. *Intermetallics* **2020**, *120*, 106741. [CrossRef]
10. Utt, D.; Stukowski, A.; Able, K. Grain boundary structure and mobility in high-entropy alloys: A comparative molecular dynamics study on a Σ11 symmetrical tilt grain boundary in face-centered cubic CuNiCoFe. *Acta Mater.* **2020**, *186*, 11–19. [CrossRef]
11. Li, J.; Fang, Q.; Liu, B.; Liu, Y.; Liu, Y. Mechanical behaviors of AlCrFeCuNihigh-entropy alloys under uniaxial tension via molecular dynamics simulation. *RSC Adv.* **2016**, *6*, 76409–76419. [CrossRef]
12. Wang, Z.; Li, J.; Fang, Q.; Liu, B.; Zhang, L. Investigation into nanoscratchingmechanical response of AlCrCuFeNi high-entropy alloys using atomic simulations. *Appl. Surf. Sci.* **2017**, *416*, 470–481. [CrossRef]
13. Hirel, P. Atomsk: A tool for manipulating and converting atomic data files. *Comput. Phys. Commun.* **2015**, *197*, 212–219. [CrossRef]
14. Plimpton, S. Fast Parallel Algorithms for Short-Range Molecular Dynamics. *J. Comput. Phys.* **1995**, *117*, 1–19. [CrossRef]
15. Li, Q.-J.; Sheng, H.; Ma, E. Strengthening in multi-principal element alloys with local-chemical-order roughened dislocation pathways. *Nat. Commun.* **2019**, *10*, 1–11. [CrossRef] [PubMed]
16. Stukowski, A. Visualization and analysis of atomistic simulation data with OVITO-the open visualization tool. *Model. Simul. Mater. Sci. Eng.* **2010**, *18*, 015012. [CrossRef]
17. Honeycutt, J.D.; Andersen, H.C. Molecular dynamics study of melting and freezing of small Lennard-Jones clusters. *J. Phys. Chem.* **1987**, *91*, 4950–4963. [CrossRef]
18. Stukowski, A. Structure identification methods for atomistic simulations of crystalline materials. *Model. Simul. Mater. Sci. Eng.* **2012**, *20*, 045021. [CrossRef]
19. Stukowski, A.; Bulatov, V.V.; Arsenlis, A. Automated identification and indexing of dislocations in crystal interfaces. *Model. Simul. Mater. Sci. Eng.* **2012**, *20*, 085007. [CrossRef]
20. Yan, S.; Zhou, H.; Xing, B.; Zhang, S.; Li, L.; Qin, Q.H. Crystal plasticity in fusion zone of a hybrid laser welded Al alloys joint: From nanoscale to macroscale. *Mater. Des.* **2018**, *160*, 313–324. [CrossRef]
21. Yan, S.; Xing, B.; Qin, Q.-H. Effect of Interface on the Deformation of Aluminium Bicrystal: Atomistic Simulation Study. *MATEC Web Conf.* **2016**, *82*, 02010. [CrossRef]
22. Liang, Y.; Yang, X.; Ming, K.; Xiang, S.; Liu, Q. In situ observation of transmission and reflection of dislocations at twin boundary in CoCrNi alloys. *Sci. China Technol. Sci.* **2021**, *64*, 407–413. [CrossRef]
23. Deng, H.W.; Xie, Z.M.; Zhao, B.L.; Wang, Y.K.; Wang, M.M.; Yang, J.F.; Zhang, T.; Xiong, Y.; Wang, X.P.; Fang, Q.F.; et al. Tailoring mechanical properties of a CoCrNi medium-entropy alloy by controlling nanotwin-HCP lamellae and annealing twins. *Mater. Sci. Eng.* **2019**, *744*, 241–246. [CrossRef]

Article

Initial Microstructure Effects on Hot Tensile Deformation and Fracture Mechanisms of Ti-5Al-5Mo-5V-1Cr-1Fe Alloy Using In Situ Observation

Mingzhu Fu [1], Suping Pan [1,2,*], Huiqun Liu [1,3,*] and Yuqiang Chen [4,*]

[1] School of Materials Science and Engineering, Central South University, Changsha 410083, China; fumingzhu@csu.edu.cn
[2] Advanced Research Center, Central South University, Changsha 410083, China
[3] State Key Laboratory of Powder Metallurgy, Central South University, Changsha 410083, China
[4] Hunan Engineering Research Center of Forming Technology and Damage Resistance Evaluation for High Efficiency Light Alloy Components, Hunan University of Science and Technology, Xiangtan 411201, China
* Correspondence: pan-su-ping@163.com (S.P.); liuhuiqun@163.com (H.L.); yqchen1984@163.com (Y.C.)

Abstract: The hot tensile deformation and fracture mechanisms of a Ti-5Al-5Mo-5V-1Cr-1Fe alloy with bimodal and lamellar microstructures were investigated by in situ tensile tests under scanning electron microscopy (SEM) and electron backscatter diffraction (EBSD). The results show that the main slip deformation modes are prismatic slip ($\{1\bar{1}00\}<11\bar{2}0>$) and pyramidal slip ($\{1\bar{1}01\}<11\bar{2}0>$) under tension at 350 °C. In the bimodal microstructure, several parallel slip bands (SBs) first form within the primary α (α_P) phase. As the strain increases, the number of SBs in the α_P phase increases significantly and multislip systems are activated to help further coordinate the increasing deformation. Consequently, the microcracks nucleate and generally propagate along the SBs in the α_P phase. The direction of propagation of the cracks deflects significantly when it crosses the α_P/β interface, resulting in a tortuous crack path. In the lamellar microstructure, many dislocations pile up at the coarse-lath α (α_L) phase near the grain boundaries (GBs) due to the strong fencing effect thereof. As a result, SBs develop first; then, microcracks nucleate at the α_L phase boundary. During propagation, the cracks tend to propagate along the GB and thus lead to the intergranular fracture of the lamellar microstructure.

Keywords: Ti alloy; microstructure evolution; hot tensile deformation; fracture mechanism

Citation: Fu, M.; Pan, S.; Liu, H.; Chen, Y. Initial Microstructure Effects on Hot Tensile Deformation and Fracture Mechanisms of Ti-5Al-5Mo-5V-1Cr-1Fe Alloy Using In Situ Observation. *Crystals* 2022, 12, 934. https://doi.org/10.3390/cryst12070934

Academic Editors: Pavel Lukáč and Shouxun Ji

Received: 16 May 2022
Accepted: 23 June 2022
Published: 1 July 2022

Publisher's Note: MDPI stays neutral with regard to jurisdictional claims in published maps and institutional affiliations.

Copyright: © 2022 by the authors. Licensee MDPI, Basel, Switzerland. This article is an open access article distributed under the terms and conditions of the Creative Commons Attribution (CC BY) license (https://creativecommons.org/licenses/by/4.0/).

1. Introduction

Near-β-titanium alloys have been widely used in aerospace applications as important structural materials, such as in aeroengine compressor disks and turbo blades, due to the excellent corrosion resistance, high-temperature mechanical properties, and creep performance [1–3]. The damage and fracture behaviors of titanium alloy subject to high-temperature environments are very important for the safety of aerospace systems [4,5]. The improvement in high-temperature performance and service security of titanium alloy is always a focus among researchers exploring both the deformation and fracture behaviors of titanium alloys at high temperature [6–8].

Recently, the microstructural evolution and deformation mechanism of titanium alloys during thermal deformation processes have been studied: these were found to be closely associated with the size, volume, morphology, and distribution of the α phase [9,10]. Luo et al. [11] discovered that the flow stress of Ti-6Al-4V alloy increases with the volume fraction of equiaxed α phase but decreases with α-grain size. Wang et al. [12] found that the effects of microstructure on the deformation mechanism of TG6 alloy were mainly determined by the morphology and size of the α_L and α colonies (small α phases lying parallel to the same orientation within the initial β grain). Lin et al. [13] revealed the effects of initial microstructures on hot tensile deformation behaviors and fracture characteristics of Ti-6Al-4V alloy. They found that the equiaxed α phases can prevent the formation and coalescence of microvoids, which is beneficial to improving the ductility. Jiang et al. [14] suggested that significant dynamic softening could occur during uniaxial hot tensile straining of the lamellar microstructure. This was mainly induced by the globularization of α_L phases accompanied by the formation of high-angle grain boundaries (HAGBs). Although many mechanisms have been proposed, there is still no consensus on what the fundamental effect of microstructure characteristics on the hot deformation behavior of titanium alloy is. This could be due to the ex situ observation methods performed by previous researchers which made it difficult to obtain the original information concerning the deformation behavior of alloy directly.

In situ scanning electron microscopy (SEM) is an emerging technique used to characterize the deformation mechanism of materials and has been successfully applied to study the deformation behavior of titanium alloys at room temperature. Huang et al. [15] observed the in situ tensile behavior of Ti-6Al alloy with extra-low interstitial at room temperature and found that there were various coordinated deformation mechanisms such as crystalline orientation rotation, slip transmission, and deformation twinning. Wang et al. [16] investigated the fracture behavior of a new metastable β titanium alloy (Ti-5Cr-4Al-4Zr-3Mo-2W-0.8Fe) at room temperature. They stated that the deformation behavior of Ti-54432 alloy was strongly dependent on dislocation slip, and the deflection of the crack as it propagated; however, there is still a lack of the detailed information concerning the microstructure effect on the crack propagation behavior and slip activation of titanium alloys at high temperature which might be obtained by using an in situ SEM method.

Ti-55511 alloy has been widely used in aircraft casings, engine fan disks, etc., which requires long-term service at 350 °C [17,18]. In this study, the effects of bimodal and lamellar microstructures on the mechanical deformation behavior of Ti-55511 alloy at 350 °C were investigated by in situ SEM. Using electron backscatter diffraction (EBSD), the activation of slip system in deformation region was studied in detail. The results would benefit the microstructure design and deformation mechanism of titanium alloys for high-temperature application.

2. Materials and Methods

2.1. Materials

The as-received material in this study was a forged TC18 alloy provided by BAOTI Group Ltd. (Baoji, China) as cuboid (98.0 mm in length, 20.0 mm in width, and 7.0 mm in height). The experimental alloy had the chemical composition (wt.%) given in Table 1. The α+β/β transition temperature ($T_{\alpha+\beta \to \beta}$) of this alloy is about 865 °C.

Table 1. The chemical composition of Ti-55511 alloy (all in wt %).

Element	Ti	Al	Mo	V	Fe	Cr
Wt. %	83.37	5.07	4.81	4.74	1.06	0.95

The heat treatment methods of Ti-55511 alloy are shown in Figure 1. To obtain a bimodal microstructure, the sample was heated at 830 °C (below $T_{\alpha+\beta \to \beta}$) for 2 h. Then, it was furnace-cooled (FC) to 750 °C and held for 2 h followed by air cooling (AC). As a result, the equiaxed primary α_P phases were distributed in the β matrix (Figure 2a). To acquire a lamellar microstructure which was characterized by coarse original β grains and coarse α_L phase at GBs (Figure 2b), the sample was held at 895 °C (above $T_{\alpha+\beta \to \beta}$) for 2 h and then FC to 750 °C. It was also kept at 750 °C for 2 h followed by AC. The volume fractions of α phase in bimodal and lamellar microstructures were measured to be 39.6% and 11.3%, respectively.

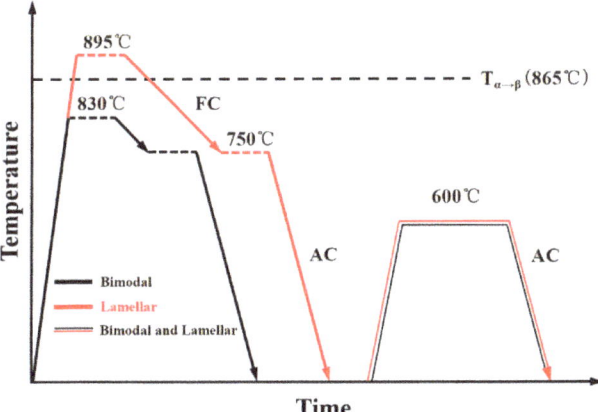

Figure 1. Heat treatment routes for bimodal and lamellar samples.

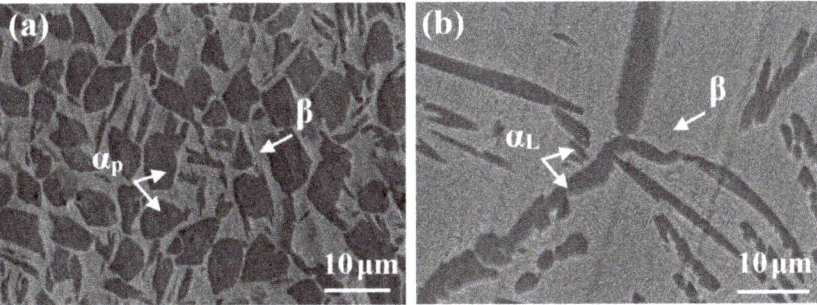

Figure 2. Microstructures after heat treatment (**a**) bimodal microstructure; (**b**) lamellar microstructure.

Both bimodal and lamellar microstructures were finally aged at 600 °C for 8 h to obtain a dispersed distribution of fine secondary α ($α_S$) phase embedded in the β matrix. The yield stress ($σ_{0.2}$) and tensile strength ($σ_b$) values of bimodal samples are 1098 MPa and 1133 MPa, respectively, significantly lower than that of lamellar samples ($σ_{0.2}$ = 1293 MPa, $σ_b$ = 1309 MPa). On the contrary, the elongation (δ) of the bimodal sample at room temperature is 17.0%, which is significantly higher than that of the lamellar sample (δ = 6.0%).

2.2. In Situ Tensile Testing

High-temperature in situ tensile tests were performed on the additional Mini-MTS (Liweiauto Ltd., Hangzhou, China) loading system equipped in a TESCAN MIRA3 SEM (TESCAN Brno,s.r.o, Brno, Czech Republic). The in situ mechanical test bench and heating device are illustrated in Figure 3a. The Mini-MTS Test Control software was used to control the loading speed of 1.5 μm/s and the test temperature of 350 °C. To ensure a uniform temperature distribution, the samples were held for 10 min at the test temperature before stretching. The loading system allows for several interruptions during the tensile test for SEM imaging. By controlling the displacement during in situ stretching, the evolution of the microstructure was observed. The dimensional size of the sample with a thickness of 1 mm is displayed in Figure 3b. The sample surface was polished to a smooth mirror surface with a mixture of SiO_2 suspension and 30% H_2O_2. The in situ tensile tests were repeated three times for each group of experimental conditions, and typical results are provided.

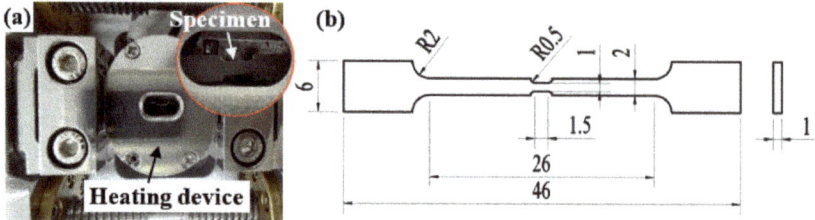

Figure 3. (**a**) Mechanical test bench and heating device; (**b**) Geometry and dimensions of in situ tensile test sample at 350 °C (units: mm).

2.3. EBSD Analysis

Electron backscattered diffraction (EBSD) measurements were performed on the samples obtained by in situ tensile testing using an AZtec system (Oxford Instruments Group, Oxford, UK) coupled with a Hitachi-Regulus 8230 cold-field emission SEM (Hitachi High-Technologies Corporation, Tokyo, Japan). The operating voltage used was 20 kV when optimizing the quality of the diffraction patterns. The EBSD samples were electropolished using a solution of 8% perchloric acid ($HClO_4$) and 92% CH_4O at −25 °C. A step size of 0.2 µm was used to collect data covering an area of 60 µm × 80 µm. The smaller step size of 0.06 µm was used for microstructural analysis of the 20 µm × 16 µm region. The EBSD data were collected by Channel 5 (HKL-Tango) and ATEX [19].

3. Results

3.1. The Stress–Displacement Curves during In Situ Testing

Figure 4 shows the stress–displacement curves of the bimodal and lamellar microstructures during in situ tensile testing at 350 °C. The drops in the curve indicate that the loading is suspended for SEM imaging during the tensile process. To observe the microstructure evolution during hot stretch, different stages of plastic deformation of the samples were marked, respectively (A, B, and F for the bimodal microstructure, and A' and F' for the lamellar microstructure). This indicates that, compared to the lamellar microstructure, the bimodal microstructure results in a lower yield strength ($\sigma_{0.2}$) and tensile strength (σ_b) but a larger tensile displacement (L_{max}).

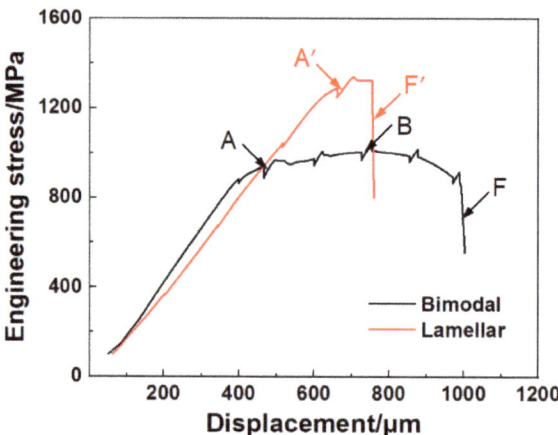

Figure 4. The stress–displacement curve of sample stretching in situ with bimodal and lamellar microstructure at 350 °C.

As shown in Table 2, $\sigma_{0.2}$ and σ_b of specimens with a lamellar microstructure are 37.3% and 28.5% higher than those with bimodal microstructure, respectively. On the contrary, L_{max} of specimens with a bimodal microstructure is 28.4% larger than that of specimens with a lamellar microstructure.

Table 2. The in situ tensile mechanical properties of bimodal and lamellar microstructure at 350 °C.

Samples	$\sigma_{0.2}$ (MPa)	σ_b (MPa)	L_{max} (μm)
Bimodal	930	1005	1012
Lamellar	1277	1291	788

3.2. Microstructural Evolution in the Bimodal Microstructure

Figure 5a illustrates the in situ SEM images of bimodal microstructure at position A. As shown in Figure 5c, some parallel SBs begin to appear within the α_P phase. These SBs generally form at an angle of 41° to 49° with the direction of the applied tension [20]. Some of them pass through the α_P interface and gradually penetrate the β matrix (Figure 5d). It is worth noting that microcracks first form within the region of the α_P phase rather than in the β matrix during the early stage of plastic deformation. As shown in Figure 5b–d, a few microcracks, evolved from the deep SBs in the α_P phase, extend macroscopically perpendicular to the tensile direction, which indicates that high stress concentration arises rapidly at the α_P phase during hot deformation.

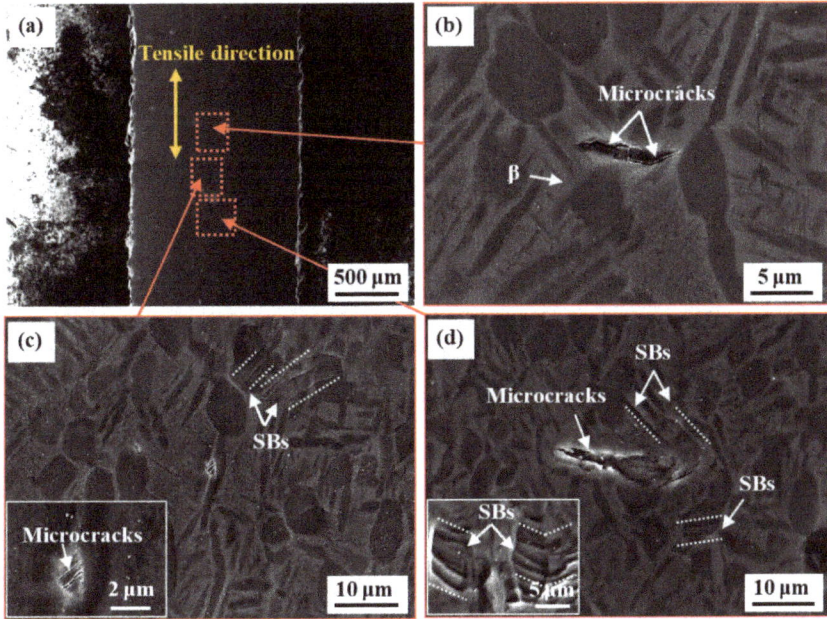

Figure 5. (a) In situ SEM images of bimodal microstructure at A (the displacement of 467 μm); (b) microcracks nucleated at α phase; (c) microcracks and SBs formed within the α_P phase; (d) magnified image showing SBs passing through the interface between the α_P and β phases.

Figure 6 presents the SEM images of bimodal microstructure at position B. As shown in Figure 6a, significant necking is found on the sample. With the increase in strain, the microcracks in Figure 6b widened significantly compared with position A (Figure 5b). Moreover, SBs in the α_P phases became more noticeable when they continued by crossing the α_P interface and extended into the β matrix. From Figure 6c, the deformation became more severe and SBs occurred in most regions composed of α_P phases. In addition, a macrocrack formed at the edge of the sample which exhibited a trend to connect the SBs in the α_p phases ahead of the crack tip and gradually grew into the center of the sample. As illustrated in Figure 6d, at the center of the sample, SBs in the α_P phases deepened significantly and gradually changed into wave-like shapes due to the increasing distortion in that local region. Moreover, many microcracks were also found to nucleate in these deep SB regions.

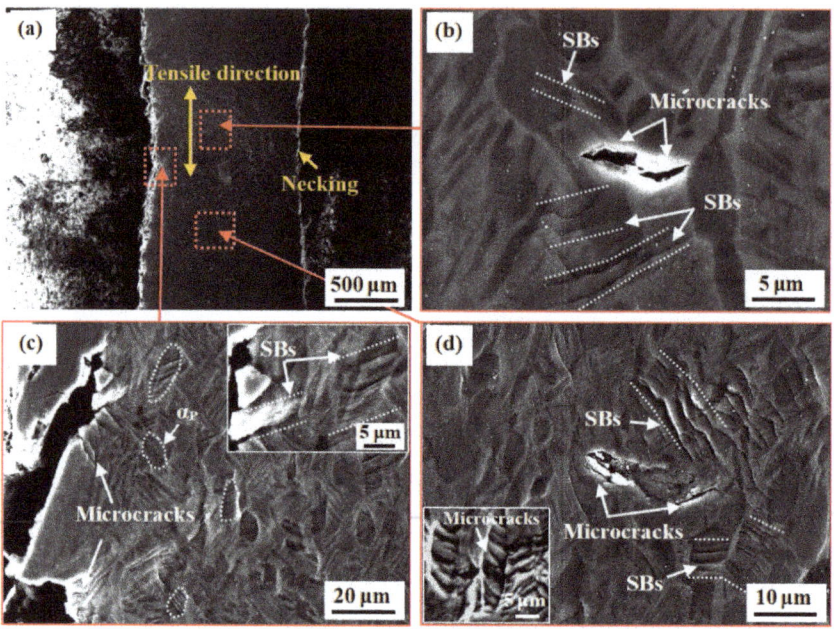

Figure 6. (**a**) In situ SEM images of bimodal microstructure at B (the displacement of 726 μm); (**b**) microcracks become deepened and widened; (**c**) magnified image showing macrocracks nucleated at the edge of the sample; (**d**) distortion in SB regions and microcracks at α_P/β interface.

Figure 7 shows in situ SEM images of bimodal microstructure at position F (after fracture). The sample was shear-fractured at about 45° to the tensile direction. Its fracture surface is relatively rough and significant necking is observed (Figure 7a). As shown in Figure 7b, many microcracks formed at the region close to the fracture surface which nucleated and generally propagated along the SBs in the α_P phase. Moreover, the crack propagation direction always deflected as it crossed the α/β interface, leading to a microscopically tortuous path. This could undoubtedly increase the total length of the crack path and therefore consume more energy before it fractured [21].

Figure 7c illustrates that the SBs in each α_P phase had a strong tendency to interconnect. During this process, the direction of SBs changed significantly at the α_P interface. Moreover, a dendritic crack formed on the edge of the sample (Figure 7d) whose path deflected or branched frequently during its progress to the center of the sample at about 45° to the tensile direction. Furthermore, many microcracks can be observed near the fractured area (Figure 7f) which implies that the sample underwent severe plastic deformation before fracture.

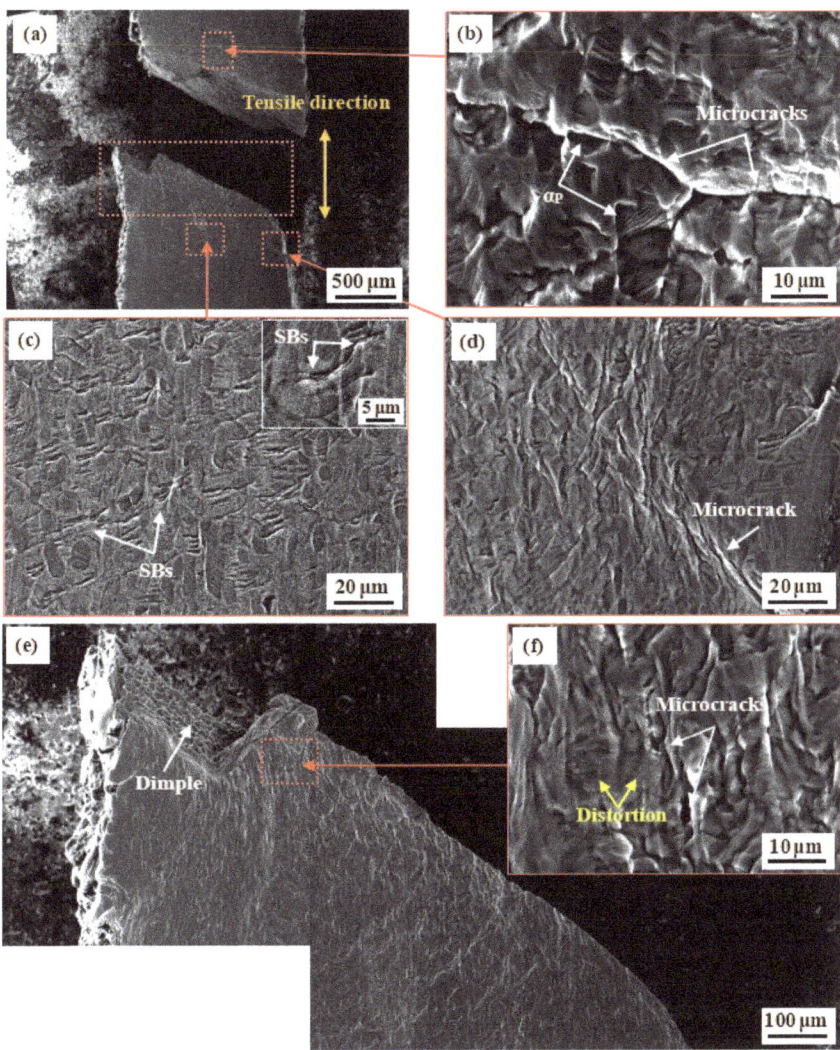

Figure 7. (**a**) In situ SEM images of bimodal microstructure sample at F (the displacement of 1012 μm); (**b**) a crack propagated along the SBs in the $α_P$ phase; (**c**) SBs in each $α_P$ phase tended to interconnect; (**d**) magnified image showing microcracks formed on the edge of the sample; (**e**) SEM images showing the rough fracture surface of the sample; (**f**) magnified image showing many microcracks existed near the fracture area.

3.3. Microstructural Evolution in the Lamellar Microstructure

Figure 8 presents SEM images of the lamellar microstructure during in situ stretching at position A'. As shown in Figure 8b,d, many parallel SBs appeared first in the $α_L$ phases near GB, while they can hardly be observed in the β matrix. As illustrated in Figure 8c, a microcrack developed in a long, coarse $α_L$ phase at a trigeminal GB. This was akin to the findings of Wang et al. [16] from the in situ tensile testing of Ti-54432 alloy at room temperature, which indicated that microcracks readily nucleated at trigeminal GBs due to the significant stress concentration therein.

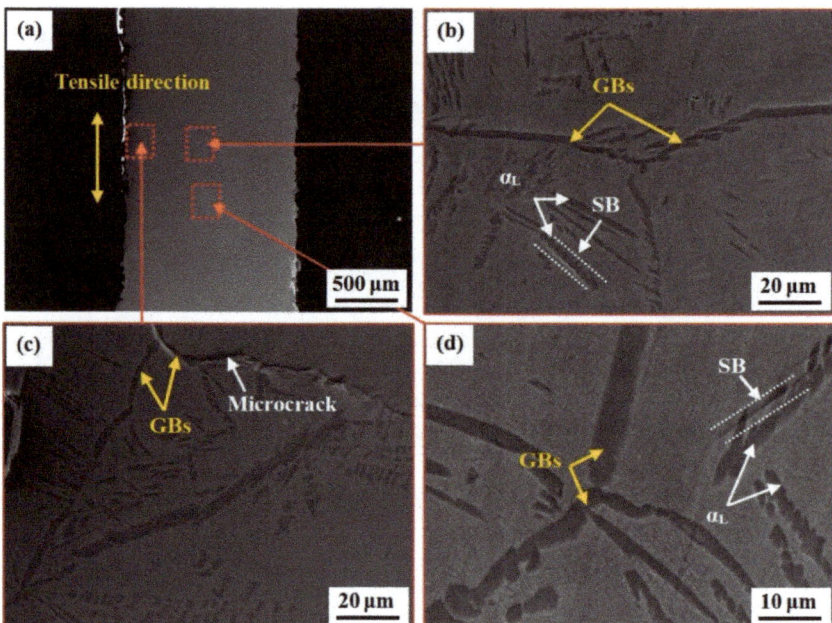

Figure 8. (a) In situ SEM images of lamellar microstructure at A' (the displacement of 661 μm); (b,d) magnified image showing SBs develop in the α_L phase near GB; (c) magnified image showing microcracking in a long, coarse α_L phase at a trigeminal GB.

Figure 9 illustrates SEM images of a lamellar sample during stretching in situ at position F'. Differing from bimodal samples, no obvious necking can be observed until the fracture of the lamellar sample (Figure 9a). This indicates that the plastic deformation therein is not evident. This can be further proved from Figure 9b,c that only slight distortion and certain microcracks are illustrated in the β matrix in the region adjacent to the fracture surface which is significantly smaller than that in specimens with a bimodal microstructure.

Figure 9d shows that, at the region near the fracture surface, a microcrack just initiated at and propagated along the large α_L phase at GB. The main crack of the lamellar microstructure grew almost only along the GB which led to the intergranular fracture of the sample (Figure 9e,f). The magnified image in Figure 9f clearly indicates that a microcrack nucleated and grew only along the α_L phase at GB. Moreover, there were many fine dimples and tearing ridges at the fracture surface which might be caused by the plastic deformation of the α_L phase at GB during the final fracture process. This explains the minor elongation of the lamellar microstructure at the elevated temperature.

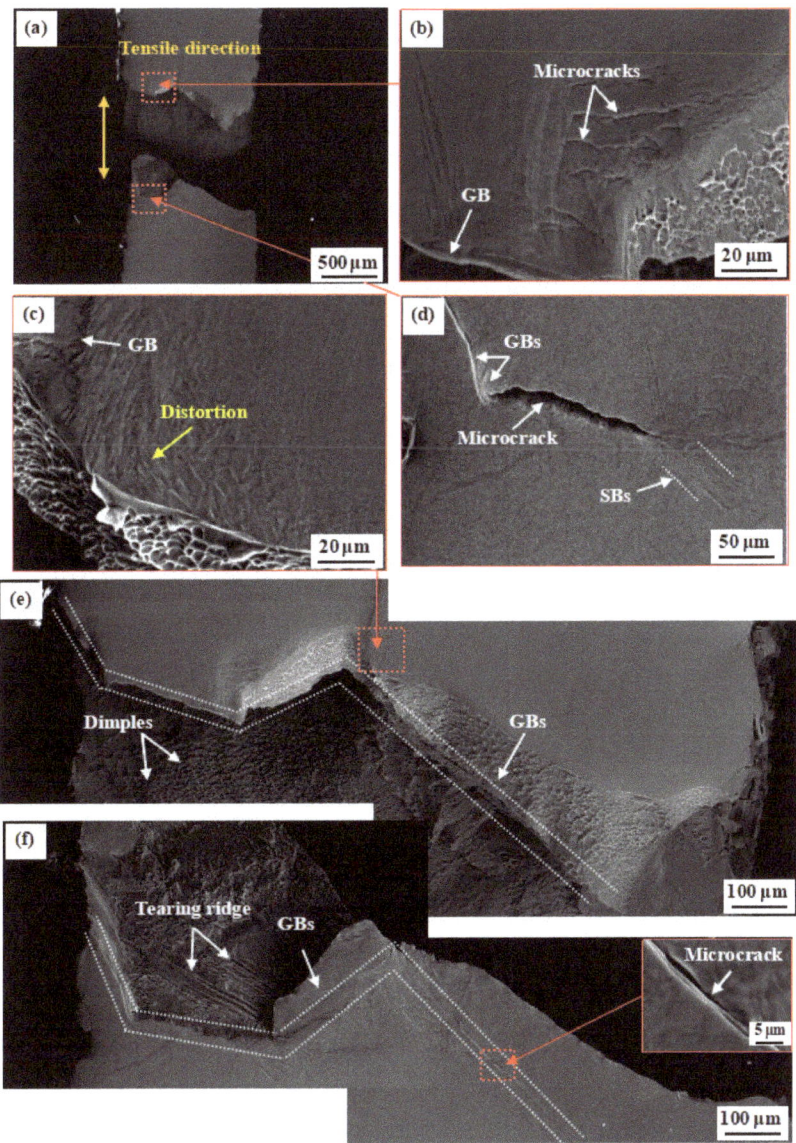

Figure 9. (**a**) In situ SEM images of lamellar microstructure at F' (the displacement of 788 µm); (**b**,**c**) microcracks and distortion occurred near the fracture area; (**d**) magnified image showing crack initiation occurred at GB; (**e**,**f**) SEM images showing the fracture surface of the sample.

4. Discussion

4.1. Activation of Slip Systems

To identify the activated slip systems of the alloy during hot deformation, EBSD of the bimodal microstructure after in situ stretching was analyzed and the results are illustrated in Figure 10.

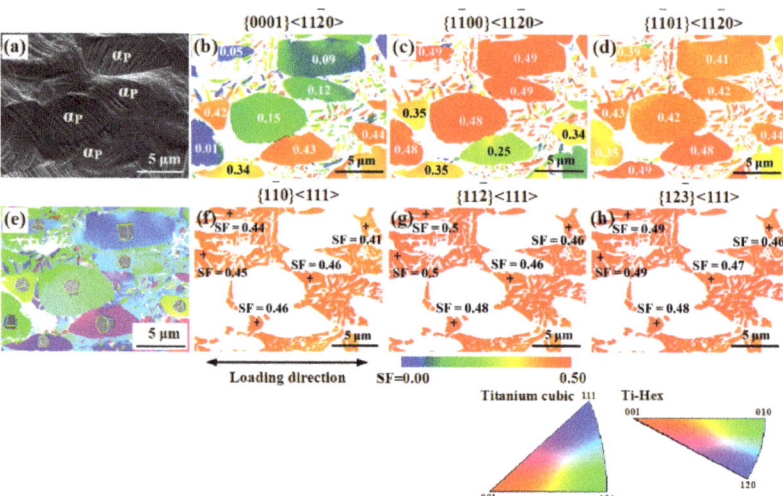

Figure 10. EBSD images of bimodal microstructure after stretching in situ; (**a**) SEM image showing many parallel SBs in the α_P phase; (**b**) SF map of basal slip for α; (**c**) SF map of prismatic slip for α; (**d**) SF map of pyramidal slip for α; (**e**) the inverse pole figure map; (**f**) SF map of β for $\{1\bar{1}0\}<111>$ slip; (**g**) SF map of β for $\{11\bar{2}\}<111>$ slip; (**h**) SF map of β for $\{12\bar{3}\}<111>$ slip.

In Figure 10a, many SBs are generated within regions composed of α_P phases. This indicates that dislocation slip is the dominant deformation mechanism of the α phase at the applied temperature [22]. Our previous study indicated that pyramidal slip is difficult to take place within the α_P phase which is more likely to slip along its basal or prismatic plane at room temperature [23]. In the present study, Figure 10b–d show that prismatic slip ($\{1\bar{1}00\}<11\bar{2}0>$) and pyramidal slip ($\{1\bar{1}01\}<11\bar{2}0>$) were prevalent in the α_P phase compared with basal slip ($\{0001\}<11\bar{2}0>$). According to the study by Lecomte et al. [24], the most common slip system is the pyramidal system for α phases in the range of 150–300 °C, and prismatic slip is the common mode of deformation at temperatures above 300 °C. The fact that basal slip is not the main slip system is the result of the combined influence of the hcp crystal structure properties of the α phase and the applied temperature. Turner et al. [25] stated that due to the changes in lattice parameters at an elevated temperature, the c/a ratio of the α phase with hcp crystals might be less than 1.6333. This leads to the (0001) basal plane no longer being the only close-packed plane, whereas the prismatic slip and pyramidal slip become the main mode of deformation [25].

As presented in Figure 10f–h, there are no significant differences in the Schmidt factor (SF) values among the main slip systems in the β phase [26,27], i.e., $\{1\bar{1}0\}<111>$, $\{11\bar{2}\}<111>$, and $\{12\bar{3}\}<111>$. Thus, a variety of slip systems can be activated simultaneously and the long-range slip along a single slip system might not be able to take place in the β matrix. Consequently, it is difficult to observe significant SBs along some specific crystal planes in the β matrix. Due to its excellent coordinated deformation capacity, the microcrack can seldomly initiate in the β matrix [28].

To explore the slip characteristics in the bimodal microstructure during hot deformation, a EBSD analysis of a large number of α_P grains were carried out for the sample in situ stretched at 350 °C, which allowed a statistical analysis of the nature (basal, prismatic, and pyramidal) and distribution of the slip systems in α_P.

According to the statistical results in Figure 11a, the SF values of prismatic slip in α_P mostly concentrated at the range of 0.35–0.50. Therefore, this relatively large SF value distribution in the α_P phase might account for the main slip system of prismatic slip during in situ stretching at 350 °C. Figure 11b–g present several typical SB morphologies in α_P phases. As shown in Figure 11b, multislip systems were activated in grain 1. Combined with Figure 11h,i, it can be determined that the specific slip systems activated were the (001)[1$\bar{2}$0] of basal slip and (0$\bar{1}$0)[2$\bar{1}$0] of prismatic slip. The activation of the multislip system helped to coordinate further the overall deformation of the polycrystal. Grains 2 and 3 formed slip steps on the surface due to the repeated dislocation slipping along certain planes since the mobility of dislocations was enhanced significantly at a higher temperature. During high-temperature in situ stretching, microcracks tend to initiate at SBs in the α_P phase, probably due to the increasing demand for stress relief [29].

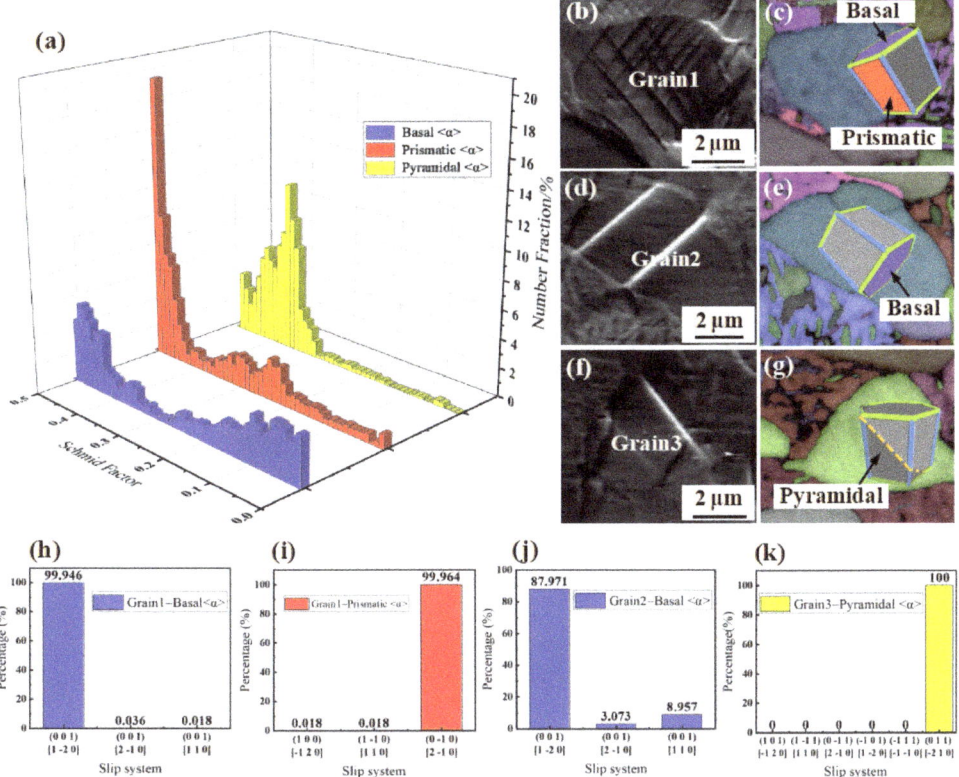

Figure 11. (a) SF histogram of α phase with basal <α>, prismatic <α>, and pyramidal <α> slip systems for bimodal microstructure after stretching in situ; (b,c) Multislip activation of Grain 1 in the α_P phase; (d,e) Basal slip activation of grain 2 in the α_P phase; (f,g) Pyramidal slip activation of Grain 3 in the α_P phase; (h) The (001)[1$\bar{2}$0] basal slip system activation for Grain 1; (i) The (0$\bar{1}$0)[2$\bar{1}$0] prismatic slip system activation for Grain 1; (j) The (001)[1$\bar{2}$0] basal slip system activation for Grain 2; (k) The (011)[$\bar{2}$10] slip system activation for Grain 3.

By calculating SF values, it is possible to reveal qualitatively the initiation trend in various dislocation slip mechanisms in specimens with a lamellar microstructure. Figure 12 shows the SF distribution in the observation region. Figure 12b implies that the SF value of basal slip in the α_L phase is less than 0.10, namely, the basal slip system is hard-orientated and difficult to activate. In Figure 12c,d, the SFs of prismatic and pyramidal slip systems are soft-orientated and easy to activate with relatively large values (SF $\geq$ 0.37). Therefore, in specimens with a lamellar microstructure, the α_L phase was more inclined to activate prismatic slip and pyramidal slip.

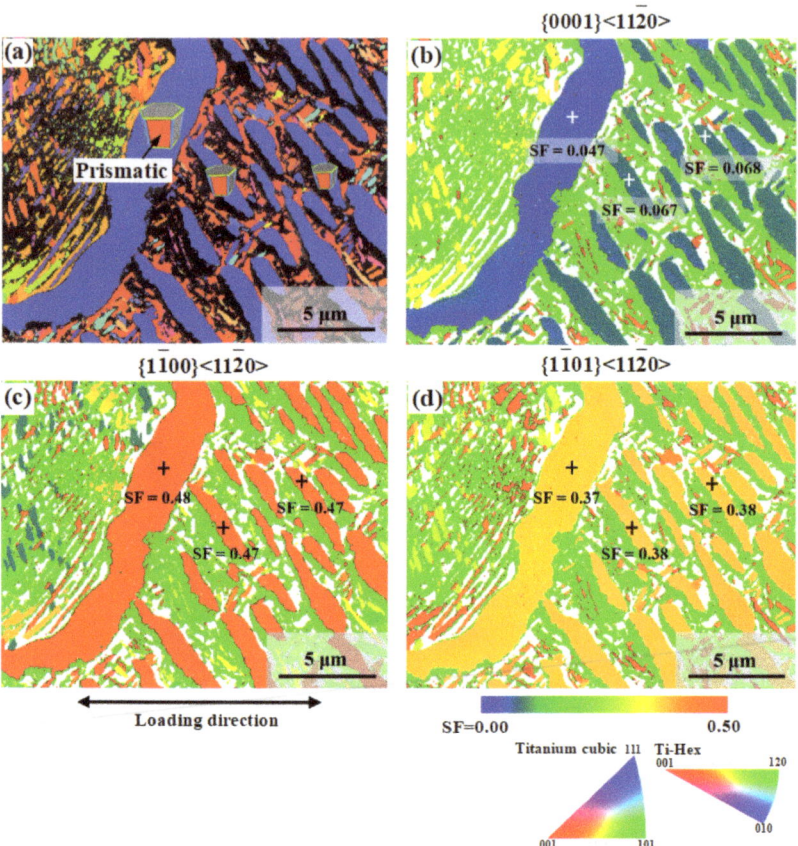

Figure 12. EBSD images of lamellar microstructure after stretching in situ; (**a**) the inverse pole figure map; (**b**) SF map of basal slip for α; (**c**) SF map of prismatic slip for α; (**d**) SF map of pyramidal slip for α.

Figure 13 shows statistics of slip deformation modes and SF distributions of the α_L phase in specimens with a lamellar microstructure. As shown in Figure 13a, the SF values of prismatic slip were mainly concentrated between 0.45 and 0.50, which dominated the slip activities during the in situ tensile process at 350 °C. In Figure 13b–g, two typical slip systems were activated in the α_L phase: (100)[$\bar{1}$20] of prismatic slip (grain 4) and (011)[$\bar{2}$10] of pyramidal slip (grain 5). During plastic deformation at elevated temperature, to coordinate deformation, slip cannot only be activated in the optimal orientation of the grain, but also in the relatively more difficult orientation [30].

Due to the limited slip systems of the α phase, once the plastic deformation in the α_L phase accumulated to a certain extent and no new plastic deformation mechanism (e.g., mechanical twinning and phase transformation) was generated to coordinate the deformation, microcrack nucleation would occur at SBs in the α_L phase due to the pile-up of a large number of dislocations therein [31].

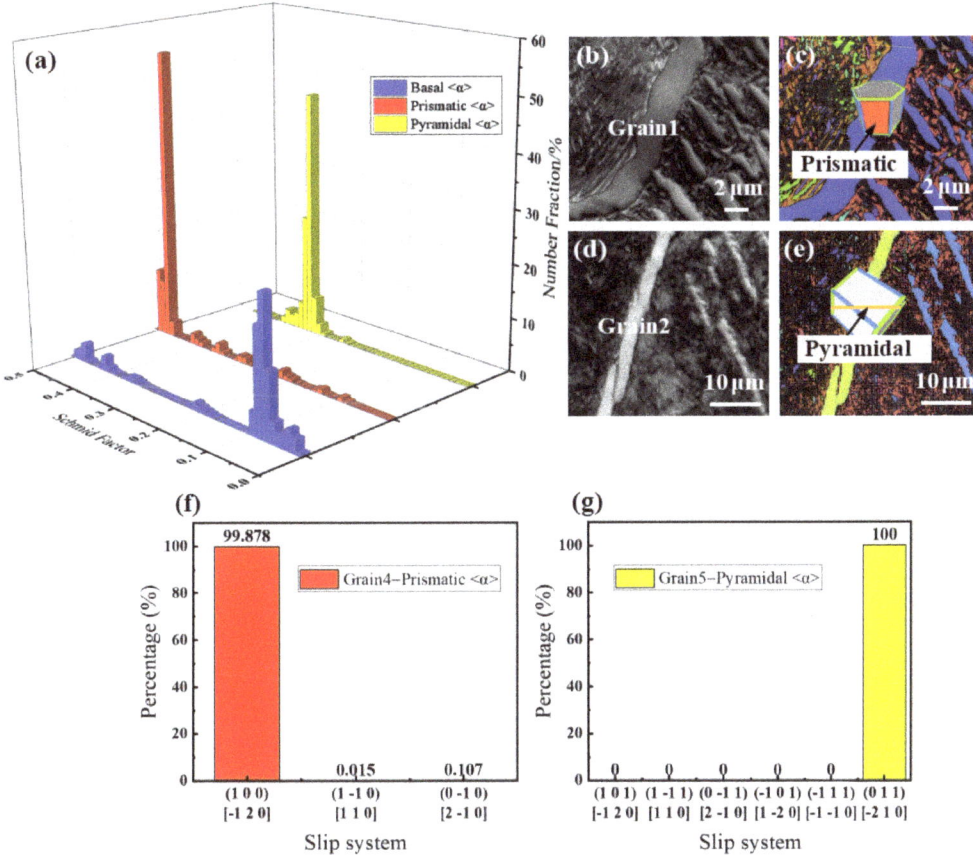

Figure 13. (a) SF histogram of α phase with basal <α>, prismatic <α>, and pyramidal <α> slip systems for specimens with a lamellar microstructure after stretching in situ; (**b**,**c**) Prismatic slip activation of grain 4 in the α$_L$ phase; (**d**,**e**) Pyramidal slip activation of grain 5 in the α$_L$ phase; (**f**) The (100)[$\bar{1}$20] of prismatic slip system activation for grain 4; (**g**) The (011)[$\bar{2}$10] of pyramidal slip system activation for grain 5.

4.2. Characteristics of Deformation and Fracture

The in situ observation of microstructure evolution of bimodal and lamellar microstructures during in situ stretching at 350 °C infers that the deformation and fracture behavior of Ti-55511 alloy are closely related to slip characteristics in the α phase.

Figure 14 shows the schematic diagram of the deformation mechanisms and microstructure evolution of bimodal and lamellar microstructures at 350 °C. At the initial stage of bimodal microstructural deformation (Figure 14b), several parallel SBs along certain slip planes formed within the α$_P$ phase. Since the α$_P$ phase has a limited deformation coordination ability [29], the formation of SBs inside the α$_P$ phase can be deemed a good way of releasing the local stress concentration therein. As the deformation further progresses, some slip systems (focused on prismatic slip and pyramidal slip) were activated to further coordinate the rising deformation during which multislip occurred. Dislocation climbing might occur at an elevated temperature [32,33], which weakened the hindering effect of the α$_P$/β interface on dislocations to a significant extent. Thus, with the further increase in strain, the SBs gradually cut across the α$_P$/β interface and extended into the β matrix or adjacent α$_P$ phase. During this process, the SB deflected significantly to accommodate the slip systems between the neighboring grains. Meanwhile, the SBs in adjacent α_P phases exhibited a strong tendency to bridge with each other because of the significant stress concentration present in the region between SBs.

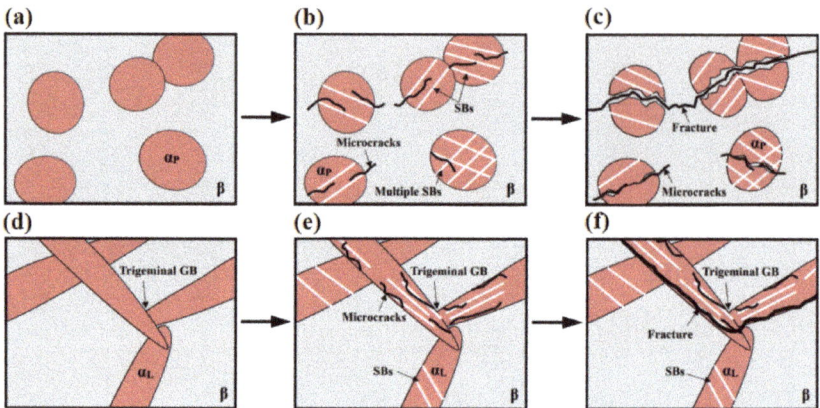

Figure 14. Schematic diagram showing the microstructure evolution of Ti-55511 alloy during stretching in situ; (**a–c**) bimodal microstructure evolution; (**d–f**) lamellar microstructure evolution.

With the cyclic-loading processes, microcracks first initiated within the region of the α_P phase due to the repeated dislocation slipping along certain slip planes (Figure 14c). Subsequently, several microcracks could evolve from the deep SBs in the α_P phase. Like the deflection of SB, once the microcracks passed through the phase interface, the propagation direction would also change, resulting in a tortuous crack propagation path. Then, many microcracks in the α_P phase interconnected along the 45° direction under the strong shearing force in that direction. This finally results in the fracture of the sample. Therefore, as shown in Figure 7, the crack propagation path in the bimodal microstructure is evidently tortuous with frequently deflected or branched cracks. This fracture mechanism endowed the sample with excellent plasticity.

In regions with a lamellar microstructure (Figure 14d), the α_L phases are mainly distributed at GBs and separate the β matrix into many relatively isolated regions, which produce a strong fencing effect on the dislocation in β. The dislocations in β do not readily cross regions consisting of the α_L phase and enter neighboring β grains. This makes it difficult for the long-range slip of dislocations and large plastic deformation to develop within the β matrix (Figure 14e). Due to this fencing effect of the α_L phase, the SBs, dominated by prismatic slip and pyramidal slip, first appear in the α_L phase near the GB due to the significant stress concentration therein, especially for the trigeminal GB where deformation coordination is hard to achieve. When the deformation is insufficient to accommodate the plastic strain, microcracks easily nucleate at the α_L interface boundary at GB. Once the cracks nucleate, they will propagate along the GBs which are thought to be low-energy channels that facilitate crack propagation (Figure 14f). This accounts for the lower ductility of specimens with a lamellar microstructure compared with those having a bimodal microstructure.

5. Conclusions

In this study, the effects of bimodal and lamellar microstructures on the mechanical deformation behavior of Ti-55511 titanium alloy at 350 °C were investigated. The main findings can be summarized as follows:

(1) Multislip systems are activated in the α_P phase to adapt to the plastic strain during the in situ tensile process of bimodal microstructural evolution. The slip modes are dominated by prismatic slip and pyramidal slip.

(2) During in situ stretching at 350 °C, there is a strong bridging tendency of SBs in adjacent α_P phases towards coordinated deformation and alleviation of the stress concentration.
(3) The SBs in the α_P phase are the preferred crack nucleation sites due to their limited deformation ability. They are also likely to connect with the main crack during crack propagation.
(4) Once the crack crosses the α_P/β phase boundary, the crack always deflects significantly, which gives rise to a tortuous crack path and endows specimens with a bimodal microstructure with excellent plasticity.
(5) During the stretching of specimens with a lamellar microstructure, high-density dislocations are concentrated in the large α_L phase region at GB. Microcracks readily initiate and propagate along the α_L phase boundary at GB, leading to the ductile intergranular fracture of the lamellar microstructure.

Author Contributions: Conceptualization, M.F. and S.P.; methodology, S.P., H.L. and M.F.; software, M.F.; formal analysis, M.F., S.P., H.L. and Y.C.; investigation, M.F. and H.L.; resources, S.P., H.L. and Y.C.; data curation, M.F. and Y.C.; writing—original draft, M.F. and Y.C.; supervision, Y.C., S.P., H.L. and M.F. All authors have read and agreed to the published version of the manuscript.

Funding: This research was funded by the State Key Laboratory of Powder Metallurgy (Grant No. 10500-621022001), Central South University, Changsha, China.

Institutional Review Board Statement: Not applicable.

Informed Consent Statement: Informed consent was obtained from all subjects involved in the study.

Data Availability Statement: The data presented in this study are available on request from the corresponding author.

Conflicts of Interest: The authors declare no conflict of interest. The funders had no role in the design of the study; in the collection, analyses, or interpretation of data; in the writing of the manuscript, or in the decision to publish the results.

References

1. Huang, H.L.; Li, D.; Chen, C.; Li, R.D.; Zhang, X.Y.; Liu, S.C.; Zhou, K.C. Selective laser melted near-beta titanium alloy Ti-5Al-5Mo-5V-1Cr-1Fe: Microstructure and mechanical properties. *J. Cent. South Univ.* **2021**, *28*, 1601–1614. [CrossRef]
2. Yadav, P.; Saxena, K.K. Effect of heat -treatment on microstructure and mechanical properties of Ti alloys: An overview. *Mater. Today Proc.* **2020**, *26*, 2546–2557. [CrossRef]
3. Antunes, R.A.; Salvador, C.A.F.; de Oliveira, M.C.L. Materials Selection of Optimized Titanium Alloys for Aircraft Applications. *Mater. Res.* **2018**, *21*, 979–987. [CrossRef]
4. Pan, S.P.; Liu, H.Q.; Chen, Y.Q.; Zhang, X.Y.; Chen, K.W.; Yi, D.Q. α_S dissolving induced mechanical properties decay in Ti-55511 alloy during uniaxial fatigue. *Int. J. Fatigue* **2020**, *132*, 105372–105388. [CrossRef]
5. Boyer, R.R.; Cotton, J.D.; Mohaghegh, M.; Schafrik, R.E. Materials considerations for aerospace applications. *Mrs Bull.* **2015**, *40*, 1055–1065. [CrossRef]
6. Pan, S.P.; Liu, H.Q.; Chen, Y.Q.; Chi, G.F.; Yi, D.Q. Lamellar α fencing effect for improving stress relaxation resistance in Ti-55511 alloy. *Mater. Sci. Eng. A* **2021**, *808*, 140945–140960. [CrossRef]
7. Zhang, H.; Shao, H.; Shan, D.; Wang, K.X.; Cai, L.L.; Yin, E.H.; Wang, Y.L.; Zhuo, L.C. Influence of strain rates on high temperature deformation behaviors and mechanisms of Ti-5Al-5Mo-5V-3Cr-1Zr alloy. *Mater. Charact.* **2021**, *171*, 110794–110802. [CrossRef]
8. Zhou, D.Y.; Gao, H.; Guo, Y.H.; Wang, Y.; Dong, Y.C.; Dan, Z.H.; Chang, H. High-Temperature Deformation Behavior and Microstructural Characterization of Ti-35421 Titanium Alloy. *Materials* **2020**, *13*, 3623. [CrossRef]
9. Pei, W.; Chen, F.H.; Eckert, J.; Pilz, S.; Scudino, S.; Prashanth, K.G. Microstructural evolution and mechanical properties of selective laser melted Ti-6Al-4V induced by annealing treatment. *J. Cent. South Univ.* **2021**, *28*, 1068–1077.
10. Sun, H.; Yu, L.M.; Liu, Y.C.; Zhang, L.Y.; Liu, C.X.; Li, H.J.; Wu, J.F. Effect of heat treatment processing on microstructure and tensile properties of Ti-6Al-4V-10Nb alloy. *Trans. Nonferrous Metals Soc. China* **2019**, *29*, 59–66. [CrossRef]
11. Luo, J.; Ye, P.; Li, M.Q.; Liu, L.Y. Effect of the alpha grain size on the deformation behavior during isothermal compression of Ti-6Al-4V alloy. *Mater. Des.* **2015**, *88*, 32–40. [CrossRef]
12. Wang, T.; Guo, H.Z.; Wang, Y.W.; Peng, X.N.; Zhao, Y.; Yao, Z.K. The effect of microstructure on tensile properties, deformation mechanisms and fracture models of TG6 high temperature titanium alloy. *Mater. Sci. Eng. A* **2011**, *528*, 2370–2379. [CrossRef]
13. Lin, Y.C.; Jiang, X.Y.; Shuai, C.J.; Zhao, C.Y.; He, D.G.; Chen, M.S.; Chen, C. Effects of initial microstructures on hot tensile deformation behaviors and fracture characteristics of Ti-6Al-4V alloy. *Mater. Sci. Eng. A* **2018**, *711*, 293–302. [CrossRef]

14. Jiang, Y.Q.; Lin, Y.C.; Jiang, X.Y.; He, D.G.; Zhang, X.Y.; Kotkunde, N. Hot tensile properties, microstructure evolution and fracture mechanisms of Ti-6Al-4V alloy with initial coarse equiaxed phases. *Mater. Charact.* **2020**, *163*, 110272–110279. [CrossRef]
15. Huang, S.X.; Zhao, Q.Y.; Lin, C.; Wu, C.; Zhao, Y.Q.; Jia, W.J.; Mao, C.L. In-situ investigation of tensile behaviors of Ti-6Al alloy with extra low interstitial. *Mater. Sci. Eng. A* **2021**, *809*, 140958–140969. [CrossRef]
16. Wang, J.; Zhao, Y.; Zhou, W.; Zhao, Q.; Huang, S.; Zeng, W. In-situ investigation on tensile deformation and fracture behaviors of a new metastable β titanium alloy. *Mater. Sci. Eng. A* **2021**, *799*, 140187–140198. [CrossRef]
17. Jin, H.X.; Wei, K.X.; Li, J.M.; Zhou, J.Y.; Peng, W.J. Advances in titanium alloys for aerospace applications. *Trans. Nonferrous Metals Soc. China* **2015**, *25*, 280–292.
18. Wang, Q.J.; Liu, J.R.; Yang, R. Present situation and prospect of high temperature titanium alloy. *J. Aeronaut. Mater.* **2014**, *34*, 1–26.
19. Beausir, B.; Fundenberger, J.J. Analysis Tools For Electron And X-ray Diffraction, ATEX-Software. Université de Lorraine-Metz. 2017. Available online: www.atex-software.eu (accessed on 1 January 2022).
20. Qin, L.; Wang, J.; Wu, Q.; Guo, X.Z.; Tao, J. In-situ observation of crack initiation and propagation in Ti/Al composite laminates during tensile test. *J. Alloys Compd.* **2017**, *712*, 69–75. [CrossRef]
21. Wang, H.; Xin, S.W.; Zhao, Y.Q.; Zhou, W.; Zeng, W.D. Plane strain fracture behavior of a new high strength Ti-5Al-3Mo-3V-2Zr-2Cr-1Nb-1Fe alloy during heat treatment. *Mater. Sci. Eng. A* **2020**, *797*, 140080–140091. [CrossRef]
22. Tan, C.S.; Sun, Q.Y.; Xiao, L.; Zhao, Y.Q.; Sun, J. Cyclic deformation and microcrack initiation during stress controlled high cycle fatigue of a titanium alloy. *Mater. Sci. Eng. A* **2018**, *711*, 212–222. [CrossRef]
23. Pan, S.P.; Fu, M.Z.; Liu, H.Q.; Chen, Y.Q.; Yi, D.Q. In Situ Observation of the Tensile Deformation and Fracture Behavior of Ti-5Al-5Mo-5V-1Cr-1Fe Alloy with Different Microstructures. *Materials* **2021**, *14*, 5794. [CrossRef]
24. Lecomte, J.S.; Philippe, M.J.; Klimanek, P. Plastic deformation of a Ti-6%Al-4%V alloy with a strong transverse-type crystallographic α-texture at elevated temperatures. *Mater. Sci. Eng. A* **1997**, *234–236*, 869–872. [CrossRef]
25. Turner, G.I.; White, J.S. Mechanical hysteresis in beryllium single crystals oriented for (10$\bar{1}$0) <11$\bar{2}$0> slip. *J. Phys. F Met. Phys.* **1973**, *3*, 926–932. [CrossRef]
26. Semiatin, S.L.; Bieler, T.R. Effect of texture and slip mode on the anisotropy of plastic flow and flow softening during hot working of Ti 6Al 4V. *Metall. Mater. Trans. A* **2001**, *32*, 1787–1799. [CrossRef]
27. Banerjee, D.; Williams, J.C. Microstructure and slip character in titanium alloys. *Def. Sci. J.* **2014**, *36*, 191–206. [CrossRef]
28. Demulsant, X.; Mendez, J. Microstructural effects on small fatigue crack initiation and growth in Ti6A14V alloys. *Fatigue Fract. Eng. Mater. Struct.* **1995**, *18*, 1483–1497. [CrossRef]
29. Kacher, J.; Eftink, B.P.; Cui, B.; Robertson, I.M. Dislocation interactions with grain boundaries. *Curr. Opin. Solid State Mater. Sci.* **2014**, *18*, 227–243. [CrossRef]
30. Bridier, F.; Villechaise, P.; Mendez, J. Analysis of the different slip systems activated by tension in a α/β titanium alloy in relation with local crystallographic orientation. *Acta Mater.* **2005**, *53*, 555–567. [CrossRef]
31. Tan, C.S.; Sun, Q.Y.; Zhang, G.J.; Zhao, Y.Q. High-cycle fatigue of a titanium alloy: The role of microstructure in slip irreversibility and crack initiation. *J. Mater. Sci.* **2020**, *55*, 12476–12487. [CrossRef]
32. Rabeeh, B.M.; Rokhlin, S.I.; Soboyejo, W.O. Microplasticity and fracture in a Ti-15V-3Cr-3Al-3Sn alloy. *Scr. Mater.* **1996**, *35*, 1429–1434. [CrossRef]
33. Ning, Y.Q.; Yao, Z.K.; Li, H.; Cuo, H.Z.; Tao, Y.; Zhang, Y.W. High temperature deformation behavior of hot isostatically pressed P/M FGH4096 superalloy. *Mater. Sci. Eng. A* **2010**, *527*, 961–966. [CrossRef]

Article

Corrosion Mechanism and the Effect of Corrosion Time on Mechanical Behavior of 5083/6005A Welded Joints in a NaCl and NaHSO₃ Mixed Solution

Yuqiang Chen [1,*], Hailiang Wu [1], Xiangdong Wang [2], Xianghao Zeng [2], Liang Huang [3], Hongyu Gu [3] and Heng Li [1,*]

1. Hunan Engineering Research Center of Forming Technology and Damage Resistance Evaluation for High Efficiency Light Alloy Components, Hunan University of Science and Technology, Xiangtan 411201, China
2. Zhuzhou CRRC Times Electric Co., Ltd., Zhuzhou 412001, China
3. Zhuzhou Times Metals Manufacturing Co., Ltd., Zhuzhou 412001, China
* Correspondence: 1030109@hust.edu.cn (Y.C.); liheng771212@163.com (H.L.)

Abstract: The effect of corrosion time on the mechanical behavior of 5083/6005A welded joints in a 3.5% NaCl + 0.01 mol/L NaHSO$_3$ solution was evaluated via scanning electron microscopy (SEM), polarization curve analysis, and X-ray photoelectron spectroscopy (XPS). The prediction model of fatigue life after corrosion was established based on the experimental results and the theory of fracture mechanics, and the formula for the effect of corrosion time on lifespan was determined. The results show that with increasing corrosion time, the corrosion of the sample becomes increasingly severe, and the elongation and fatigue life of the 5083/6005A welded joints decrease significantly. The corrosion resistance of the 5083/6005A welded joints decreases with increasing corrosion time because the corrosive medium promotes the destruction of the oxide film and thereby reduces the corrosion resistance. The corrosion products of the 5083/6005A welded joints are Al(OH)$_3$ and AlCl$_3$.

Keywords: 5083/6005A welding joint; mechanical properties; corrosion mechanism; corrosion model; fatigue life prediction model

Citation: Chen, Y.; Wu, H.; Wang, X.; Zeng, X.; Huang, L.; Gu, H.; Li, H. Corrosion Mechanism and the Effect of Corrosion Time on Mechanical Behavior of 5083/6005A Welded Joints in a NaCl and NaHSO₃ Mixed Solution. *Crystals* **2022**, *12*, 1150. https://doi.org/10.3390/cryst12081150

Academic Editor: Cyril Cayron

Received: 19 June 2022
Accepted: 9 August 2022
Published: 16 August 2022

Publisher's Note: MDPI stays neutral with regard to jurisdictional claims in published maps and institutional affiliations.

Copyright: © 2022 by the authors. Licensee MDPI, Basel, Switzerland. This article is an open access article distributed under the terms and conditions of the Creative Commons Attribution (CC BY) license (https://creativecommons.org/licenses/by/4.0/).

1. Introduction

Due to their low density, good formability, and excellent corrosion resistance, 5083 and 6005A aluminum alloys have been widely used for preparing critical components in high-speed trains, such as body structures, electrical cabinets, and lifting lugs [1], which are usually joined by melt inert-gas welding (MIG) [2,3]. In practice, these components of high-speed trains are exposed to diverse service environments, such as humidity and salt spray. The 5083/6005A aluminum alloy welded joints are prone to corrosion, causing a sharp decline in fatigue life and premature fracture of aluminum alloy welded joints [4]. Thus, it is highly desirable and necessary to assess the effect of corrosion on the mechanical behavior of the 5083/6005A welded joints.

It is generally recognized that aluminum alloys are vulnerable to chloride (Cl$^-$) corrosion, especially in the marine atmosphere polluted by industry. The deposition of industrial pollutants such as SO$_2$ in the environment further accelerates the anodic dissolution process of metals. Therefore, many scholars have explored the effect of aluminum alloy corrosion on mechanical behavior in NaCl solutions [5–7], NaHSO$_3$, and NaCl mixed solutions [8]. These studies found that Cl$^-$ and SO$_4^{2-}$ have significant effects on the mechanical properties of the aluminum alloy after corrosion [9–11]. Mishra's study found that the corrosion of aluminum alloy in 3.5% NaCl solution reduces the alloy's mechanical tensile properties and fatigue life, and the impact on fatigue properties is particularly significant [10]. Aluminum alloy exposed to a solution containing Cl$^-$ rapidly generates corrosion pits on its surface, and cracks are induced to propagate under the action of stress [11]. Ma et al. [12] found that, after a 7075 aluminum alloy welded joint was corroded by 3.5% NaCl solution, there were multiple crack sources in the fracture after fatigue fracture, which mainly appeared in the corrosion pits, causing a significant reduction in the fatigue life of the aluminum alloy. In addition, when $NaHSO_3$ is added to NaCl solution, HSO_3^- in the solution is ionized to H^+ and SO_3^{2-}, and SO_3^{2-} is readily oxidized to SO_4^{2-}, which promotes further ionization of HSO_3^-, leading to a decrease in solution pH [8]. In an acidic environment, the lower the pH value, the faster the corrosion rate of aluminum alloys, and the more significant the effect on the mechanical properties of aluminum alloys [13]. Ge et al. [14] estimated the effects of adding different concentrations of $NaHSO_3$ to 3.5% NaCl solution on the mechanical properties of 2024 and 7075 aluminum alloys. They found that the tensile strength and elongation decreased with increasing HSO_3^- concentration, and the corrosion rates of the two aluminum alloys increased. However, the fatigue behavior of dissimilar aluminum alloy welded joints after corrosion in mixed solutions has rarely been reported, especially the fatigue behavior of 5*xxx* and 6*xxx* aluminum alloy welded joints after corrosion in NaCl and $NaHSO_3$ mixed solution. Thus, the fatigue mechanism of 5*xxx* and 6*xxx* aluminum alloy welded joints after corrosion remains unclear.

In the present work, the effect of corrosion time on the mechanical behavior of 5083/6005A welded joints in 3.5% NaCl + 0.01 mol/L $NaHSO_3$ solution was studied. The corrosion mechanism was revealed, and the corrosion model was established by measuring the corrosion products of the 5083/6005A welded joints. The fatigue life prediction model of the 5083/6005A welded joints after corrosion was established based on the fatigue test results after corrosion. This study aims to provide a reference for the safe application of 5083/6005A welded joints in high-speed trains.

2. Materials and Methods

2.1. Materials

The 5083/6005A welded joint specimens were provided by Zhuzhou Times Metal Manufacturing Co., Ltd. (Zhuzhou, China) The welding joint material was 6.0 mm 5083-H111 and 6005A-T6 aluminum alloy plate, and the welding wire was ER5356. Under a flow of 12 L/min of argon, a Fronius TPS4000 MIG welding machine was used to weld the plate perpendicular to the rolling direction. The chemical compositions of these materials determined by SPECTRO BLUE SOP type inductively coupled plasma optical emission spectroscopy (ICP-OES) (SPECTRO Analytical Instruments, Kleve, Germany) are listed in Table 1.

Table 1. Chemical compositions of the 5083 and 6005A aluminum alloys and the ER5356 filler metal (wt. %).

Material	Si	Fe	Cu	Mn	Mg	Zn	Cr	Ti	Al
5083	0.09	0.20	0.01	0.75	4.98	0.02	0.09	0.05	Balance
6005A	0.50	0.19	0.01	0.26	0.71	0.02	0.16	0.06	Balance
ER5356	0.12	0.12	0.08	0.15	4.90	0.12	0.11	0.12	Balance

2.2. Corrosion Tests

Figure 1a shows the welding joint dimensions of the 5083/6005A specimen, and Figure 1b illustrates a schematic diagram of the corrosion testing, in which the size of the service device was 200 × 100 × 250 mm. The corrosion solution was 3.5% NaCl + 0.01 mol/L $NaHSO_3$, and the corrosion times were 3, 7, 15, and 30 days.

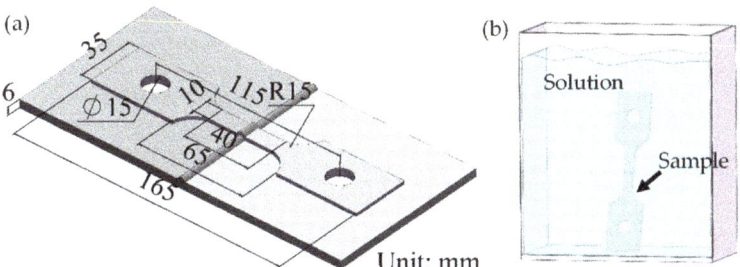

Figure 1. (a) Typical 5083/6005A welded joint; (b) Schematic representation a of the corrosion test.

2.3. Tensile and Fatigue Tests

The tensile tests and fatigue tests were conducted on an MTS-322 servo-hydraulic test machine (maximum load-carrying capacity of ±500 kN) (MTS System Corporation, Eden Prairie, MN, USA) at room temperature. Sinusoidal loading was used with a frequency (f) of 75 Hz, a stress ratio (R) of 0.1, and maximum stress (σ_{max}) of 100 MPa in the fatigue tests. These tests were conducted according to GB/T 228.1-2010, GB/T 3075-2008, and HB 5287-96. Three replicate specimens were tested.

2.4. Electrochemical Measurement

The electrochemical performance of the 5083/6005A welded joints was investigated using a CHI760e electrochemical workstation (CH Instruments Ins., Shanghai, China) with a classic three-electrode system. The electrochemical measurements were conducted in accordance with standard GB/T 24196-2009. The working, counter, and reference electrodes were the 5083/6005A welded joints, a platinum slice, and a saturated calomel electrode (SCE), respectively. The electrolyte was the 3.5 wt.% NaCl aqueous solution, and the tested area for each working electrode was 10 mm × 10 mm. The polarization curves were recorded by scanning the potential from −1.2 V (versus SCE) to −0.3 V at 0.5 mV s^{-1}. Three replicate specimens were tested for each electrochemical test, and a typical polarization curve is reported.

2.5. Characterization Techniques

The surface morphology and corrosion products of the 5083/6005A welded joints were observed by SEM (Hitachi, Tokyo, Japan) and XPS (ThermoFisher-VG Scientific, Waltham, MA, USA), in which the Hitachi SU3500 was used for scanning electron microscopy at an operating voltage of 25 kV, and an ESCALAB 250Xi spectrometer was used for XPS observations with Al Kα radiation as an excitation source.

3. Results and Discussion

3.1. Effect of Corrosion Time on Mechanical Properties

The mechanical properties and stress–strain curves of the 5083/6005A welded joints after being corroded by the 3.5% NaCl + 0.01 mol/L NaHSO$_3$ solution for different times are demonstrated in Figure 2. Figure 2a shows that the yield strength (σ_s), tensile strength (σ_b), and elongation of the uncorroded 5083/6005A welded joints were 125 MPa, 204 MPa, and 11.29%, respectively. After 3, 7, 15, and 30 days of corrosion, the σ_s values of the 5083/6005A welded joints were 126 MPa, 125 MPa, 125 MPa, and 121 MPa, respectively; the σ_b values were 205 MPa, 203 MPa, 200 MPa, and 196 MPa, respectively; and the elongation values were 9.31%, 8.96%, 8.27%, and 7.96%, respectively. The results indicate that corrosion greatly affected the elongation of the 5083/6005A welded joints but had little influence on σ_b and σ_s. As the corrosion time increased, the elongation of the 5083/6005A welded joints decreased, while σ_b and σ_s decreased slightly. Compared with that of the uncorroded 5083/6005A welded joints, the elongation of the 5083/6005A welded joints after corrosion for 3, 7, 15, and 30 days decreased by 17.5%, 20.6%, 26.7%, and 29.5%, respectively.

The results show that the elongation of the 5083/6005A welded joints decreased greatly once they were corroded, but the elongation reduction rate decreased with increasing corrosion time.

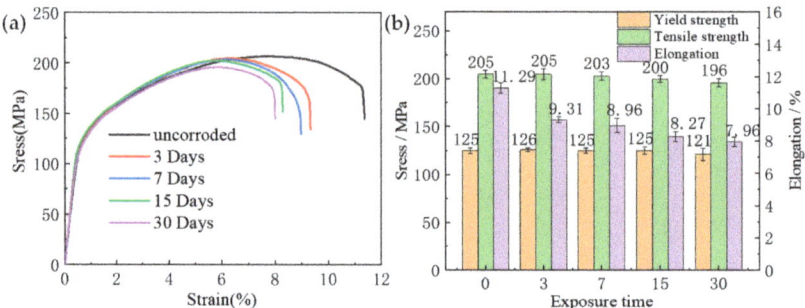

Figure 2. Mechanical properties and stress–strain curves of the 5083/6005a welded joints after corrosion in the 3.5% NaCl + 0.01 mol /L NaHSO$_3$ solution for different times: (**a**) stress–strain curve; (**b**) mechanical properties.

The fatigue properties of the 5083/6005A welded joints corroded for different times are shown in Table 2 and Figure 3: after 3, 7, 15, and 30 days of corrosion, the fatigue life decreased by 84.5%, 90.7%, 93.1%, and 95.7%, respectively. Corrosion had a significant effect on the fatigue life of the 5083/6005A welded joints, and the fatigue life decreased with increasing corrosion time.

Table 2. Fatigue life of 5083/6005A welded joints.

Specimen	Uncorroded	3 Days	7 Days	15 Days	30 Days
N_f/ cycles	10^7	1,159,257	884,098	805,323	449,833
	6,804,918	907,457	672,835	353,091	267,661

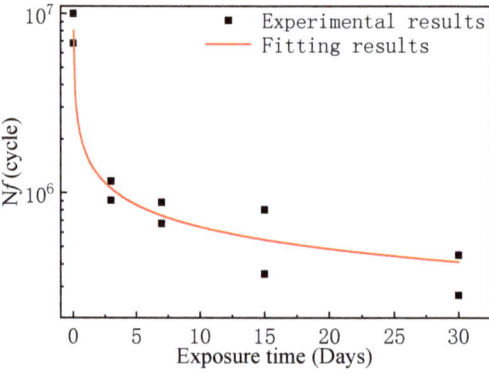

Figure 3. The effect of corrosion time on the fatigue life, obtained by fitting the results in Table 2.

In the fatigue process, stress concentration occurs at the corrosion pits due to alternating stress, resulting in crack nucleation at the corrosion pits [15]. As is shown in Figure 4a, there were many corrosion pits on the surface of the 5083/6005A welded joint after 30 days of corrosion, among which some of them formed clusters with Al(Fe,Mn)Si inclusions contained (Figure 4b,c). Thus, the fatigue life of the 5083/6005A welded joints decreased sharply after corrosion.

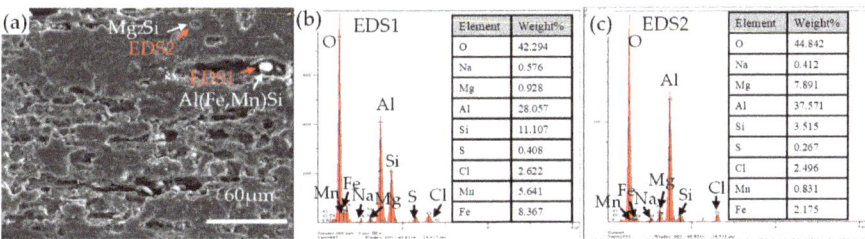

Figure 4. (**a**) SEM of a 5083/6005A welded joint corroded for 30 days; (**b**) EDS1 results; (**c**) EDS2 results.

Figure 5 shows SEM images of fatigue fractures and fracture locations of the 5083/6005A welded joints after corrosion for different times: the fatigue fracture position of the 5083/6005A welded joints after corrosion was mostly on the 6005A side, because the corrosion resistance of aluminum alloy 5xxx is better than that of 6xxx aluminum alloy [8]. Therefore, after the same corrosion time, there were more corrosion pits on one side of 6005A. Previous research indicated that when pits become larger and deeper or form clusters, a higher stress concentration is caused in the fatigue process, resulting in accelerated fatigue initiation and crack propagation and reduced fatigue life [15]. Moreover, the crack propagation region was found to decrease with increasing corrosion time, which may be driven by the decrease in the ductility of the samples (Figure 2). Besides this, the number of corrosion pits at the crack source also gradually increased.

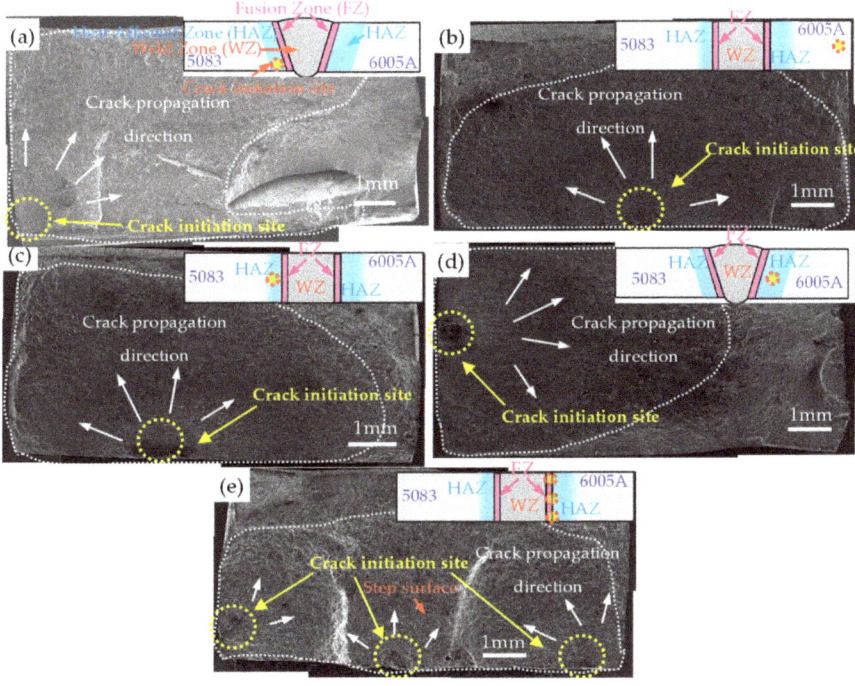

Figure 5. SEM images of fatigue fractures of the 5083/6005A welded joints after corrosion in the 3.5% NaCl + 0.01 mol/L NaHSO$_3$ solution: (**a**) uncorroded; (**b**) 3 days; (**c**) 7 days; (**d**) 15 days; (**e**) 30 days.

Figures 6 and 7 show two typical SEM fractographies of the 5083/6005A welded joints. The crack source is marked with a yellow dotted line, the crack growth area is marked with a white dotted line, and the directions of crack growth are indicated by white arrows. Figure 6a illustrates that the 5083/6005A welded joint fracture cracks occurred at the corrosion pit after three days of corrosion, and then extended radially. In Figure 6a, fatigue striations can be observed in the crack propagation region over a distance of about 0.77 μm. The upper side of the fracture was the final fracture region in which many dimples were distributed, as illustrated in Figure 6d.

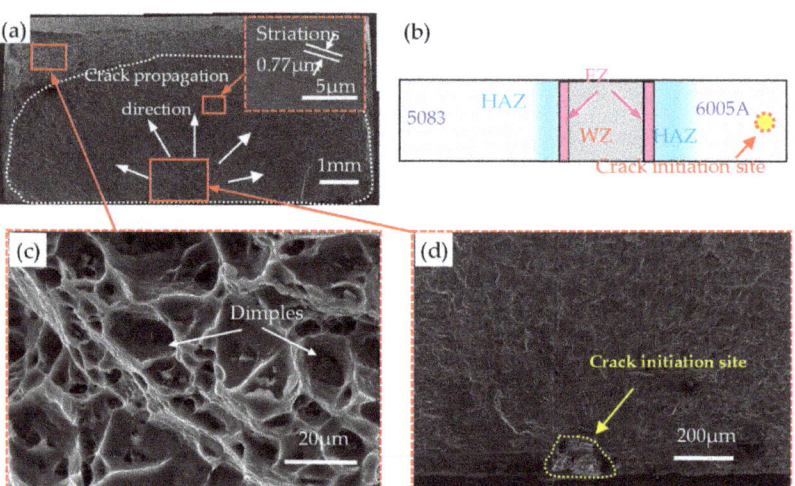

Figure 6. SEM images of fatigue fracture of the 5083/6005A welded joints corroded for three days: (**a**) overall fracture diagram; (**b**) schematic diagram of fracture location; (**c**) dimples; (**d**) crack source.

Figure 7a shows that the fracture surface of the 5083/6005A welded joints corroded for 30 days was irregular and there were three crack sources, with cracks extending radially at each crack source; because the three crack sources were in different planes, when they extended to the same position, they merged into a main crack and continued to extend, leading to the formation of a stepped plane. In Figure 7a, fatigue striations can be seen in the three crack source propagation regions, at spacing of 0.78 μm, 0.91 μm, and 0.83 μm, respectively. The upper end of the fracture was the final fracture region, which was composed of many small dimples, as illustrated in Figure 7b. Figure 7d–f shows rock-candy-type fracture features easily observed at the fracture surface close to crack initiation sites. This indicates that intergranular fracture dominated the early stage of the crack propagation process.

3.2. Effect of Corrosion Time on Corrosion Resistance

Figure 8 shows the polarization curves of the 5083/6005A welded joints in the 3.5% NaCl solution after different corrosion times. The results of the corrosion current density (J_{corr}) calculated by Tafel extrapolation are presented in Table 3. The corrosion potential can reflect the corrosion tendency from a thermodynamic perspective, and the corrosion current density can be used to assess the corrosion rate from the reaction kinetics. Figure 8a and Table 3 indicate that with increasing corrosion time, the corrosion potential (E_{corr}) of the 5083 aluminum alloy matrix of the 5083/6005A welded joints decreased continuously, while J_{corr} increased gradually at the beginning of corrosion and reached its peak value between 15 and 30 days of corrosion, then it decreased continuously. The fast dissolution of the matrix at the early stage of corrosion is probably due to the galvanic cell reaction between the 5083 aluminum alloy matrix and the second phase. Meanwhile, after a certain corrosion time (15 to 30 days), most of the second phase on the surface of the 5083/6005A welded joints had dissolved, resulting in a trend of J_{corr} reduction as evinced by the polarization curve. Compared with that of the uncorroded 5083 aluminum alloy matrix, the E_{corr} of the 5083 aluminum alloy matrix was decreased by 15.3% after 30 days of corrosion. These results show that with increasing corrosion time, the tendency to corrosion of the 5083 aluminum alloy matrix increased, but the rate of corrosion decreased.

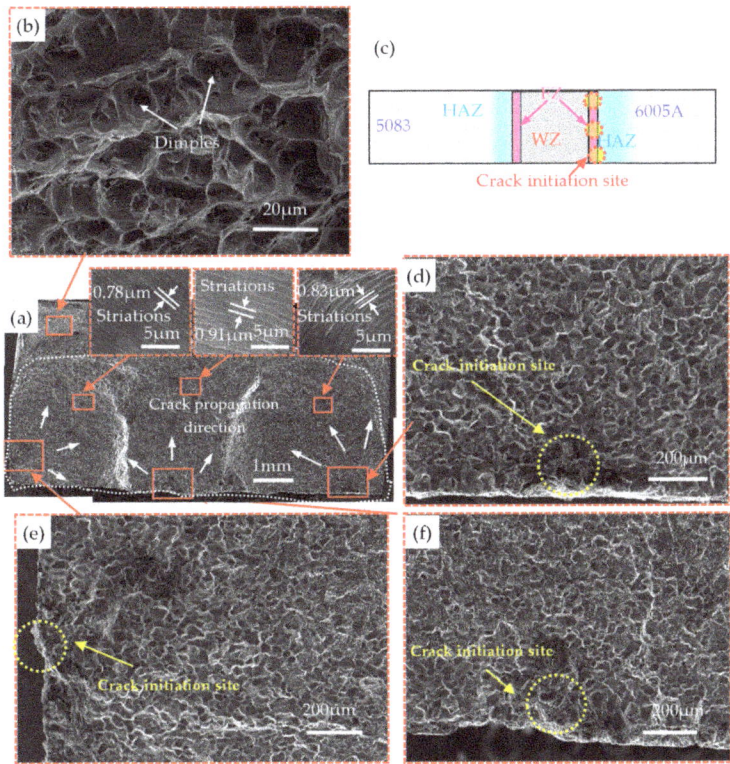

Figure 7. SEM images of fatigue fracture of the 5083/6005A welded joints corroded for 30 days: (**a**) overall fracture diagram; (**b**) dimples; (**c**) fracture location diagram; (**d**) crack source 1; (**e**) crack source 2; (**f**) crack source 3.

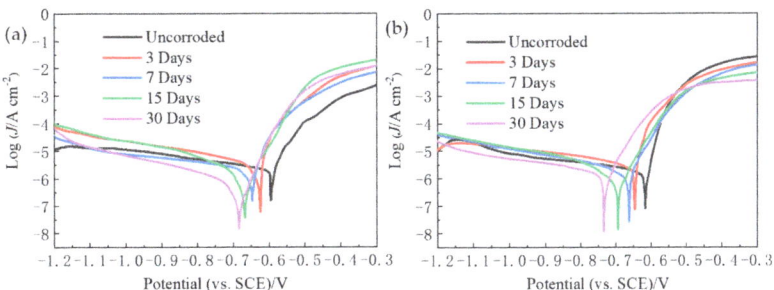

Figure 8. Polarization curves of the 5083/6005a welded joints after corrosion by the 3.5% + 0.01 mol/L $NaHSO_3$ solution: (**a**) 5083 aluminum alloy matrix; (**b**) 6005A aluminum alloy matrix.

Table 3. Test results of polarization curves in different environments for the 5083/6005A welded joints.

Experimental Conditions	5083		6005A	
	E_{corr} (vs. SCE)/ (V)	J_{corr}/ ($\mu A \cdot cm^{-2}$)	E_{corr} (vs. SCE)/ (V)	J_{corr}/ ($\mu A \cdot cm^{-2}$)
Uncorroded	−0.594	2.0	−0.617	1.95
After being corroded for 3 days	−0.626	4.11	−0.647	2.89
After being corroded for 7 days	−0.650	4	−0.663	2.50
After being corroded for 15 days	−0.668	5.01	−0.693	2.18
After being corroded for 30 days	−0.685	0.812	−0.731	1.38

As can be seen from Figure 8b and Table 3, the E_{corr} value of the 5083/6005A welded joint 6005A aluminum alloy matrix decreased with increasing corrosion time, while the J_{corr} value did not change significantly. The E_{corr} and J_{corr} values of 6005A aluminum alloy were lower than those of the 5083 aluminum alloy matrix.

3.3. Analysis of Corrosion Products

The XPS analysis results of the 5083 matrix and 6005A matrix of the 5083/6005A welded joints after 7 days of corrosion are demonstrated in Figure 9. The XPS survey spectrum shown in Figure 9a proves the existence of Mg, Na, O, N, C, Cl, and Al on the 5083/6005A welded joints, among which the C and N elements are mainly due to dissolved carbon dioxide and nitrogen dioxide from the air in the solution, while the Mg, Fe, Al, and Cl elements are due to the corrosion products formed by the second phase and the matrix of the aluminum alloy after corrosion. The presence of elemental S was not detected in the XPS spectrum due to the low content of $NaHSO_3$ in the solution, which resulted in almost no corrosion products containing S on the surface of the 5083/6005A welded joints after corrosion.

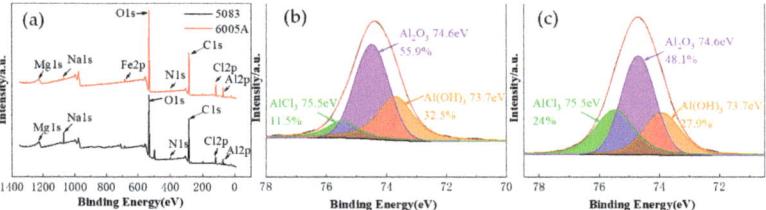

Figure 9. XPS spectra of the 5083/6005A welded joints corroded in the 3.5% NaCl + 0.01 mol/L $NaHSO_3$ solution for 7 days: (**a**) 5083 and 6005A survey spectra; (**b**) 5083 Al2p spectrum; (**c**) 6005A Al2p spectrum.

The spectrum of Al2p in Figure 10b,c is composed of three peaks with respective binding energies of 75.7 eV, 74.5 eV, and 73.9 eV [16,17] (the corresponding substances are $AlCl_3$, Al_2O_3, and $Al(OH)_3$). Since Al_2O_3, the main component of the oxide film, is not a corrosion product formed in the corrosion process, the corrosion products of the 5083/6005A welded joints are $Al(OH)_3$ and $AlCl_3$ [18].

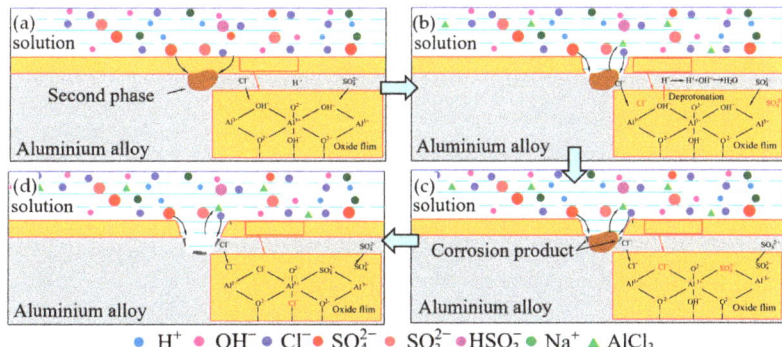

Figure 10. Corrosion model of the 5083/6005A welded joints in a corrosion solution: (**a**) initial stage; (**b**) pitting stage; (**c**) corrosion pit expansion stage; (**d**) second phase shedding stage.

3.4. Corrosion Mechanism and Model

Sakairi et al. [19] found that in the stage of oxide film formation, the oxide film consists of a gel-like structured hydrated oxide, which contains a large amount of bound water. After dehydration, an oxide film with better structure and stronger protection is formed. The oxide film of an aluminum alloy contains various chemical bonds, including HO-Al-OH, O-Al-O, and so on. However, the morphology and thickness of the oxide film in these places are greatly changed due to the presence of intermetallic compounds, forming a defective oxide film [20].

Hoar suggested that chloride ion adsorption on the alloy surface can migrate to the oxide film in a certain way, and then arrive through the oxide film and metal interface. This leads to blunt damage and promotes the dissolution of the active interface of the metal matrix, thus causing local corrosion, and aggressive ion adsorption results in a decrease in the oxide film's surface tension [21,22].

The corrosion mechanism of 6xxx and 5xxx aluminum alloys is related to the iron-rich phase and Mg–Si intermetallic compounds [23]. According to the energy spectrum analysis in Figure 4, the second phase is mainly composed of Al, Mn, Fe, and Si elements. Therefore, the second phase may be Al(Fe,Mn)Si. During the corrosion process, the iron-rich phase acts as the cathode phase, and the aluminum matrix acts as the soluble anode [23].

Therefore, in 3.5% NaCl + 0.01 mol/L NaHSO$_3$ solution, corrosive ions readily migrate to the oxide film, leading to the gradual dissolution of the 5083/6005A aluminum alloy welds. The aluminum matrix reacts as the anode (1) and the second-phase particles react as the cathode (2), gradually reacting (3) and finally forming Al(OH)$_3$.

$$Al - 3e^- \rightarrow Al^{3+} \tag{1}$$

$$O_2 + 2H_2O + 4e^- \rightarrow 4OH^- \tag{2}$$

$$Al^{3+} + 3OH^- \rightarrow Al(OH)_3 \tag{3}$$

The solution contains HSO$_3^-$, which is electrolyzed into SO$_3^{2-}$ and H$^+$ in water (4). As SO$_3^{2-}$ is easily oxidized to SO$_4^{2-}$ in air, a reaction occurs (5).

$$HSO_3^- \rightarrow H^+ + SO_3^{2-} \tag{4}$$

$$2SO_3^{2-} + O_2 \rightarrow 2SO_4^{2-} \tag{5}$$

In the solution containing Cl$^-$ and SO$_4^{2-}$, these ions compete with OH$^-$ to adsorb on the surface of the oxide film. OH$^-$ in the oxide film is gradually replaced, and with increasing corrosion time, the oxide film gradually becomes sparse and breaks [24].

The XPS analysis results imply that the corrosion products of the 5083/6005A welded joints in the 3.5% NaCl + 0.01 mol/L NaHSO$_3$ solution are Al(OH)$_3$ and AlCl$_3$. Therefore, the reaction (6) occurs when the concentration of Al(OH)$_3$ is high in the corrosion pit.

$$Al(OH)_3 + 3Cl^- \rightarrow AlCl_3 + 3OH^- \tag{6}$$

Based on the above analysis, the corrosion model of the 5083/6005A welded joints in the 3.5% NaCl + 0.01 mol/L NaHSO$_3$ solution was established as shown in Figure 10. At the initial stage of corrosion, Cl$^-$ and SO$_4^{2-}$ are adsorbed on the oxide film, and the defective oxide film formed near the Al(Fe,Mn)Si phase is rapidly destroyed under the action of Cl$^-$ and SO$_4^{2-}$, resulting in the corrosion of the aluminum alloy matrix of the 5083/6005A welded joints near the Al(Fe,Mn)Si phase. OH$^-$ in the oxide film not near the Al(Fe,Mn)Si phase is gradually replaced by Cl$^-$ and SO$_4^{2-}$, as shown in Figure 10a,b. With the progress of corrosion, the aluminum alloy matrix near the 5083/6005A welding joint with its Al(Fe,Mn)Si phase is corroded more severely, accompanied by the generation of many corrosion products. However, the surface of the corrosion pit is only covered with a small amount of corrosion products, because NaCl is the main component in the 3.5% NaCl + 0.01 mol/L NaHSO$_3$ solution, and the main corrosion product is AlCl$_3$. AlCl$_3$ is easily soluble in water, and Al2(SO$_4$)$_3$ will cover the surface of the aluminum alloy. No elemental S could be detected in the XPS spectra, suggesting that the content of Al2(SO$_4$)$_3$ is very small. Therefore, few corrosion products are present in the pit, as shown in Figure 10c. At the later stage of corrosion, the aluminum alloy matrix around the Al(Fe,Mn)Si phase is completely corroded, and then the Al(Fe,Mn)Si phase spalls, causing the formation of corrosion pits, as shown in Figure 10d.

3.5. Prediction Model of Fatigue Life after Corrosion

A surface corrosion pit on a 5083/6005A welded joint is equivalent to an initial crack on a semi-elliptical surface, so the fatigue life was predicted based on the methods of fracture mechanics. A Cartesian coordinate system was established at the center of the corrosion pit as the origin O, with the width (w) direction as the x-axis and the thickness direction (t) as the y-axis. A schematic representation of the equivalent semi-elliptic crack is shown in Figure 11. The intersection points of the corrosion pit and the positive axes of the two axes are A (0, a_i) and B (b_i, 0), respectively, as shown in Figure 12a–f. In the model with multiple corrosion pits, it is necessary to superimpose the long and short axes of multiple equivalent semi-ellipses into the long and short axes of an equivalent semi-ellipse crack, the long and short axes of which are a_p and b_p, respectively.

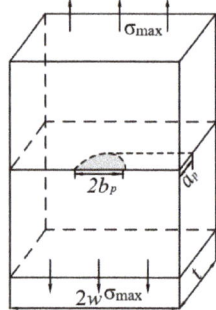

Figure 11. Schematic representation of the equivalent semi-elliptic crack.

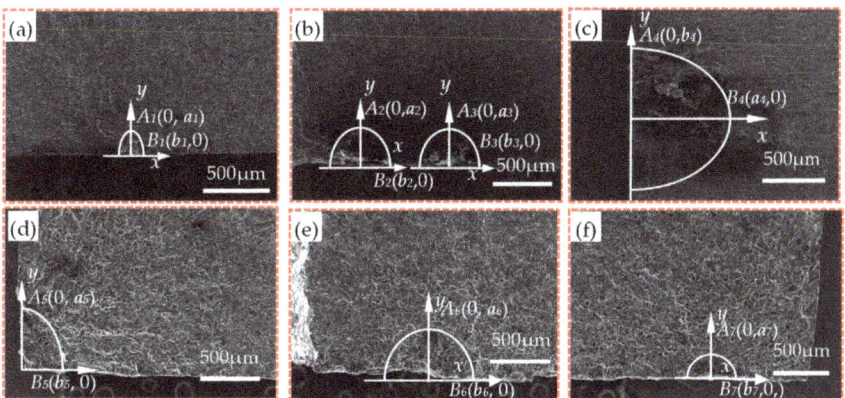

Figure 12. Equivalent model of a corrosion pit: (**a**) 3 days; (**b**) 7 days; (**c**) 15 days; (**d**), (**e**) and (**f**) 30 days.

Many scholars have previously established fatigue life prediction models by combining the crack growth rate formula with the initial crack length and the critical crack length when fracture occurs [25,26]. The rate of crack growth da/dN is commonly used to indicate the crack growth performance of metal materials. Herein, the fatigue life prediction of the 5083/6005A was made based on Paris law [27] as given by:

$$\frac{da}{dN} = C(\Delta K)^n \qquad (7)$$

where a denotes the crack depth, N is the fatigue life, C and n are the material parameters, and ΔK is the applied stress intensity factor range.

The crack can only extend when it is opened, so the crack opening function f_{op} is introduced as follows [28]:

$$\begin{cases} f_{op} = \frac{\sigma_{op}}{\sigma_{max}} = \frac{K_{op}}{K_{max}} = \begin{cases} A_0 + A_1 R + A_2 R^2 + A_3 R^3 & R \geq 0 \\ A_0 + A_1 R & R < 0 \end{cases} \\ A_0 = (0.825 - 0.34\alpha + 0.05\alpha^2)[\cos(\frac{\pi \sigma_{max}}{2\sigma_0})]^{\frac{1}{\alpha}} \\ A_1 = (0.415 - 0.071\alpha)\frac{\sigma_{max}}{\sigma_0} \\ A_2 = 1 - A_0 - A_1 - A_3 \\ A_3 = 2A_0 + A_1 - 1 \end{cases} \qquad (8)$$

where K_{OP} is the K value when mating surfaces of the crack make contact; α is the plane stress constraint factor, such that $\alpha = 1$ denotes plane stress conditions; and σ_0 is the flow stress, such that $\sigma_0 = 1.15(\sigma_s + \sigma_b)/2$.

When $K < K_{OP}$, crack growth is suppressed, since the crack is closed. In this case, the fatigue crack propagation behavior is found to depend on the effective stress intensity factor range (ΔK_{eff}) rather than the nominally applied value [29,30].

$$\Delta K_{eff} = K_{max} - K_{OP} \qquad (9)$$

The stress intensity factor K_{max} of a Type-I equivalent crack can be expressed as follows:

$$K_{max} = \frac{\sigma_{max}\sqrt{\pi a}}{E(k)} \bullet F_I(\frac{a_p}{b_p}, \frac{a_p}{t}, \frac{b_p}{w}, \theta) \qquad (10)$$

where σ_{max} is the maximum stress, and $E(k)$ is the complete elliptic integral of the second kind. F_I represents the correction function for the crack, and the expression of F_I [31] is:

$$F_I = [M_1 + M_2(\frac{a_p}{t})^2 + M_3(\frac{a_p}{t})^4]g_1 f_\theta f_w \tag{11}$$

When $a_p/b_p \leq 1$, the specific expressions of each coefficient in the formula are given as follows:

$$\begin{cases} M_1 = 1.13 - 0.09(\frac{a_p}{b_p}) \\ M_2 = -0.54 + \frac{0.89}{0.2+(\frac{a_p}{b_p})} \\ M_3 = 0.5 - \frac{1.0}{0.65+(\frac{a_p}{b_p})} + 14(1-\frac{a_p}{b_p})^{24} \\ g_1 = 1 + [0.1 + 0.35(\frac{a_p}{t})^2](1-\sin\theta)^2 \\ f_\theta = [(\frac{a_p}{b_p})^2 \cos^2\theta + \sin^2\theta]^{\frac{1}{4}} \\ f_w = [\sec(\frac{\pi b_p}{2w}\sqrt{\frac{a_p}{t}})]^{\frac{1}{2}} \end{cases}, \frac{a_p}{b_p} \leq 1 \tag{12}$$

When $a_p/b_p > 1$, the specific expressions of each coefficient in the formula are as follows:

$$\begin{cases} M_1 = \sqrt{\frac{b_p}{a_p}}(1 + 0.04\frac{b_p}{a_p}) \\ M_2 = 0.2(\frac{b_p}{a_p})^4 \\ M_3 = -0.11(\frac{b_p}{a_p})^4 \\ g_1 = 1 + [0.1 + 0.35(\frac{b_p}{a_p})(\frac{a_p}{t})^2](1-\sin\theta)^2 \\ f_\theta = [(\frac{b_p}{a_p})^2 \sin^2\theta + \cos^2\theta]^{\frac{1}{4}} \\ f_w = [\sec(\frac{\pi b_p}{2w}\sqrt{\frac{a_p}{t}})]^{\frac{1}{2}} \end{cases}, \frac{a_p}{b_p} > 1 \tag{13}$$

where θ denotes the angular location.

The complete elliptic integral of the second kind is expressed as follows:

$$\begin{cases} E(k) = [1 + 1.464(\frac{a_p}{b_p})^{1.65}]^{\frac{1}{2}} & \frac{a_p}{b_p} \leq 1 \\ E(k) = [1 + 1.464(\frac{b_p}{a_p})^{1.65}]^{\frac{1}{2}} & \frac{a_p}{b_p} > 1 \end{cases} \tag{14}$$

Under the action of external force, the stress intensity factor K at the tip of a Type-I crack increases with crack propagation, and when K reaches the critical value K_{IC}, the sample fractures. When fracturing occurs, the critical crack length is as follows:

$$a_c = \frac{1}{\pi}(\frac{K_{IC} \bullet E(k)}{F_I \bullet \sigma_{max}})^2 \tag{15}$$

After the transformation of Formula (7), it is integrated from the initial crack depth a_p to the critical crack depth at fracture a_c, which is shown below.

$$N = \int_{a_p}^{a_c} \frac{da}{C(\Delta K_{eff})^n} \tag{16}$$

The material crack growth performance parameters are $n = 7.664$ and $C = 4 \times 10^{-13}$ mm/cycle [32]. The fatigue load parameters are $R = 0.1$ and $\sigma_{max} = 100$ MPa. The fracture toughness K_{IC} of the 6005A side of the 5083/6005A welded joints is 51 Mpa·m$^{1/2}$ [33]. The geometric shape parameters of the sample, σ_0 values, geometric parameters of corrosion pits, experimental results of fatigue life, and predicted results are listed in Table 4 (it should be pointed out that the values of t and w for the samples after corrosion for 15 days were exchanged since the crack propagated along the transverse, rather than thickness, direction of the plate).

Table 4. Parameters of the fatigue life prediction model for the 5083/6005A welded joints.

Sample	a_p/μm	b_p/μm	t/mm	w/mm	θ	σ_0/MPa	Experimental Fatigue Life (Average)/Cycle	Predicted Fatigue Life/Cycles
3 days	198	91	6	5	0.125π	190.3	1,033,357	1,075,621
7 days	673	541	6	5	0.125π	188.6	778,467	852,645
15 days	721	547	10	3	0.1875π	187.0	579,207	604,837
30 days	1076	934	6	5	0.125π	182.8	338,747	309,234

Based on Formula (17), the effect of corrosion time on the fatigue life of the 5083/6005A welded joints can be obtained by fitting the experimental results (Figure 13).

$$Nf = 1.7 \times 10^6 \bullet t^{-0.418} \tag{17}$$

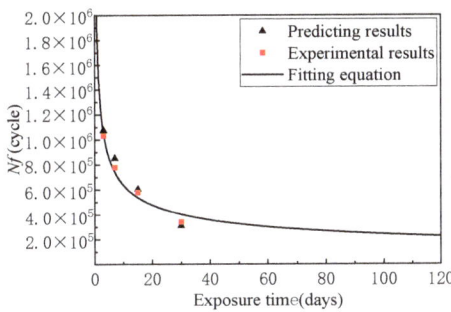

Figure 13. The effect of corrosion time on the fatigue life of the 5083/6005A welded joints.

4. Conclusions

1. With increasing corrosion time, the elongation and fatigue life of the 5083/6005A welded joints continued to decrease, and the corrosion phenomenon became more severe.

2. After 3, 7, 15, and 30 days of corrosion in the solution, the elongation of the 5083/6005A welded joint specimens was decreased by 17.5%, 20.6%, 26.7%, and 29.5%, respectively, and the fatigue life was decreased by 84.5%, 90.7%, 93.1%, and 95.7%, respectively.

3. After 30 days of corrosion, the tendency to corrosion of the 5083/6005A welded joint specimens increased, while the rate of corrosion decreased.

4. The corrosion products of the 5083/6005A welded joints in the 3.5% NaCl + 0.01 mol/L NaHSO$_3$ solution were Al(OH)$_3$ and AlCl$_3$.

5. The formula describing the effect of corrosion time on the fatigue life of the 5083/6005A welded joints is $N_f = 1.7 \times 10^6 \bullet t^{-0.418}$.

Author Contributions: Conceptualization, H.W. and Y.C.; methodology, H.W., H.L. and Y.C.; software, H.W. and H.L.; formal analysis, H.W. and Y.C.; investigation, X.Z. and X.W.; resources, H.G., and L.H.; data curation, H.W. and Y.C.; writing—original draft, H.W. and Y.C.; supervision, H.W., Y.C., X.Z., X.W., H.G., H.L. and L.H. All authors have read and agreed to the published version of the manuscript.

Funding: This research was funded by the National Natural Science Foundation of China (Grant nos. 52075166 and U21A20130), the Natural Science Foundation of Hunan Province (2020JJ5171), the High-level Talent Gathering-Innovative Talent Project of 2021 Hunan (2021RC5010), and China Postdoctoral Science Foundation (2022M712642).

Institutional Review Board Statement: Not applicable.

Informed Consent Statement: Not applicable.

Data Availability Statement: The data presented in this study are available on request from the corresponding author.

Conflicts of Interest: The authors declare no conflict of interest. The funders had no role in the design of the study; in the collection, analyses, or interpretation of data; in the writing of the manuscript; or in the decision to publish the results.

References

1. Fan, L.; Ma, J.J.; Zou, C.X.; Gao, J.; Wang, H.S.; Sun, J.; Guan, Q.M.; Wang, J.; Tang, B.; Li, J.S.; et al. Revealing foundations of the intergranular corrosion of 5xxx and 6xxx Al alloys. *Mater. Lett.* **2020**, *271*, 127767. [CrossRef]
2. Chen, Y.Q.; Xu, J.B.; Pan, S.P.; Li, N.B.; Ou, C.G.; Liu, W.H.; Song, Y.F.; Tan, X.R.; Liu, Y. Effect of T6I6 treatment on dynamic mechanical behavior of Al-Si-Mg-Cu cast alloy and impact resistance of its cast motor shell. *J. Cent. South Univ.* **2022**, *29*, 924–936. [CrossRef]
3. Arifurrahman, F.; Budiman, B.A.; Aziz, M. On the lightweight structural design for electric road and railway vehicles using fiber reinforced polymer composites-a review. *Int. J. Sustain. Transp.* **2018**, *1*, 21–29. [CrossRef]
4. Lu, W.; Ma, C.P.; Gou, G.Q.; Fu, Z.H.; Sun, W.G.; Che, X.L.; Chen, H.; Gao, W. Corrosion fatigue crack propagation behavior of A7N01P-T4 aluminum alloy welded joints from high-speed train underframe after 1.8 million km operation. *Mater. Corros.* **2021**, *72*, 879–887. [CrossRef]
5. Chanyathunyaroj, K.; Phetchcrai, S.; Laungsopapun, G.; Rengsomboon, A. Fatigue characteristics of 6061 aluminum alloy subject to 3.5% NaCl environment. *Int. J. Fatigue* **2020**, *133*, 105420. [CrossRef]
6. Wang, L.W.; Liang, J.M.; Li, H.; Cheng, L.J.; Cui, Z.Y. Quantitative study of the corrosion evolution and stress corrosion cracking of high strength aluminum alloys in solution and thin electrolyte layer containing Cl^-. *Corros. Sci.* **2021**, *178*, 109076. [CrossRef]
7. Shen, L.; Chen, H.; Che, X.L.; Xu, L.D. Corrosion—Fatigue crack propagation of aluminum alloys for high-speed trains. *Int. J. Mod. Phys. B* **2017**, *31*, 1744009. [CrossRef]
8. Zhang, Z.; Xu, Z.L.; Sun, J.; Zhu, M.T.; Yao, Q.; Zhang, D.J.; Zhang, B.W.; Xiao, K.; Wu, J.S. Corrosion behaviors of AA5083 and AA6061 in artificial seawater, effects of Cl^-, HSO_3^- and temperature. *Int. J. Electrochem. Sci.* **2020**, *15*, 1218–1229. [CrossRef]
9. Zhao, Q.Y.; Guo, C.; Niu, K.K.; Zhao, J.B.; Huang, Y.H.; Li, X.G. Long-term corrosion behavior of the 7A85 aluminum alloy in an industrial-marine atmospheric environment. *J. Mater. Res. Technol.* **2021**, *12*, 1350–1359. [CrossRef]
10. Mishra, R.K. Study the effect of pre-corrosion on mechanical properties and fatigue life of aluminum alloy 8011. *Mater. Today Proc.* **2020**, *25*, 602–609. [CrossRef]
11. Ghosh, R.; Venugopal, A.; Rao, G.S.; Narayanan, P.R.; Pant, B.; Cherian, R. Effect of temper condition on the corrosion and fatigue performance of AA2219 aluminum alloy. *J. Mater. Eng. Perform.* **2018**, *27*, 423–433. [CrossRef]
12. Ma, Q.N.; Shao, F.; Bai, L.Y.; Xu, Q.; Xie, X.K.; Shen, M. Corrosion fatigue fracture characteristics of FSW 7075 aluminum alloy joints. *Materials* **2020**, *13*, 4196. [CrossRef] [PubMed]
13. Seetharaman, R.; Ravisankar, V.; Balasubramanian, V. Corrosion performance of friction stir welded AA2024 aluminium alloy under salt fog conditions. *Trans. Nonferrous Met. Soc. China* **2015**, *25*, 1427–1438. [CrossRef]
14. Ge, F.; Zhang, L.; Tian, H.Y.; Yu, M.D.; Liang, J.M.; Wang, X. Stress corrosion cracking behavior of 2024 and 7075 high-strength aluminum alloys in a simulated marine atmosphere contaminated with SO_2. *J. Mater. Eng. Perform.* **2020**, *29*, 410–422. [CrossRef]
15. Weber, M.; Eason, P.D.; Özdeş, H.; Tiryakioğlu, M. The effect of surface corrosion damage on the fatigue life of 6061-T6 aluminum alloy extrusions. *Mater. Sci. Eng. A* **2017**, *690*, 427–432. [CrossRef]
16. Wang, B.; Zhang, L.W.; Su, Y.; Mou, X.L.; Xiao, Y.; Liu, J. Investigation on the corrosion behavior of aluminum alloys 3A21 and 7A09 in chloride aqueous solution. *Mater. Des.* **2013**, *50*, 15–21. [CrossRef]
17. Yang, X.K.; Zhang, L.W.; Zhang, S.Y.; Liu, M.; Zhou, K.; Mu, X.L. Properties degradation and atmospheric corrosion mechanism of 6061 aluminum alloy in industrial and marine atmosphere environments. *Mater. Corros.* **2017**, *68*, 529–535. [CrossRef]
18. Li, Z.; Zhang, Z.; Chen, X.G. Microstructure, elevated-temperature mechanical properties and creep resistance of dispersoid-strengthened Al-Mn-Mg 3xxx alloys with varying Mg and Si contents. *Mater. Sci. Eng. A* **2017**, *708*, 383–394. [CrossRef]
19. Sakairi, M.; Otani, K.; Kaneko, A.; Seki, Y.; Nagasawa, D. Analysis of chemical compositions and morphology of surface films formed on 3003 aluminum alloy by immersion in different cation containing model tap waters. *Surf. Interface Anal.* **2013**, *45*, 1517–1521. [CrossRef]
20. Dong, C.; Kui, X.; Lin, X.; Sheng, H. Characterization of 7A04 aluminum alloy corrosion under atmosphere with chloride ions using electrochemical techniques. *Rare Met. Mater. Eng.* **2011**, *40*, 275–279.
21. Hoar, T.P.; Mears, D.C.; Rothwell, G.P. The relationships between anodic passivity, brightening and pitting. *Corros. Sci.* **1965**, *5*, 279–289. [CrossRef]
22. Hoar, T.P. The production and breakdown of the passivity of metals. *Corros. Sci.* **1967**, *7*, 341–355. [CrossRef]

23. Kumar, K.K.; Kumar, A.; Satyanarayana, M. Effect of friction stir welding parameters on the material flow, mechanical properties and corrosion behavior of dissimilar AA5083–AA6061 joints. *Proc. Inst. Mech. Eng. Part C* **2022**, *236*, 2901–2917. [CrossRef]
24. Chen, Y.Q.; Zhang, H.; Pan, S.P.; Song, Y.F.; Liu, W.H. Effects of service environment and pre-deformation on the fatigue behaviour of 2524 aluminium alloy. *Arch. Civ. Mech. Eng.* **2020**, *20*, 5. [CrossRef]
25. Nan, Z.Y.; Ishihara, S.; Goshima, T. Corrosion fatigue behavior of extruded magnesium alloy AZ31 in sodium chloride solution. *Int. J. Fatigue* **2008**, *30*, 1181–1188. [CrossRef]
26. Mohabeddine, A.; Correia, J.A.F.O.; Montenegro, P.A.; Castro, J.M. Fatigue crack growth modelling for cracked small-scale structural details repaired with CFRP. *Thin Wall Struct.* **2021**, *161*, 107525. [CrossRef]
27. Chen, Y.Q.; Zhang, H.; Song, W.W.; Pan, S.P.; Liu, W.H.; Zhu, B.W.; Song, Y.F.; Zhou, W. Acceleration effect of a graphite dust environment on the fatigue crack propagation rates of Al alloy. *Int. J. Fatigue* **2019**, *126*, 20–29. [CrossRef]
28. Newman, J.C. A crack opening stress equation for fatigue crack growth. *Int. J. Fatigue* **1984**, *24*, R131–R135. [CrossRef]
29. Chen, Y.Q.; Tang, Z.H.; Pan, S.P.; Liu, W.H.; Song, Y.F.; Zhu, B.W.; Liu, Y.; Wen, Z.L. The fatigue crack behavior of 7N01-T6 aluminum alloy in different particle environments. *Arch. Civ. Mech. Eng.* **2020**, *20*, 129. [CrossRef]
30. Chen, Y.Q.; Tang, Z.H.; Pan, S.P.; Liu, W.H.; Song, Y.F.; Liu, Y.; Zhu, B.W. The fatigue crack growth behaviour of 2524-T3 aluminium alloy in an Al_2O_3 particle environment. *Fatigue Fract. Eng. Mater. Struct.* **2020**, *43*, 2376–2389. [CrossRef]
31. Newman, J.C.; Raju, I.S. Stress-intensity factor equations for cracks in three-dimensional. In Proceedings of the Fourteenth National Symposium on Fracture Mechanics, Los Angeles, CA, USA, 30 June 1981.
32. Rubio-González, C.; Ocaña, J.L.; Gomez-Rosas, G. Effect of laser shock processing on fatigue crack growth and fracture toughness of 6061-T6 aluminum alloy. *Mater. Sci. Eng. A* **2004**, *386*, 291–295. [CrossRef]
33. Mrowka-Nowotnik, G. Influence of precipitation strengthening process on tensile and fracture behaviour of the 6005 and 6082 alloys. *Adv. Manuf. Sci. Technol.* **2008**, *32*, 31–50.

Article

Microstructure and Mechanical Behavior of Cu–Al–Ag Shape Memory Alloys Processed by Accumulative Roll Bonding and Subsequent Annealing

Parinaz Seifollahzadeh [1], Morteza Alizadeh [1,*], Ábel Szabó [2], Jenő Gubicza [2,*] and Moustafa El-Tahawy [2,3]

[1] Department of Materials Science and Engineering, Shiraz University of Technology, Modarres Blvd., Shiraz 71557-13876, Iran
[2] Department of Materials Physics, Eötvös Loránd University, P.O. Box 32, H-1518 Budapest, Hungary
[3] Department of Physics, Faculty of Science, Tanta University, Tanta 31527, Egypt
* Correspondence: alizadeh@sutech.ac.ir (M.A.); jeno.gubicza@ttk.elte.hu (J.G.); Tel.: +98-713-7278491 (M.A.); +36-1-372-2876 (J.G.)

Abstract: Ultrafine-grained Cu/Al/Ag composites were processed by an accumulative roll bonding (ARB) technique from pure copper and aluminum sheets and a silver powder. The Al content was fixed to 11 wt.% while the silver concentration was 1, 2, or 3 in wt.%. The ARB-processed samples were heat treated at different temperatures between 750 and 1050 °C for 60 min and then quenched to room temperature (RT) for producing Cu–Al–Ag alloys. The effect of the addition of different Ag contents and various heat treatment temperatures on the structural evolution was investigated. The ARB-processed samples were composed of Cu and Al layers with high dislocation density and fine grain size (a few microns). During heat treatment of the ARB-processed samples, new intermetallic phases formed. For the lowest Ag content (1 wt.%), the main phase was a brittle simple cubic Al_4Cu_9, while for higher Ag concentrations (2 and 3 wt.%), the quenched samples contain mainly an orthorhombic β1-$AlCu_3$ martensite phase. The martensite phase consisted of very fine lamellas with a thickness of one micron or less. The heat treatment increased the microhardness and the strength of the samples at RT due to the formation of a fine-grained hard martensite phase. For 2 and 3% Ag, the highest martensite phase content was achieved at 850 and 950 °C, respectively. The annealed and quenched samples exhibited good shape memory behavior at RT.

Keywords: Cu–Al–Ag alloy; martensite; accumulative roll bonding (ARB); shape memory alloy (SMA); microstructure; mechanical properties

Citation: Seifollahzadeh, P.; Alizadeh, M.; Szabó, Á.; Gubicza, J.; El-Tahawy, M. Microstructure and Mechanical Behavior of Cu–Al–Ag Shape Memory Alloys Processed by Accumulative Roll Bonding and Subsequent Annealing. *Crystals* **2022**, *12*, 1167. https://doi.org/10.3390/cryst12081167

Academic Editors: Yang Zhang and Yuqiang Chen

Received: 25 July 2022
Accepted: 14 August 2022
Published: 19 August 2022

Publisher's Note: MDPI stays neutral with regard to jurisdictional claims in published maps and institutional affiliations.

Copyright: © 2022 by the authors. Licensee MDPI, Basel, Switzerland. This article is an open access article distributed under the terms and conditions of the Creative Commons Attribution (CC BY) license (https:// creativecommons.org/licenses/by/ 4.0/).

1. Introduction

In recent years, composites with ultrafine-grained (UFG) microstructures have attracted significant attention due to their improved mechanical properties [1–3]. UFG microstructure can be fabricated by severe plastic deformation (SPD) methods such as equal channel angular pressing (ECAP) [4,5], high-pressure torsion (HPT) [6,7], repetitive corrugation and straightening (RCS) [8], and accumulative roll bonding (ARB) [9–11]. Among all of these methods, ARB is the easiest way to apply in an industrial environment since it can be performed by conventional rolling apparatuses. During ARB processing, rolling, cutting, and stacking steps are applied consecutively on a sheet in order to achieve grain refinement without reduction in the thickness of the workpiece [12]. If a sandwich-like material consisting of different metallic layers is processed by ARB, a layered composite structure forms [1]. Post-processing heat treatment of layered composite structures obtained by ARB can yield multiphase alloys [13].

UFG Cu–based alloys with different compositions have already been produced by ARB in the literature [14–18]. The as-processed alloys have advantageous properties such as excellent wear resistance, high mechanical strength, and good corrosion resistance, which make them potential candidates as raw materials in electronic and automotive industries [19,20]. In addition, some Cu–based compositions exhibit a shape memory effect (SME). SME occurs due to martensitic thermoelastic transformation [21,22]. Binary Cu–Al and Cu–Zn alloys are the two main systems of Cu–based shape memory alloys (SMAs). In both systems, the main phase must be the β-phase for exhibiting SME. However, it has been reported that binary Cu–Al alloys exhibit a weak SME; therefore, attempts were made for improving SME by adding a third alloying element [23]. For example, Zn and Mn have been added to Cu–Al alloys, and then processed by ARB [24,25]. It was reported that the type and concentration of alloying elements influence the microstructure evolution during ARB and the properties of Cu–based SMAs. It has also been shown that the addition of Ag to Cu–based SMAs fabricated by casting not only yields a higher degree of martensitic transformation but also improves the ductility and increases the martensitic transformation temperature [23,26]. Namely, Cu–Al alloys containing Ag exhibit martensitic transformation temperatures well above 200 °C, while the most commonly used Cu–Ni–Ti and Cu–Al–Ni SMAs can be used only below 200 °C. For obtaining the best SME in Cu–Al–Ag alloys, the silver content and the heat treatment conditions must be optimized. For this purpose, experiments must be conducted for studying the phase evolution in Cu–Al–Ag SMAs during thermal cycling between martensite finish (M_f) and austenite finish (A_f) temperatures.

The classical methods for the fabrication of SMAs are casting and powder metallurgy. In casting methods, oxidation or evaporation of some elements can occur, resulting in a change in the alloy composition. To avoid this effect, the application of vacuum or shielding gases is necessary. Moreover, casting results in a coarse-grained material, and additives are needed to refine the grain size. Powder metallurgy leads to porosity inside the samples at the end of processing [24]. Thus, in order to obtain fine-grained SMAs with high strength, SPD-processing routes are suggested to be used instead of the conventional casting technique. ARB is a non-expensive SPD method that can result in a fine-grained material with no or little porosity. However, this is a two-step method since first a layered composite of the constituents (e.g., Cu and Al) forms and then the ARB-processed sandwich material must be heat treated in order to obtain SMA.

In the present work, UFG multilayered Cu–Al–Ag materials with three different Ag contents were fabricated by ARB. The as-processed multilayered samples were subjected to heat treatment processes in order to produce Cu–Al–Ag SMAs for investigating the effect of different heat treatments and Ag contents on the martensite phase evolution in these alloys. Immediately after ARB, the UFG microstructure in the major Cu phase was studied by X-ray line profile analysis (XLPA) in order to reveal the microstructure evolution due to SPD processing. The microstructure and the phase composition were investigated by scanning electron microscopy (SEM), electron backscatter diffraction (EBSD), and X-ray diffraction (XRD) for the ARB-processed and the subsequently heat-treated specimens. The effect of the Ag content and the annealing temperature on the phase composition and the mechanical behavior is discussed in detail.

2. Materials and Methods

2.1. Starting Materials and ARB Processing of Multi-Layered Composites

Cu/Al/Ag composites with three different compositions were produced by ARB. In this process, pure copper (>99.9 wt.%) and aluminum (>99 wt.%) in sheet form and pure silver powder (>99 wt.%) with a particle size of <5 μm were used as starting materials. The initial thicknesses of the Cu and Al sheets were 0.5 and 0.3 mm, respectively. These sheets and the Ag powder were ARB-processed for 1, 5, and 9 cycles [27] in order to obtain layered composites. The silver powder was placed between the Cu and Al layers before ARB. The samples have a constant amount of Al with three different Ag concentrations as Cu/11 wt.% Al/x wt.% Ag (x = 1, 2, or 3). It should be noted that the chemical composition of the as-processed composites depends on the thicknesses and the numbers of the Cu and

Al sheets as well as the fraction of the Ag powder. These data before the first ARB cycle are listed in Table 1.

Table 1. Dimensions, chemical composition, and number of layers before the first cycle of ARB-processing.

Starting Materials	Composition (wt.%)	Weight (g)	Number of Layers	Dimensions of the Sheets (mm^3)/ Particle Size of the Powder (µm)
Cu sheets	Pure Cu (>99.9)	80.46	4	150 × 30 × 0.5
Al sheets	Pure Al (>99)	10.935	3	150 × 30 × 0.3
Ag powder	Pure Ag (>99)	5, 10, and 15	-	<5

A surface preparation process was performed on Cu and Al sheets for a better bonding between the layers. The following steps were carried out:

- Degreasing the sheets with acetone in order to remove surface contaminations.
- Scratching the sheets with a stainless-steel brush in order to promote cold welding between the layers during the ARB process.
- Sectioning the sample into two halves.
- Stacking them again, repeating the ARB process up to nine passes.

The diameters of the brush and its wires were 100 mm and 0.3 mm, respectively. During the preparation of the samples for ARB, oxidation of the sheets may occur. To avoid this effect, attention must be paid to the time gap between the surface preparation and the next ARB cycle. In this research, the time gap was less than 10 min. The Cu layers were considered as the matrix and the Al layers and the silver powder were placed between them. The rolling speed, the load, and the diameter of the rolls were 10 rpm, 30 tons, and 30 cm, respectively. In this study, the following coding is used for labeling the as-prepared samples. The code consists of one letter and two numbers. The letter (C) stands for the word "composite" denoting the ARB-processed samples, the first number shows the amount of Ag in wt.% and the last number indicates the number of ARB cycles. For example, C11 is the composite with 1 wt.% Ag processed by one cycle of ARB and C35 is the composite sample with 3 wt.% Ag after the 5th ARB cycle. It should be noted that the presence of silver powder hinders the development of bonding between the Cu and Al layers. To overcome this problem, the magnitude of cross-section reduction was increased in the first ARB cycle. Namely, three rolling steps were applied; during the first rolling step the thickness reduction was 65%, while in the next two steps, the thickness reduction during each rolling was 50%. These three rolling steps were considered as the first cycle of ARB. During the subsequent ARB processing, one cycle corresponds to 50% thickness reduction as usual.

2.2. Alloy Fabrication from the ARB-Processed Layered Composites via Heat Treatment

For obtaining alloys from the ARB-processed layered composites, the samples processed by 9 passes were heat treated at temperatures between 750 and 1050 °C for 60 min, and then quenched into an ice/water mixture to produce a martensite phase. According to the Cu–Al phase diagram [28], Cu with 11 wt.% Al can reach the β-phase region in this temperature range; therefore, we selected this range for our study. It should be noted that the adding of Ag may change slightly the temperatures of interest. The martensite phase plays an important role in SMEs. In this regard, martensite start temperature (M_s) is a very important parameter. It is essential to find alloys with high transformation temperatures for high-temperature applications. Most Cu-based shape memory alloys, with Cu–Al–Ni being the most common one, can be used at temperatures up to 200 °C. It has been shown that for Cu–Al–Ag alloys containing different amounts of Ag (fabricated by casting), M_s is well above 200 °C [26]. Among all concentrations of Ag, 2 and 3 wt.% yielded the highest M_s. Thus, three different Ag concentrations below 3 wt.% were chosen in the present study. A coding system was also used for the annealed samples. In this system, the letter A stands for the word "alloy" (or "annealed samples"). The number next to it reveals the wt.% of Ag constituent, the second number shows the annealing temperature in °C, and the last one indicates the holding time in minutes (60 min for all studied specimens). For example, a sample with the code of "A2–950–60" indicates the alloy with 2 wt.% Ag, which was heat treated at 950 °C for 60 min.

2.3. Characterization of the Phase Composition and the Microstructure

The phase composition of the ARB-processed composites and the heat-treated samples was determined by XRD using a powder diffractometer (type: Smartlab, manufacturer: Rigaku) with Bragg–Brentano geometry and a D/Tex detector (applying CuKα radiation with a wavelength of λ = 0.15418 nm). The diffractograms were evaluated with the PDXL2 program using the ICDD-2018 database.

The microstructure of the Cu phase in the ARB-processed samples was studied by XLPA. The X-ray diffraction patterns were measured by a high-resolution rotating anode diffractometer (type: RA-MultiMax9, manufacturer: Rigaku) using CuKα_1 radiation (wavelength, λ = 0.15406 nm). The X-ray diffraction peak profiles were evaluated by the convolutional multiple whole profile (CMWP) fitting procedure [29]. In this method, the diffraction pattern is fitted by the sum of a background spline and the convolution of the theoretical line profiles related to crystallite size and dislocations. The first eight reflections of Cu were used in the evaluation. These peaks can be found in the diffraction angle (2θ) range between 40 and 155°. The area-weighted mean crystallite size ($<x>_{area}$) and the dislocation density (ρ) were determined by the CMWP fitting evaluation procedure of the diffraction patterns. The value of $<x>_{area}$ is calculated as $<x>_{area} = m.\exp(2.5\sigma^2)$, where m is the median and σ^2 is the lognormal variance of the crystallite size distribution. As an example, Figure 1 shows the CMWP fitting for the sample containing 1% Ag and ARB-processed for 9 cycles. More details about the XLPA evaluation can be found in [30].

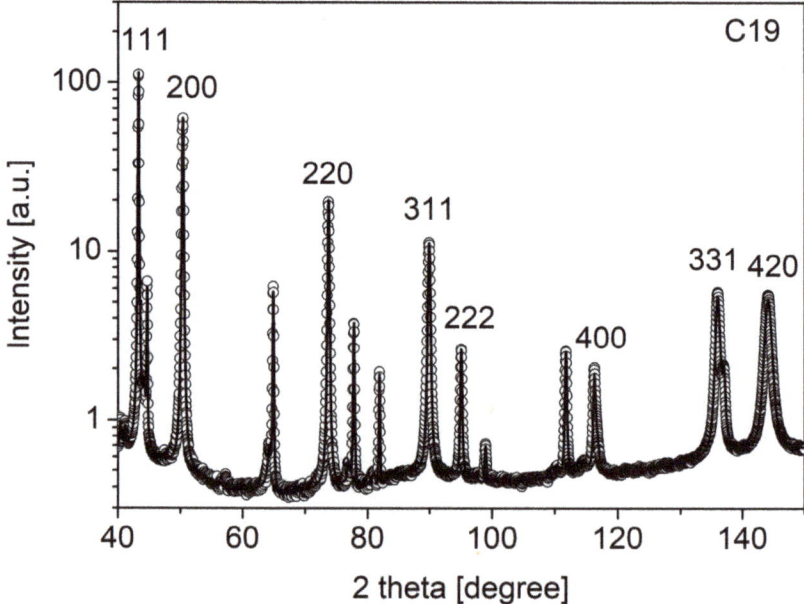

Figure 1. CMWP fitting for the sample containing 1% Ag and processed by 9 passes of ARB. The open circles and the solid line represent the measured and the calculated patterns in the case of the best fitting. The indices of the fitted Cu peaks are indicated in the figure. The non-indexed peaks are related to the Al phase.

The ARB-processed and the annealed microstructures were studied on the cross sections of the samples by SEM using an FEI Quanta 3D microscope (manufacturer: Thermo Fisher Scientific, Waltham, MA, USA). Each surface was mechanically polished with 1200, 2500, and 4000 grit SiC abrasive papers, and then the polishing was continued with a colloidal silica suspension (OP-S) first with a particle size of 1 micron and then 40 nm. Finally, the surface was electropolished at 5 V and 0.5 A using an electrolyte D2 from Stuers. EBSD images were taken with a step size between 35 and 250 nm, depending on the magnification, and evaluated using OIM software (manufacturer: TexSem Laboratories). The grain size was taken as the size of the volumes bounded by high-angle grain boundaries (HAGBs) with misorientation angles higher than 15°. Only those areas were taken as grains that contained at least 4 pixels. The area-weighted mean grain size was used for the characterization of the microstructure. Energy dispersive X-ray spectroscopy (EDS) was applied for the analysis of the chemical composition using the same electron microscope (FEI Quanta 3D).

2.4. Mechanical Characterization

The Vickers microhardness of the samples was measured by a Wolpert Wilson hardness tester (manufacturer: Buehler, Düsseldorf, Germany) in accordance with ASTM E384 standard. The measurements were carried out with a load of 100 g and a loading time of 10 s. In order to increase the reliability of hardness measurement, several indents were placed in different parts of the samples and their average was used as the final hardness number.

The strength, ductility, and shape memory behavior of the samples were studied by tension using an STM-50 universal testing machine (manufacturer: Santam, Tehran, Iran). For this experiment, 1/5 miniaturized JIS-5 tensile test specimens were prepared by a wire cut machine. The longitudinal axis of the tensile samples was parallel to the rolling direction of ARB processing. The length, width, gauge length, and gauge width of the tensile test specimens were 50, 15, 10, and 5 mm, respectively. During the tensile tests, a constant cross-head velocity of 0.5 mm/min was applied, which corresponded to an initial strain rate of about 10^{-3} s^{-1}.

3. Results and Discussion

3.1. Microstructure of the ARB-Processed and the Subsequently Annealed Samples

3.1.1. Microstructure of the Sandwich-like Specimens Obtained by ARB

Figure 2 shows the XRD patterns obtained for the samples with three different Ag contents processed by nine ARB cycles. In all samples, besides the main Cu phase, the peaks of face-centered cubic (FCC) Al and Ag are visible in the diffractograms. The reflections of Ag are very weak due to its small volume fraction. It can be concluded that no reaction occurred between the constituents during ARB owing to the low temperature of the ARB process. Although the temperature of the ARB samples usually increases from room temperature (RT) to about 100 °C due to the friction between the strips and the rolls in non-lubricant conditions [31], this temperature is not enough for a considerable diffusion of the alloying elements into the Cu matrix during ARB processing. Therefore, the phase composition of the ARB-processed multilayered Cu–Al–Ag materials is far from equilibrium, since the phase diagram predicts the existence of intermetallic compounds for the studied compositions even at low temperatures. For instance, if the Ag content is neglected then for Cu–11 wt.% Al composition between RT and 100 °C, an ordered FCC Cu_3Al phase must form according to the equilibrium phase diagram.

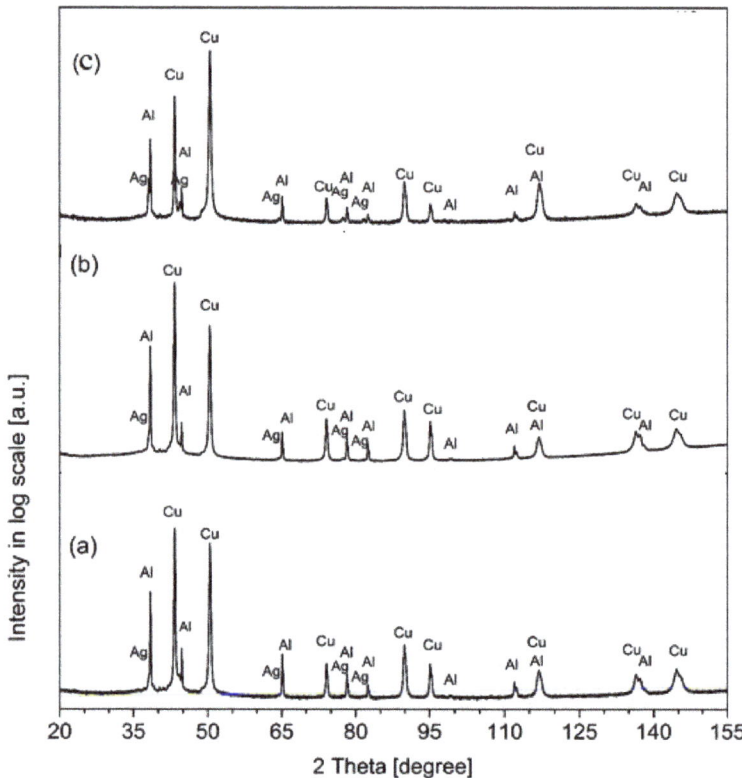

Figure 2. XRD patterns for the samples containing (**a**) 1%, (**b**) 2%, and (**c**) 3% Ag processed by 9 cycles of ARB.

In order to characterize the microstructure of the ARB-processed samples, the XRD peaks of the main Cu phase were evaluated for the crystallite size and the dislocation density using the XLPA method. Table 2 shows the effect of the ARB cycles and the Ag contents on the crystallite size ($<x>_{area}$) and the dislocation density (ρ). The crystallite size varied between ~65 and ~103 nm while the dislocation density was about 7×10^{14} m^{-2} for the different samples. It is obvious that the crystallite size is small and the dislocation density is high even after the first cycle of ARB, which is not in accordance with the expected gradual development of the microstructure versus the number of ARB cycles. The observed fast microstructural refinement can be attributed to the addition of Ag powder to the Cu–Al system and the corresponding modification of the ARB process. Namely, the presence of silver powder between Cu and Al layers hinders the development of bonding between them [32]. To achieve a proper bonding strength between the layers, either the number of rolling cycles or the magnitude of cross-section reduction should be increased. In this study, three rolling steps were used during the first cycle of ARB as described in Section 2.1. Since several rolling steps were required to achieve a proper bonding strength between the layers due to the presence of the silver powder, the density of dislocations became high and the crystallite size had a low value even in the early stage of ARB (i.e., immediately after the first cycle). Therefore, the crystallite size and the dislocation density only slightly changed with increasing the number of ARB cycles. The XLPA results also suggest that the Ag content has only a marginal effect on the microstructural parameters. Most probably, the silver particles at the interfaces of the Cu and Al layers cannot influence the microstructure inside the Cu layers considerably. Since the number of ARB cycles has no significant effect on the microstructural parameters determined by XLPA, an SEM study on the grain structure was performed only for the highest applied ARB cycle (e.g., for nine cycles). These results are shown in the next paragraphs.

Table 2. The crystallite size (<x>area) and the dislocation density (ρ) for the ARB-processed samples as obtained by XLPA.

Sample	<x>$_{area}$ (nm)	ρ (10^{14} m^{-2})
C11	103 ± 12	9.2 ± 1.0
C15	91 ± 10	6.4 ± 0.8
C19	70 ± 9	7.4 ± 0.9
C21	87 ± 10	6.8 ± 0.8
C25	93 ± 10	7.3 ± 0.9
C29	80 ± 9	6.4 ± 0.8
C31	65 ± 8	6.2 ± 0.7
C35	83 ± 9	6.8 ± 0.8
C39	67 ± 8	7.4 ± 0.9

The backscattered SEM images in Figure 3 illustrate the microstructures obtained on the cross section of the samples containing different Ag contents and processed for nine cycles of ARB. The bright and dark areas correspond to Cu and Al phases, respectively. It can be seen that the samples contain thick Cu and thin Al layers or their fragments, which are elongated in the rolling direction (direction RD). After the first ARB cycle, the samples consist of four Cu and three Al layers. The number of layers increases with increasing the number of ARB cycles and at the end of the ninth cycle, the composites contain 1792 layers. However, it should be noted that the determination of the number of layers in the micrographs is difficult due to their fragmentation and deformation caused by ARB straining. With increasing the number of cycles, the thickness of the layers decreases (not studied here). The thickness reduction may be different for the various layers, depending on their mechanical properties [24]. Some ARB processing parameters, such as the load, also affect the thickness reduction [33]. Figure 3 revealed that there is a broad thickness distribution for the layers of the copper matrix.

Figure 4a shows a large Cu layer fragment from the center of the backscattered SEM image in Figure 3a obtained on the ARB-processed sample C19. The contrast differences inside the Cu layer suggest a fine-grained microstructure. Indeed, the inverse pole figure (IPF) map in Figure 4b taken on the same area reveals an average grain size of a few microns on the top and bottom part of the layer (zone "b" in Figure 4a) while the UFG microstructure with an average grain size of several hundreds of nanometers exists in the middle of the Cu layer (indicated by the letter "a" in Figure 4a). The much smaller grain size in the middle region may be caused by a recrystallization due to the severe deformation during ARB. Indeed, former studies suggested that in metals with low and medium stacking fault energies (SFEs), recrystallization is the dominant mechanism of grain refinement during ARB, leading to nanostructured or UFG microstructures [31,34]. However, grain subdivision is the dominant mechanism in metals with high SFEs [31,32]. Since the values of SFEs for copper and aluminum are 78 and 166 mJ/m^2, respectively [35], it is suggested that recrystallization and grain subdivision are the dominant mechanisms for copper and aluminum, respectively. In the grain subdivision mechanism, dislocations are formed at the beginning of SPD processing, which are clustered into dislocation cell walls at a later stage of deformation [36]. In the present material, Cu and Al co-exist; therefore, both mechanisms contributed to the structural refinement. It should be noted that in ARB processing, in addition to the high applied strain, other factors also contribute to the structural refinement [2,37,38]. For instance, friction between the rolls and the samples, as well as scratch brushing of the surface, can increase the dislocation density and refine the grains. In addition, the interfaces between the layers are strong obstacles against dislocation motion, thereby inducing a more intense dislocation multiplication and thus contributing to grain refinement. The difference between the thermal expansion coefficients of Cu, Al, and Ag can also yield dislocation generation if the increase in temperature during ARB is considered [37].

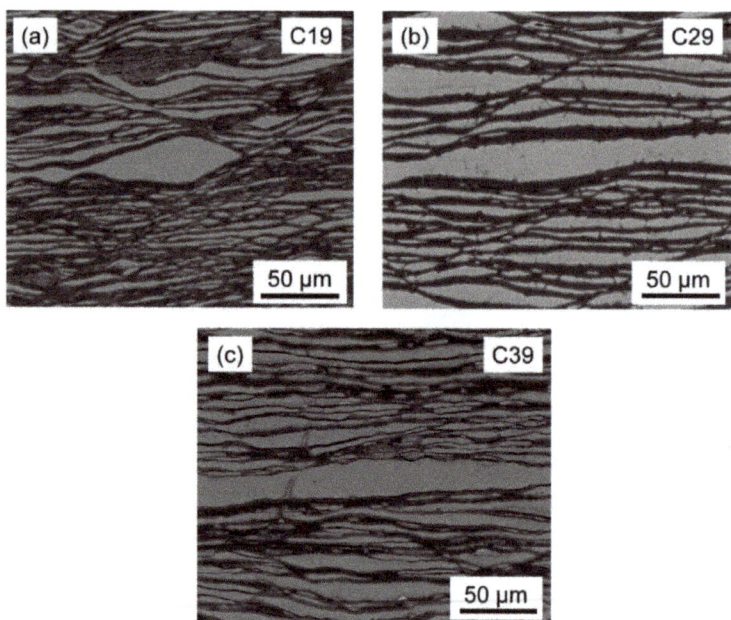

Figure 3. Backscattered SEM image of the microstructure obtained on the cross-section of the composites containing (**a**) 1 wt.%, (**b**) 2 wt.%, and (**c**) 3 wt.% Ag processed for 9 passes of ARB (denoted as C19, C29, and C39, respectively). The bright and dark areas correspond to Cu and Al phases, respectively.

Inhomogeneities in the grain structure were found not only for specimen C19 but also for the samples containing 2% and 3% Ag and processed by 9 cycles of ARB, as shown in Figure 4c,d. The average grain size was in the range of 2–5 µm for the three samples with different Ag contents and ARB-processed for nine cycles. It was also found that most grains in the Cu matrix are elongated in the rolling direction. It is worth noting that the crystallite size obtained by XLPA is much smaller than the grain size determined by microscopic methods such as EBSD, which is a usual effect in SPD-processed metallic materials [19]. This difference can be explained by the sensitivity of the XLPA method on misorientations inside the grains since volumes even with very low orientation differences (a few tenths of degrees) scatter X-rays incoherently. Therefore, the XLPA method measures the size of subgrains or dislocation cells in severely deformed metals rather than the grain size. The microstructure in the minor Al phase in the ARB-processed samples was not studied either XLPA or EBSD due to its very low fraction.

3.1.2. XRD Study of the Evolution of the Phase Composition during Annealing of the ARB-Processed Samples

The ARB-processed samples were heat treated at different temperatures for 60 min in order to induce the formation of the β-phase with good SME. In the selection of the annealing conditions, it was considered that Cu–based SMAs exhibit their shape memory features in the zone of the β-phase. The shape memory characteristics of these alloys depend mainly on the properties of the β-phase. During cooling of the β-phase from 565 °C, usually, an eutectoid decomposition of $\beta \rightarrow \alpha + \gamma_2$ occurs. However, high cooling rates are able to prevent this phase from eutectoid decomposition and enable the martensitic transformation [39]. Thus, the authors decided to study the effect of annealing at different temperatures of 750, 850, 950, and 1050 °C for 60 min on the phase composition of the present ARB-processed Cu–Al–Ag materials. Since the number of ARB cycles has no considerable effect on the phase composition of the studied samples immediately after ARB (see Section 3.1.1), the annealing experiments were performed only for the specimens processed by nine cycles of ARB.

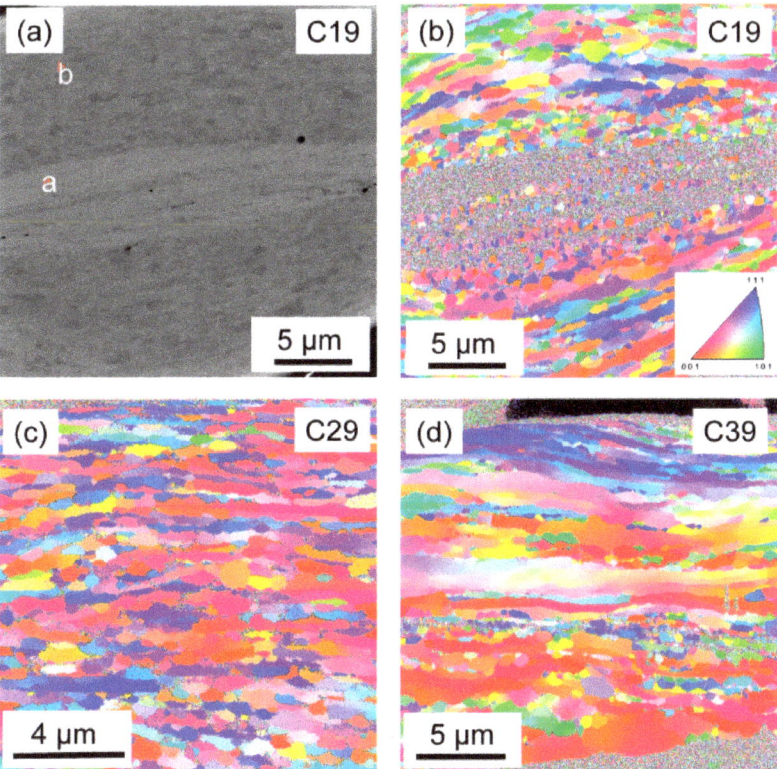

Figure 4. (**a**) Image quality (IQ) and (**b**) IPF color maps for the ARB-processed sample C19, as obtained by EBSD. In figure (a), the letters "a" and "b" indicate zones with average grain sizes of several hundreds of nanometers and a few microns, respectively. IPF color maps for the ARB-processed specimens (**c**) C29, and (**d**) C39.

The phase composition for all the three silver contents changed during the annealing of the ARB samples as revealed by XRD. For the sample containing 1% Ag and processed by nine passes of ARB, the material transformed into a single phase simple cubic Al_4Cu_9 (PDF card number: 01-075-6862), irrespective of the annealing temperature, as shown in Figure 5. It should be noted, however, that the composition of the Al_4Cu_9 phase corresponds to an Al content of 31 at. % while the nominal Al content in the studied materials is only 23 at. % (11 wt.%). Therefore, a part of Al sites in the present Al_4Cu_9 simple cubic structure must be occupied by Cu atoms; i.e., the real composition of this phase most probably deviates from Al_4Cu_9. It is also worth noting that former studies have shown that the formation probability of the Al_4Cu_9 intermetallic compound depends on the Cu content as well as the time and temperature of annealing [40]. According to the equilibrium Cu–Al phase diagram, this phase forms between 360 and 570 °C for the Al content of 11 wt.%. At the temperatures of the present heat treatments (between 750 and 1050 °C), the stable phase is the β-phase. Therefore, the Al_4Cu_9 intermetallic compound most probably formed from the β-phase when the temperature fell below 570 °C during cooling even if the samples were quenched. Since Cu–based alloys with Al_4Cu_9 as a major phase do not show SME and exhibit a low ductility (i.e., they are very brittle), this material is not suitable for shape memory applications.

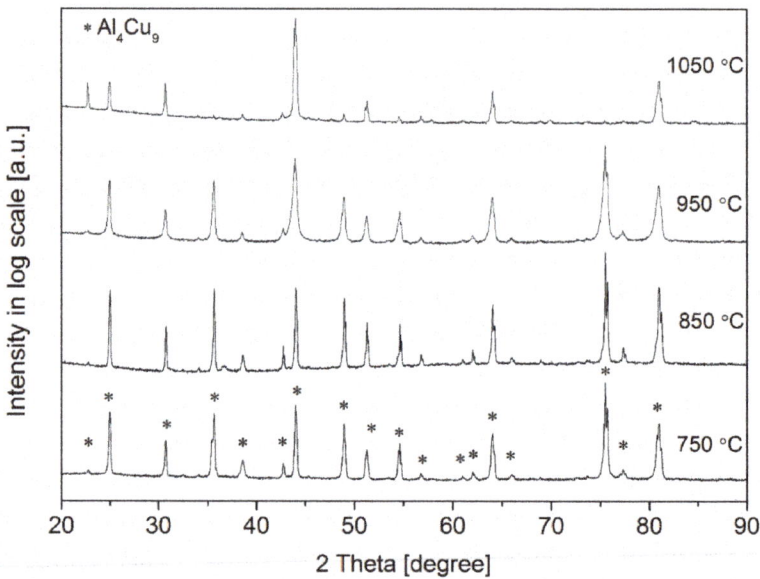

Figure 5. XRD patterns for the samples containing 1 wt.% Ag annealed at different temperatures for 60 min.

The phase compositions of the samples containing 2% Ag, processed by nine ARB passes, and subsequently annealed, are shown in Table 3. The corresponding XRD patterns are presented in Figure 6. It is revealed that the specimen heat treated at 950 °C for 60 min contains the highest fraction of the $β1$-$AlCu_3$ phase (~82%). It has been reported [39] that in Cu–based SMAs with an Al content of 11 wt.% or more, the body-centered cubic (BCC) structure transformed to a DO3-type superlattice by transferring the β to an ordered β1-phase prior to martensitic transformation. In this case, the martensite "inherits" the ordered structure. When the Al content is between 11 and 13 wt.%, β′1 martensite having a monoclinic 18R1 structure prevails. When Al content exceeds 13 wt.%, orthorhombic 2H-type γ′1 martensite forms. Our results suggest that the increase in Ag content from 1 to 2% stabilizes the β1-$AlCu_3$ phase during cooling. This effect can be attributed to the reduced diffusion rate caused by the addition of a third alloying element such as Ag [39,41,42]. The slower diffusion can hinder the decomposition of the β1-phase during cooling. In addition, large Ag solute atoms or silver particles can hinder the motion of phase boundaries, thereby impeding the development of equilibrium phase composition during quenching. According to the Cu–Al phase diagram during the cooling of the β-phase below 564 °C, an eutectoid decomposition of $β \rightarrow α + γ_2$ or $β \rightarrow α + γ_1$ occurs [39]. Silva et al. [42] suggested that Ag addition hinders these reactions and promotes $β \rightarrow β1$ ordering during cooling; therefore, the amount of the β1 phase increases in accordance with the results of the present study. It is noted that oxide phases also developed in the samples heat treated at very high temperatures (950 and 1050 °C), which can be explained by the high rate of oxidation at elevated temperatures.

Table 3. XRD intensity fractions in % of the different phases for the samples with 2% Ag processed by 9 ARB passes and subsequently annealed at different temperatures. PDF card numbers: β1-AlCu$_3$ (00-028-0005), Al$_4$Cu$_9$ (01-075-6862), γ-AlCu$_3$ (01-074-6895), CuO (01-073-6023), Cu$_2$O (01-080-7711), and CuAlO$_2$ (01-075-2356).

Sample Name	γ-AlCu$_3$	β1-AlCu$_3$	Al$_4$Cu$_9$	CuO	Cu$_2$O	CuAlO$_2$
A2-750-60	27	32	41	-	-	-
A2-850-60	10	51	39	-	-	-
A2-950-60	-	82	-	12	3	3
A2-1050-60	-	82	-	5	-	13

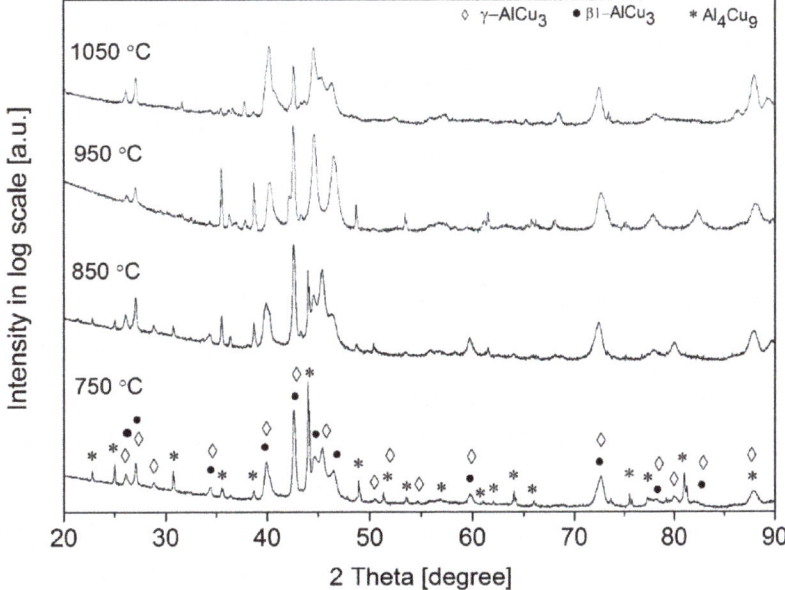

Figure 6. XRD patterns for the samples containing 2 wt.% Ag annealed at different temperatures for 60 min.

For the specimens containing 3% Ag processed by nine passes of ARB and subsequently annealed at different temperatures, the phase composition obtained after annealing is listed in Table 4. The XRD patterns of these samples are shown in Figure 7. Table 4 reveals that β1-AlCu$_3$ is the major phase for all studied annealing temperatures. It seems that with increasing Ag content, the main phase changed from Al$_4$Cu$_9$ to β1-AlCu$_3$. The highest fraction of the β1 phase was obtained at 850 °C (~95%). For the samples A3–950–60 and A3–1050–60, an FCC Cu (Al) phase was also detected. For this phase, the lattice constant was about 0.3677 ± 0.0002 nm, which is much larger than that of pure Cu (0.3615 nm). The high lattice constant was most probably caused by the solute Al and Ag atoms, which have a higher size than that of Cu. Thus, it can be concluded that the optimal annealing temperatures for obtaining the highest fraction of the β1-AlCu$_3$ phase are 950 and 850 °C for 2 and 3% Ag, respectively. Therefore, in the next section, the microstructure is investigated only for the samples annealed at 850 and 950 °C for both silver concentrations. It is worth mentioning that other researchers have also reported these temperatures as the optimum for processing Cu–Al–Ag SMAs by other methods than ARB [42].

Table 4. XRD intensity fractions in % of the different phases for the samples with 3% Ag processed by 9 ARB passes and subsequently annealed at different temperatures for various times. PDF card numbers: $\beta 1$-AlCu$_3$ (00-028-0005), CuO (01-080-1916), CuAlO$_2$ (01-075-2356), Cu (01-071-4610), and Ag (01-073-6977).

Sample Name	$\beta 1$-AlCu$_3$	CuO	Cu (Al)	Ag	CuAlO$_2$	Cu$_2$O
A3–750–60	89	8	2	-	-	1
A3–850–60	95	4	-	1	-	-
A3–950–60	79	5	11	2	3	-
A3–1050–60	66	-	12	-	22	-

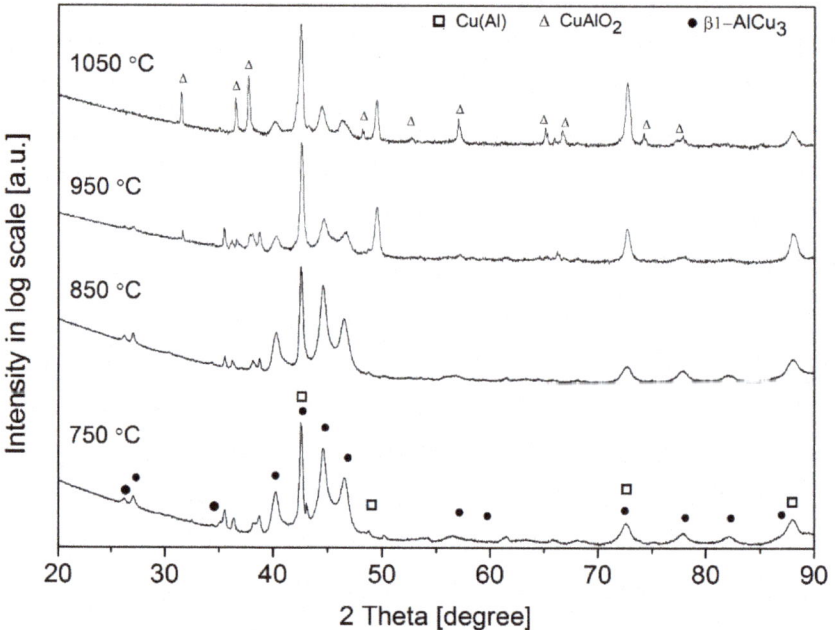

Figure 7. XRD patterns for the samples containing 3 wt.% Ag annealed at different temperatures for 60 min.

3.1.3. Microstructure Analysis of the Annealed Samples

Figure 8a,b show EBSD IPF maps for samples with 1 wt.% Ag annealed at 850 °C and 950 °C for 60 min and quenched in ice/water mixture, respectively. For both temperatures, the grain shape is rather equiaxial compared to the elongated morphology in the ARB-processed state (see Figure 3). These materials are single-phase simple cubic structures (Al$_4$Cu$_9$ type), which formed during cooling from the β-phase regime. By comparing Figure 8a,b, it can be concluded that the grain size was enhanced by increasing the annealing temperature. The average grain size was about 14 μm for sample A1–850–60, while it was about 30 μm for sample A1–950–60.

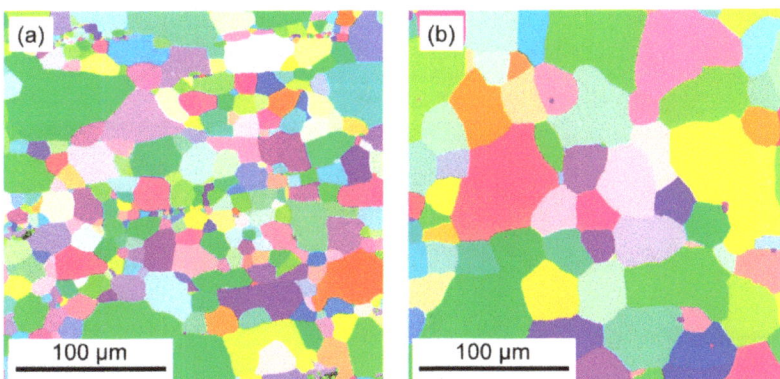

Figure 8. EBSD IPF maps for the samples with 1% Ag processed by 9 passes of ARB and then annealed at (**a**) 850 and (**b**) 950 °C for 60 min.

Figure 9a,b show EBSD IQ maps for the samples containing 2% Ag processed by nine ARB passes and then annealed at 850 and 950 °C for 60 min, respectively (samples A2–850–60 and A2–950–60). EBSD IPF color maps were not prepared for these specimens since they are multiphase materials and the β1-phase structure was not available in the database used by the indexing OIM software. Nevertheless, the IQ images are suitable for the determination of the size and morphology of grains. Namely, it can be seen that after quenching the samples below, the martensite finish temperature (M_f) spear/needle-like martensite is formed. It has been reported [43] that martensite is usually formed in two morphologies in Cu–based alloys: plates and thin needles. The martensite plates form as self-accommodation variant groups. The self-accommodation phenomenon is characterized by a zero apparent change in the shape of SMAs when they transform from the high-temperature austenite to the low-temperature martensite phase. During martensitic transformation, the crystalline lattice of the high-temperature phase (austenite) undergoes a shearing parallel to a particular crystallographic plane and with a shearing vector in a particular crystallographic direction [44]. The crystal variants of martensite have different potential orientations in which the shearing can be produced. For both samples, although the mean grain/packet size is tens of microns, the grains/packets contain lamellas with a thickness of about 3 and 1 μm for samples A2–850–60 and A2–950–60, respectively.

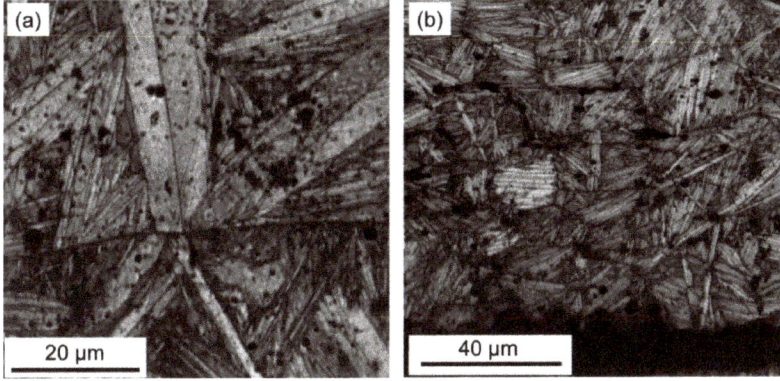

Figure 9. EBSD IQ maps for the material containing 2% Ag and processed by 9 passes of ARB and then annealed at (**a**) 850 and (**b**) 950 °C for 60 min, (samples A2–850–60 and A2–950–60, respectively).

Figure 10 shows EDS elemental maps for alloy A2–950–60 exhibiting the highest β1-phase fraction among the annealed specimens with 2% Ag content. Considerable chemical inhomogeneities cannot be observed.

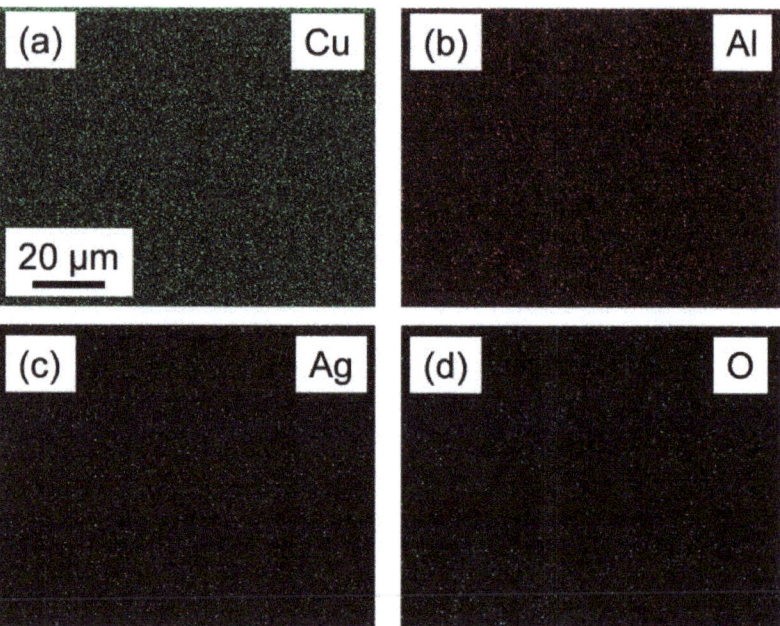

Figure 10. (**a**) Cu, (**b**) Al, (**c**) Ag, and (**d**) O elemental maps for sample A2–950–60 as obtained by SEM-EDS.

Figure 11a, b show EBSD IQ maps for the material containing 3% Ag, processed by nine passes of ARB and then annealed at 850 and 950 °C for 60 min, respectively (samples A3–850–60 and A3–950–60). It can be seen from Figure 11a that in specimen A3–850–60 the grain size is about 15 μm. These grains contain lamellas with a thickness of about 0.5 μm. From Figure 11b, it is obvious that the microstructure consists of martensite and the grain size increased to about 30 μm for sample A3–950–60 due to the increase in annealing temperature. The grains contain lamellas with a thickness between 0.1 and 0.7 μm. The lamellar microstructure inside the yellow square in Figure 11b is shown with a higher magnification in Figure 11c where a martensitic structure with different lamella sizes is visible.

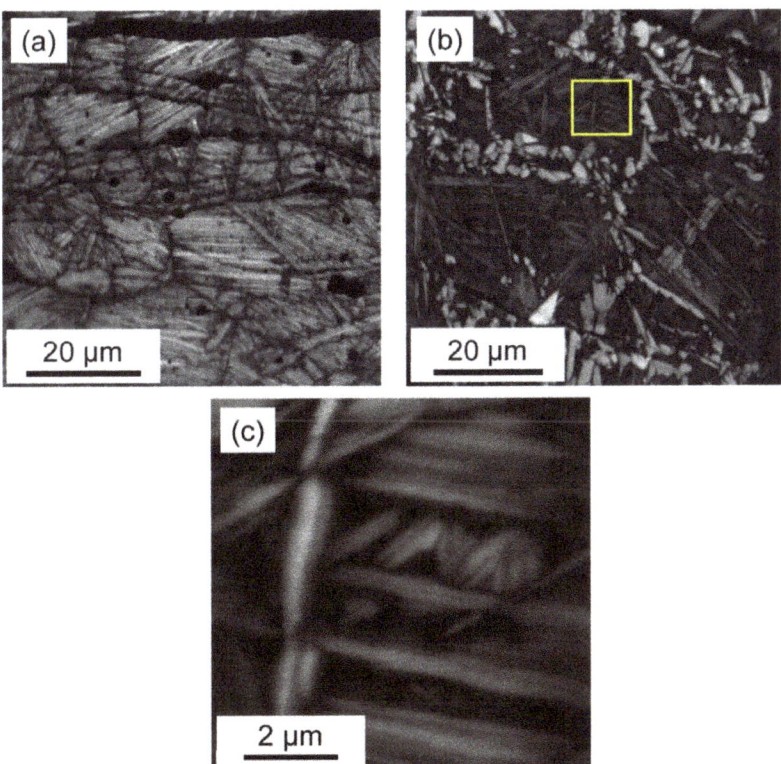

Figure 11. EBSD IQ maps for the material containing 3% Ag and processed by 9 passes of ARB and then annealed at (**a**) 850 and (**b**) 950 °C for 60 min (samples A3–850–60 and A3–950–60, respectively). The lamellar microstructure inside the yellow square in (**b**) is shown with a higher magnification in (**c**).

Figure 12 shows SEM-EDS elemental maps for sample A3–850–60, which has the highest β1-phase fraction among the annealed specimens with 3% Ag content. Considerable chemical inhomogeneities were not detected.

As it can be seen from Figures 9 and 11, the as-quenched structures for the alloys with 2 and 3 wt.% Ag are composed of a spear or needle-like grains of the β1-phase. This type of microstructure is commonly observed in as-quenched Cu-based SMAs [43]. There is a slight change in martensite needle size with increasing the Ag content from 2 to 3 wt.%, suggesting that nucleation and growth of the martensite phase are influenced by the alloying elements' concentration [45]. For instance, Ag particles can reduce the mobility of interfaces during the growth of the newly nucleated β1-phase grains, resulting in a finer microstructure. In addition, the alloying elements may have an effect on the nucleation of the martensite phase since they influence the martensite transformation temperature. For example, it has been found that with increasing the concentration of Ti and Zr, the martensite transformation temperature decreased and increased, respectively [46]. The addition of Ag to Cu–Al alloys also influences the transformation temperature. Namely, Cu–Al–Ag SMAs exhibit higher transformation temperatures than other Cu–based alloys. It was also reported that with increasing silver content, the M_s temperature changed from 370 to 257 °C [23].

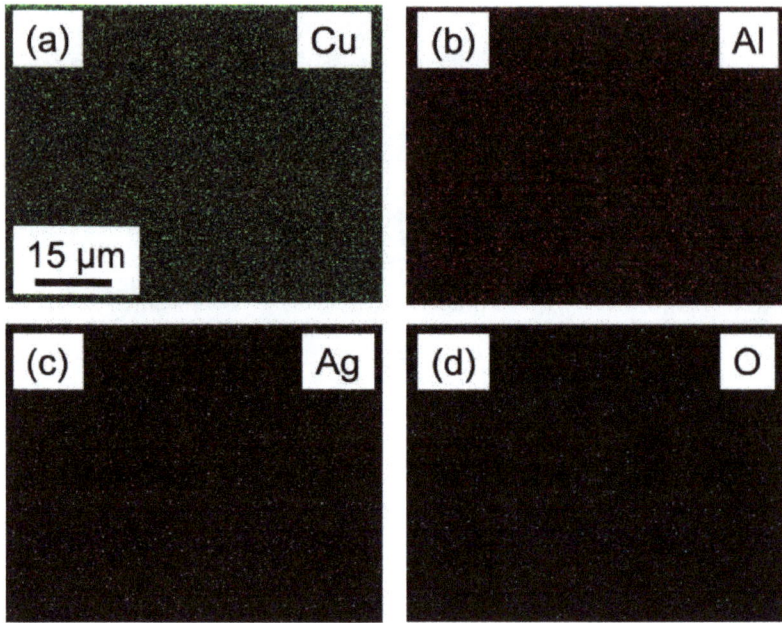

Figure 12. (**a**) Cu, (**b**) Al, (**c**) Ag, and (**d**) O elemental maps for sample A3–850–60 as obtained by SEM-EDS.

For 2 and 3% Ag concentrations, the samples heat treated at 950 and 850 °C, respectively, exhibit a compromise between high β1-phase content and low grain size. For higher temperatures (e.g., at 1050 °C), grain growth is expected to be more pronounced without increasing the amount of the β1-phase; i.e., annealing at this temperature is not beneficial for our goal (to obtain SMA with high strength). Therefore, in the study of the mechanical behavior, we focused on these optimal microstructures (i.e., samples A2–950–60 and A3–850–60).

3.2. Effect of Phase Composition and Microstructure on Mechanical Properties

3.2.1. Microhardness of the ARB-Processed and the Subsequently Annealed Samples

It is obvious that mechanical properties are structure sensitive [47]; thus, the number of ARB cycles and the heat treatment conditions should play an important role in the mechanical behavior of the studied samples. Table 5 shows the effect of ARB cycles on the Vickers microhardness of Cu–Al–Ag multilayered materials with different Ag contents. During the first cycle, the Cu and Al layers were completely separated in the samples; therefore, their microhardness values are reported individually in Table 5. Since the microhardness of Cu and Al layers did not depend on the Ag content, only a single hardness value is reported for each layer, which is valid for samples C11, C21, and C31. For five and nine cycles of ARB, the microhardness values in Table 5 characterize the whole sample. The data in Table 5 show that the microhardness increases with increasing the number of ARB cycles. Since the dislocation density saturated in the main Cu phase even after the first cycle as revealed by XLPA (see Table 2), the hardness increase is most probably caused by the refinement of the microstructure [48]. On the other hand, it is evident that the thickness of the elongated Cu and Al layers decreases gradually with increasing the number of ARB cycles (even if it was not monitored experimentally in the present study), and these interfaces are strong obstacles against dislocation motion; therefore, they considerably contribute to hardening. The Ag content has only a marginal effect on the hardness of the ARB-processed materials. Most probably, the Ag particles are located at the Cu/Al interfaces; therefore, besides the strengthening effect of interfaces, they have no significant contribution to the hardness.

Table 5. Effect of ARB cycles on the microhardness of Cu/Al/Ag samples with different Ag contents. The relative error of the hardness values is about 5%.

Sample	Cu Layers	Al Layer	C15	C19	C25	C29	C35	C39
Microhardness (HV)	55	26	101	124	109	131	108	133

Table 6 lists the microhardness values for the alloys with different Ag contents that were processed by nine cycles of ARB and then heat treated at 850 °C and 950 °C for 60 min. It is revealed that the microhardness was enhanced during the annealing of the ARB-processed samples due to the formation of hard phases such as Al_4Cu_9 and martensite. In addition, the martensite microstructure contains very fine lamellas (see Figure 11c), which also contributes to hardening. Although, it has been reported that the addition of Ag to the Cu–Al binary alloys increases their microhardness by influencing the nucleation rate and the activation energy of the eutectoid decomposition reaction [49], in the present study this effect is marginal compared to other hardening contributions (e.g., grain size).

Table 6. Effect of heat treatment on the microhardness of Cu–Al–Ag alloys. The relative error of the hardness values is about 5%.

Sample	A1–950–60	A2–850–60	A2–950–60	A3–850–60	A3–950–60
Microhardness (HV)	210	245	270	281	266

3.2.2. Tensile Testing

The tensile behavior of the materials containing 1 wt.% Ag was not studied since after annealing the samples became very brittle due to the formation of the Al_4Cu_9 phase; therefore, a reasonable stress–strain curve cannot be detected. Figure 13 shows the engineering stress–strain curves for the samples containing 2 and 3 wt.% Ag. The ultimate tensile strength of the ARB-processed multilayered samples was close to each other: ~340 and ~360 MPa for specimens C29 and C39, respectively. The elevated tensile strength is a consequence of different hardening effects, such as dislocations formed during ARB, grain refinement, and strong interfaces between the layers [33,50]. Another phenomenon that influences the tensile strength of the multilayered ARB-processed samples is the strain incompatibility. As a result of the co-existence of Cu and Al layers with different mechanical properties (elastic modulus, yield strength, etc.), a strain incompatibility phenomenon occurs between the constituents during ARB. Due to this effect, strain gradients and geometrically necessary dislocations near the interfaces can develop, resulting in an additional back-stress hardening [33,51].

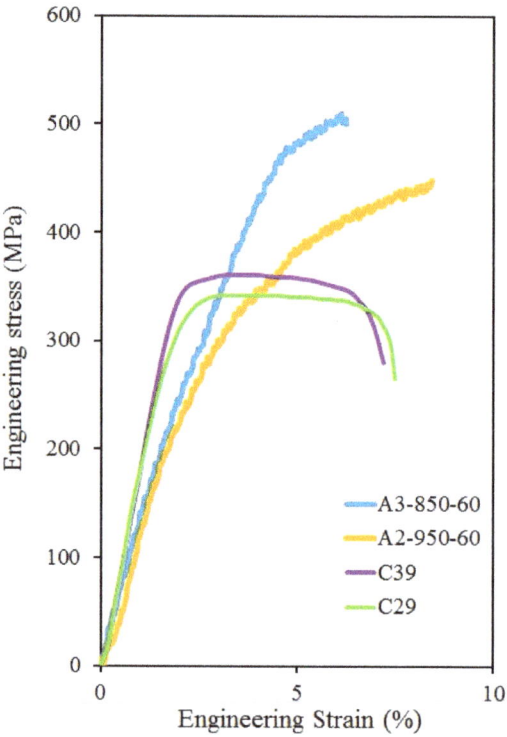

Figure 13. Tensile stress–strain curves for the samples processed for 9 ARB cycles and containing 2 and 3 wt.% Ag (specimens C29 and C39) and their counterparts heat-treated at 950 °C and 850 °C for 60 min, respectively (samples A2–950–60 and A3–850–60).

As it can be seen from Figure 13, the heat treatment at 950 °C for 60 min yielded an increase in the tensile strength for the samples containing 2 wt.% Ag and ARB-processed for nine cycles. For sample A2–950–60, the strength reached a value as high as ~460 MPa with a good ductility (about 9%). For sample with 3 wt.% Ag, a similar increase in tensile strength occurred at 850 °C. The strength of sample A3–850–60 is even higher (~520 MPa) than that for specimen A2–950–60. The improvement of the tensile strength due to annealing can be explained by the formation of a hard martensite phase with fine microstructure as discussed in Section 3.1. Former studies have also reported improvement in tensile behavior of Cu alloys with decreasing grain size [52,53]. Indeed, Cu–based SMAs with coarse grains are not appropriate for commercial applications [54–56]. The grain refinement in Cu–Al SMAs can be achieved by increasing the Al content or adding other alloying elements [57,58]. The solute atoms and the secondary phase particles can hinder the interface motion, thereby stabilizing the martensite phase exhibiting good SME [59]. It is also worth noting that the fine-grained microstructure developed during ARB had both direct and indirect effects on the microstructure and the mechanical properties of the studied Cu–Al–Ag alloys. First, this microstructure yielded a higher strength in comparison with the counterparts processed by casting [60] due to the Hall–Petch strengthening effect caused by the low grain size [61]. In addition, dislocations and grain boundaries in the ARB-processed samples are fast diffusion paths, which facilitate the formation of the martensite phase during annealing. Thus, the annealing time can be shortened, thereby minimizing the grain growth during the heat treatment (the effect of annealing time on the microstructure was not studied here). Moreover, dislocations and grain boundaries are preferred sites for nucleation of new phases during phase transformation; therefore, their high amount in the ARB-processed samples can yield a finer martensite microstructure.

3.2.3. Shape Memory Behavior of the Alloys Containing a High Fraction of the β-Phase

The shape memory behavior of the specimens containing the highest fraction of the β1-phase (i.e., for specimens A2–950–60 and A3–850–60) was characterized by the magnitude of the strain recovered during unloading of the samples tensile tested for small plastic strains (about 1%). Figure 14a, b shows the loading–unloading tensile stress–strain curves for the annealed alloys A2–950–60 and A3–850–60. According to Figure 14a, the retained plastic strain and the recovered strain for sample A2–950–60 were 0.7% and 3.0%, respectively. For sample A3–850–60, the values of retained and recovered strains were 1.2% and 3.4%, respectively. The high recovered strains for both samples can be attributed to the large β1-phase fraction (82–95% as shown in Tables 3 and 4). The slightly better shape memory behavior for sample A3–850–60 can be explained by the higher amount of the β1-phase. These two alloys showed a better SME than other Cu–based SMAs. For instance, it has been reported that Cu–Al–Ni SMAs have lower flow stress (~200 MPa) and recovered strain (~1.5%) values after being treated for SME [39]. The strength of Cu–Al–Ni SMAs was improved to about 400 MPa with the addition of 0.4 wt.% Ti; however, the recovered strain remained only ~1.5% [39]. The strength and the recovered strain obtained for a Cu–Al–Mn SMA with the composition of $Cu_{72}Al_{17}Mn_{11}$ [19] were lower than the values determined for the present ARB-processed and annealed Cu–Al–Ag sample with 3 wt.% Ag (sample A3–850–60). It was tried to improve both the strength and the recovered strain of a Cu–Al–Mn SMA by adding more Mn alloying elements, but they could not improve both properties simultaneously. For example, the alloy $(Cu_{72}Al_{17}Mn_{11})_{99.8}–B_{0.2}$ showed a strength of about 400 MPa; however, the recovered strain was only about 1%. On the other hand, the addition of a small amount of Co to Cu–Al–Mn SMA (yielding the composition of $(Cu_{72.5}Al_{17}Mn_{10.5})_{99.5}–Co_{0.5}$) increased the recovered strain to about 7%, but the strength decreased to ~150 MPa simultaneously.

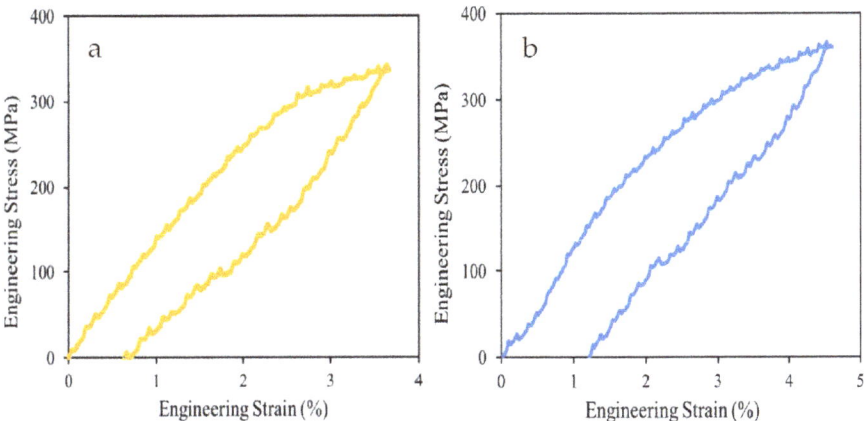

Figure 14. Engineering stress–strain curves obtained during loading and unloading in a tensile test performed for samples (**a**) A2–950–60 and (**b**) A3–850–60.

4. Conclusions

Experiments were conducted for the study of the microstructure and mechanical behavior of Cu–11 wt.% Al–x wt.% Ag (x = 1, 2, or 3) SMAs processed by ARB and subsequent annealing at different temperatures. The following conclusions were drawn from the results:

1. The ARB-processed samples contain Cu and Al layers or layer fragments since intermetallic phases were not formed due to the slow diffusion of Cu and Al at room temperature. The dislocation density in the main Cu phase was about 7×10^{14} m^{-2}, irrespective of the number of ARB cycles and the Ag content. The early saturation of the dislocation density can be attributed to the addition of Ag powder to the Cu–Al system and the corresponding modification of the ARB process. Namely, the presence of silver powder between Cu and Al layers hindered the development of bonding between them. Therefore, three rolling steps were used during the first cycle of ARB and the density of dislocations became high even after the first cycle. The grain size after nine cycles was a few microns.
2. During heat treatment of the ARB-processed samples, new intermetallic phases such as β_1-AlCu$_3$, Al$_4$Cu$_9$, and γ-AlCu$_3$ were formed. For the lowest Ag content (1 wt.%), the main phase was the brittle Al$_4$Cu$_9$, irrespective of the temperature of heat treatment. For higher Ag concentrations (2 and 3 wt.%), the annealed samples contain mainly the β_1-AlCu$_3$ phase. After 60 min of annealing, the best phase compositions were achieved at 950 and 850 °C for the samples containing 2 and 3 wt.% Ag, respectively. The martensite phase consisted of very fine lamellas with a thickness of one micron or less. Since dislocations and grain boundaries facilitate the nucleation of new phases, ARB processing must have a significant role in obtaining fine-grained martensite microstructure during annealing.
3. The heat treatment at 850 and 950 °C for 60 min increased the microhardness and the strength of the presently studied Cu–Al–Ag alloys due to the formation of fine-grained hard intermetallic phases. For the samples containing 2–3 wt.% Ag, annealing at 950 and 850 °C for 60 min after nine cycles of ARB increased the hardness from about 130 to 280 HV and the tensile strength from 340–360 to 460–520 MPa.
4. The alloys containing 2 and 3 wt.% Ag, processed by nine ARB cycles and then annealed at 950 and 850 °C for 60 min, respectively, exhibited a good SME. The recovered strain was about 3% while the tensile strength was as high as ~500 MPa. These values are outstanding among the Cu–based SMAs.

Author Contributions: Conceptualization, P.S. and M.A.; methodology, P.S., M.A. and J.G.; validation, P.S., M.A. and J.G.; formal analysis, P.S., Á.S., J.G. and M.E.-T.; investigation, P.S., Á.S. and M.E.-T.; resources, M.A. and J.G.; data curation, P.S., Á.S. and M.E.-T.; writing—original draft preparation, P.S., M.A. and J.G.; writing—review and editing, P.S., M.A. and J.G.; visualization, P.S., M.E.-T. and J.G.; supervision, M.A. and J.G.; funding acquisition, M.A. and J.G. All authors have read and agreed to the published version of the manuscript.

Funding: This research received no external funding.

Institutional Review Board Statement: Not applicable.

Informed Consent Statement: Not applicable.

Data Availability Statement: The evaluated data presented in this study are available in the tables of this paper. The raw measured data of this study are available on request from the corresponding author.

Acknowledgments: M.E.-T. thanks the cultural affairs and missions sector of Egypt for providing a postdoc fellowship to the Eötvös Loránd University, Budapest, Hungary.

Conflicts of Interest: The authors declare no conflict of interest.

References

1. Mahdavian, M.; Ghalandari, L.; Reihanian, M. Accumulative roll bonding of multilayered Cu/Zn/Al: An evaluation of microstructure and mechanical properties. *Mater. Sci. Eng. A* **2013**, *579*, 99–107. [CrossRef]
2. Eizadjou, M.; Kazemitalachi, A.; Daneshmanesh, H.; Shahabi, H.S.; Janghorban, K. Investigation of structure and mechanical properties of multi-layered Al/Cu composite produced by accumulative roll bonding (ARB) process. *Compos. Sci. Technol.* **2008**, *68*, 2003–2009. [CrossRef]

3. Ghalandari, L.; Moshksar, M. High-strength and high-conductive Cu/Ag multilayer produced by ARB. *J. Alloys Compd.* **2010**, *506*, 172–178. [CrossRef]
4. Mathis, K.; Gubicza, J.; Nam, N. Microstructure and mechanical behavior of AZ91 Mg alloy processed by equal channel angular pressing. *J. Alloys Compd.* **2005**, *394*, 194–199. [CrossRef]
5. Reihanian, M.; Ebrahimi, R.; Tsuji, N.; Moshksar, M. Analysis of the mechanical properties and deformation behavior of nanostructured commercially pure Al processed by equal channel angular pressing (ECAP). *Mater. Sci. Eng. A* **2008**, *473*, 189–194. [CrossRef]
6. Sakai, G.; Horita, Z.; Langdon, T.G. Grain refinement and superplasticity in an aluminum alloy processed by high-pressure torsion. *Mater. Sci. Eng. A* **2005**, *393*, 344–351. [CrossRef]
7. Horita, Z.; Langdon, T.G. Microstructures and microhardness of an aluminum alloy and pure copper after processing by high-pressure torsion. *Mater. Sci. Eng. A* **2005**, *410–411*, 422–425. [CrossRef]
8. Huang, J.; Zhu, Y.; Alexander, D.J.; Liao, X.; Lowe, T.C.; Asaro, R.J. Development of repetitive corrugation and straightening. *Mater. Sci. Eng. A* **2004**, *371*, 35–39. [CrossRef]
9. Saito, Y.; Utsunomiya, H.; Tsuji, N.; Sakai, T. Novel ultra-high straining process for bulk materials—Development of the accumulative roll-bonding (ARB) process. *Acta Mater.* **1999**, *47*, 579–583. [CrossRef]
10. Tsuji, N.; Ito, Y.; Saito, Y.; Minamino, Y. Strength and ductility of ultrafine grained aluminum and iron produced by ARB and annealing. *Scr. Mater.* **2002**, *47*, 893–899. [CrossRef]
11. Eizadjou, M.; Manesh, H.D.; Janghorban, K. Microstructure and mechanical properties of ultra-fine grains (UFGs) aluminum strips produced by ARB process. *J. Alloys Compd.* **2009**, *474*, 406–415. [CrossRef]
12. Borhani, E.; Jafarian, H.; Terada, D.; Adachi, H.; Tsuji, N. Microstructural Evolution during ARB Process of Al–0.2 mass% Sc Alloy Containing Al3Sc Precipitates in Starting Structures. *Mater. Trans.* **2012**, *53*, 72–80. [CrossRef]
13. Rezaei, M.R.; Toroghinezhad, M.; Ashrafizadeh, F. Analysis of Strengthening Mechanisms in an Artificially Aged Ultrafine Grain 6061 Aluminum Alloy. *J. Ultrafine Grained Nanostruct. Mater.* **2017**, *50*, 152–160. [CrossRef]
14. Takata, N.; Lee, S.-H.; Tsuji, N. Ultrafine grained copper alloy sheets having both high strength and high electric conductivity. *Mater. Lett.* **2009**, *63*, 1757–1760. [CrossRef]
15. Nomura, K.; Miwa, Y.; Takagawa, Y.; Watanabe, C.; Monzen, R.; Terada, D.; Tsuji, N. Influence of Accumulative Roll Bonding and Cold Rolling Processes on the Precipitation Strengthening Properties for Cu-Ni-P Alloy. *J. Jpn. Inst. Met.* **2011**, *75*, 509–515. [CrossRef]
16. Altenberger, I.; Kuhn, H.-A.; Gholami, M.; Mhaede, M.; Wagner, L. Ultrafine-Grained Precipitation Hardened Copper Alloys by Swaging or Accumulative Roll Bonding. *Metals* **2015**, *5*, 763–776. [CrossRef]
17. Takagawa, Y.; Tsujiuchi, Y.; Watanabe, C.; Monzen, R.; Tsuji, N. Improvement in Mechanical Properties of a Cu–2.0 mass% Ni–0.5 mass% Si–0.1 mass% Zr Alloy by Combining Both Accumulative Roll-Bonding and Cryo-Rolling with Aging. *Mater. Trans.* **2013**, *54*, 1–8. [CrossRef]
18. Kitagawa, K.; Akita, T.; Kita, K.; Gotoh, M.; Takata, N.; Tsuji, N. Structure and Mechanical Properties of Severely Deformed Cu-Cr-Zr Alloys Produced by Accumulative Roll-Bonding Process. *Mater. Sci. Forum* **2008**, *584–586*, 791–796. [CrossRef]
19. Sutou, Y.; Omori, T.; Wang, J.J.; Kainuma, R.; Ishida, K. Characteristics of Cu–Al–Mn-based shape memory alloys and their applications. *Mater. Sci. Eng. A* **2004**, *378*, 278–282. [CrossRef]
20. Otsuka, K.; Wayman, C.M. *Shape Memory Materials*; Cambridge University Press: Cambridge, UK, 1999.
21. Hartl, D.; Lagoudas, D. Thermomechanical Characterization of Shape Memory Alloy Materials. In *Shape Memory Alloys*; Springer: Berlin/Heidelberg, Germany, 2008; Volume 1, pp. 53–119. [CrossRef]
22. Mazzer, E.M.; da Silva, M.R.; Gargarella, P. Revisiting Cu-based shape memory alloys: Recent developments and new perspectives. *J. Mater. Res.* **2022**, *37*, 162–182. [CrossRef]
23. Guilemany, J.M.; Fernandez, J.; Zhang, X.M. TEM study on the microstructure of Cu–Al–Ag shape memory alloys. *Mater. Sci. Eng. A* **2006**, *438*, 726–729. [CrossRef]
24. Alizadeh, M.; Avazzadeh, M. Evaluation of Cu-26Zn-5Al shape memory alloy fabricated by accumulative roll bonding process. *Mater. Sci. Eng. A* **2019**, *757*, 88–94. [CrossRef]
25. Alizadeh, M.; Dashtestaninejad, M.K. Fabrication of manganese-aluminum bronze as a shape memory alloy by accumulative roll bonding process. *Mater. Des.* **2016**, *111*, 263–270. [CrossRef]
26. Guilemany, J.M.; Fernández, J.; Franch, R.; Benedetti, A.V.; Adorno, A.T. A new Cu-based SMA with extremely high martensitic transformation temperatures. *Le J. Phys. IV* **1995**, *5*, C2-361–C2-365. [CrossRef]
27. Alizadeh, M.; Dashtestaninejad, M.K. Development of Cu-matrix, Al/Mn-reinforced, multilayered composites by accumulative roll bonding (ARB). *J. Alloys Compd.* **2018**, *732*, 674–682. [CrossRef]
28. Swann, P.; Warlimont, H. The electron-metallography and crystallography of copper-aluminum martensites. *Acta Met.* **1963**, *11*, 511–527. [CrossRef]
29. Ribárik, G.; Gubicza, J.; Ungár, T. Correlation between strength and microstructure of ball-milled Al–Mg alloys determined by X-ray diffraction. *Mater. Sci. Eng. A* **2004**, *387–389*, 343–347. [CrossRef]
30. Gubicza, J. *X-ray Line Profile Analysis in Materials Science*; IGI Global: Hershey, PA, USA, 2014. [CrossRef]
31. Jamaati, R.; Toroghinejad, M.R.; Dutkiewicz, J.; Szpunar, J.A. Investigation of nanostructured Al/Al$_2$O$_3$ composite produced by accumulative roll bonding process. *Mater. Des.* **2012**, *35*, 37–42. [CrossRef]

32. Alizadeh, M.; Paydar, M.H.; Terada, D.; Tsuji, N. Effect of SiC particles on the microstructure evolution and mechanical properties of aluminum during ARB process. *Mater. Sci. Eng. A* **2012**, *540*, 13–23. [CrossRef]
33. Jiang, S.; Jia, N.; Zhang, H.; He, T.; Zhao, X. Microstructure and Mechanical Properties of Multilayered Cu/Ti Composites Fabricated by Accumulative Roll Bonding. *Mater. Trans.* **2017**, *58*, 259–265. [CrossRef]
34. Shaarbaf, M.; Toroghinejad, M.R. Nano-grained copper strip produced by accumulative roll bonding process. *Mater. Sci. Eng. A* **2008**, *473*, 28–33. [CrossRef]
35. Maeda, M.Y.; Quintero, J.J.H.; Izumi, M.T.; Hupalo, M.F.; Cintho, O.M. Study of Cryogenic Rolling of FCC Metals with Different Stacking Fault Energies. *Mater. Res.* **2017**, *20*, 716–721. [CrossRef]
36. Lee, S.; Saito, Y.; Tsuji, N.; Utsunomiya, H.; Sakai, T. Role of shear strain in ultragrain refinement by accumulative roll-bonding (ARB) process. *Scr. Mater.* **2002**, *46*, 281–285. [CrossRef]
37. Alizadeh, M. Strengthening mechanisms in particulate Al/B4C composites produced by repeated roll bonding process. *J. Alloys Compd.* **2011**, *509*, 2243–2247. [CrossRef]
38. Ghalandari, L.; Mahdavian, M.; Reihanian, M. Microstructure evolution and mechanical properties of Cu/Zn multilayer processed by accumulative roll bonding (ARB). *Mater. Sci. Eng. A* **2014**, *593*, 145–152. [CrossRef]
39. Al-Humairi, S.N.S. Cu-Based Shape Memory Alloys: Modified Structures and Their Related Properties. In *Recent Advancements in the Metallurgical Engineering and Electrodeposition*; Al-Naib, U.B., Vikraman, D., Karuppasamy, K., Eds.; IntechOpen: London, UK, 2020. [CrossRef]
40. Xu, H.; Liu, C.; Silberschmidt, V.V.; Pramana, S.S.; White, T.J.; Chen, Z.; Acoff, V.L. Behavior of aluminum oxide, intermetallics and voids in Cu–Al wire bonds. *Acta Mater.* **2011**, *59*, 5661–5673. [CrossRef]
41. Adorno, A.T.; Silva, R.A.G. Effect of 4 mass% Ag addition on the thermal behavior of the Cu-9 mass% Al alloy. *J. Therm. Anal.* **2003**, *73*, 931–938. [CrossRef]
42. Silva, R.A.G.; Cuniberti, A.; Stipcich, M.; Adorno, A.T. Effect of Ag addition on the martensitic phase of the Cu–10 wt.% Al alloy. *Mater. Sci. Eng. A* **2007**, *456*, 5–10. [CrossRef]
43. Kim, H.W. A study of the two-way shape memory effect in Cu–Zn–Al alloys by the thermomechanical cycling method. *J. Mater. Process. Technol.* **2004**, *146*, 326–329. [CrossRef]
44. Bhattacharya, K. *Microstructure of Martensite: Why It Forms and How It Gives Rise to the Shape-Memory Effect*; Oxford University Press: Oxford, UK, 2003.
45. Alaneme, K.K.; Okotete, E.A.; Maledi, N. Phase characterisation and mechanical behaviour of Fe–B modified Cu–Zn–Al shape memory alloys. *J. Mater. Res. Technol.* **2017**, *6*, 136–146. [CrossRef]
46. CM, W. Grain refinement of a Cu-Al-Ni shape memory alloy by Ti and Zr additions. *Trans. Japan Inst. Met.* **1986**, *27*, 584–591.
47. Alaneme, K.K.; Anaele, J.U.; Okotete, E.A. Martensite aging phenomena in Cu-based alloys: Effects on structural transformation, mechanical and shape memory properties: A critical review. *Sci. Afr.* **2021**, *12*, e00760. [CrossRef]
48. Liu, X.; Zhuang, L.; Zhao, Y. Microstructure and Mechanical Properties of Ultrafine-Grained Copper by Accumulative Roll Bonding and Subsequent Annealing. *Materials* **2020**, *13*, 5171. [CrossRef] [PubMed]
49. Adorno, A.T.; Silva, R.A.G. Isothermal decomposition kinetics in the Cu–9% Al–4% Ag alloy. *J. Alloys Compd.* **2004**, *375*, 128–133. [CrossRef]
50. Min, G.; Lee, J.-M.; Kang, S.-B.; Kim, H.-W. Evolution of microstructure for multilayered Al/Ni composites by accumulative roll bonding process. *Mater. Lett.* **2006**, *60*, 3255–3259. [CrossRef]
51. Ohsaki, S.; Kato, S.; Tsuji, N.; Ohkubo, T.; Hono, K. Bulk mechanical alloying of Cu–Ag and Cu/Zr two-phase microstructures by accumulative roll-bonding process. *Acta Mater.* **2007**, *55*, 2885–2895. [CrossRef]
52. Awan, I.Z.; Khan, A.Q. Fascinating Shape Memory Alloys. *J. Chem. Soc. Pakistan.* **2018**, *40*, 1–23.
53. Punburi, P.; Tareelap, N.; Srisukhumbowornchai, N.; Euaruksakul, C.; Yordsri, V. Correlation between electron work functions of multiphase Cu-8Mn-8Al and de-alloying corrosion. *Appl. Surf. Sci.* **2018**, *439*, 1040–1046. [CrossRef]
54. Oliveira, J.; Crispim, B.; Zeng, Z.; Omori, T.; Fernandes, F.B.; Miranda, R. Microstructure and mechanical properties of gas tungsten arc welded Cu-Al-Mn shape memory alloy rods. *J. Mater. Process. Technol.* **2019**, *271*, 93–100. [CrossRef]
55. Yang, S.; Zhang, J.; Chen, X.; Chi, M.; Wang, C.; Liu, X. Excellent superelasticity and fatigue resistance of Cu-Al-Mn-W shape memory single crystal obtained only through annealing polycrystalline cast alloy. *Mater. Sci. Eng. A* **2019**, *749*, 249–254. [CrossRef]
56. Tian, J.; Zhu, W.; Wei, Q.; Wen, S.; Li, S.; Song, B.; Shi, Y. Process optimization, microstructures and mechanical properties of a Cu-based shape memory alloy fabricated by selective laser melting. *J. Alloys Compd.* **2019**, *785*, 754–764. [CrossRef]
57. Hussain, S.; Pandey, A.; Dasgupta, R. Designed polycrystalline ultra-high ductile boron doped Cu–Al–Ni based shape memory alloy. *Mater. Lett.* **2019**, *240*, 157–160. [CrossRef]
58. Guniputi, B.N.; Murigendrappa, S. Influence of Gd on the microstructure, mechanical and shape memory properties of Cu-Al-Be polycrystalline shape memory alloy. *Mater. Sci. Eng. A* **2018**, *737*, 245–252. [CrossRef]
59. Morris, M.; Günter, S. Effect of heat treatment and thermal cycling on transformation temperatures of ductile Cu-Al-Ni-Mn-B alloys. *Scr. Met. Mater.* **1992**, *26*, 1663–1668. [CrossRef]
60. Dar, R.D.; Yan, H.; Chen, Y. Grain boundary engineering of Co–Ni–Al, Cu–Zn–Al, and Cu–Al–Ni shape memory alloys by intergranular precipitation of a ductile solid solution phase. *Scr. Mater.* **2016**, *115*, 113–117. [CrossRef]
61. Hall, E.O. The Deformation and Ageing of Mild Steel: III Discussion of Results. *Proc. Phys. Soc. Sect. B* **1951**, *64*, 747–753. [CrossRef]

Article

Mechanical Characterization and Microstructural Analysis of Hybrid Composites (LM5/ZrO$_2$/Gr)

Sunder Jebarose Juliyana [1], Jayavelu Udaya Prakash [1], Sachin Salunkhe [1,*], Hussein Mohamed Abdelmoneam Hussein [2,3] and Sharad Ramdas Gawade [4]

[1] Department of Mechanical Engineering, Vel Tech Rangarajan Dr. Sagunthala R&D Institute of Science and Technology, Chennai 600062, India
[2] Mechanical Engineering Department, Faculty of Engineering and Technology, Future University in Egypt, New Cairo 11835, Egypt
[3] Mechanical Engineering Department, Faculty of Engineering, Helwan University, Cairo 11732, Egypt
[4] Sharadchandra Pawar, College of Engineering and Technology, Someshwar, Baramati 412306, India
* Correspondence: drsalunkhesachin@veltech.edu.in

Abstract: Hybrid composites recently developed as highly effective, high-strength structural materials that are increasingly used. Aluminum matrix hybrid composites strengthened with ceramic particulates are commonly used in marine, aerospace, and defense applications because of their exceptional properties. Zirconia-reinforced composites are favored because these composites display high refractory properties, excellent abrasion resistance, and chemical resistance compared to composites of other reinforcements.For applications where lightweight and superior performance is paramount, such as parts for spacecraft, fighter aircraft, and racecars, graphite compositesare the material of choice. In this research work, an effort was made to combine the properties of zirconia and graphite by producing a unique metal matrix composite of LM5 aluminum alloy reinforced with 6% zirconium dioxide (zirconia), using the stir casting process by changing the percentage of the weight of graphite to 2%, 3%, and 4%. The test specimens were prepared and evaluated in compliance with ASTM standards to study micro- and macrohardness, and impact, tensile, and compressive strength. Microstructural studies of composites performed through optical microscopy and SEM expose the unvarying dispersal of particulates of ZrO$_2$/graphite in the aluminum matrix. The hardness, impact, and compressive strength are enhanced due to the addition of reinforcement.

Keywords: hybrid composites; characterization; zirconia; graphite; SEM; microstructure

1. Introduction

Aluminiummatrix composites (AMCs) are recognized as materials with enhanced reliability for specific engineering fields. In some instances, they substitute homogenous alloy systems and, in particular cases, similar materials in terms of efficiency and economy [1,2]. Among several light metals such as Mg, Al, and Ti used as matrices, Al and its alloys are used more extensively as the matrices for MMCs [3,4]. This is related to properties such as being lightweight, and having a high corrosion resistance and ease of fabrication, which satisfy a broad range of current and potential requirements [5,6]. LM5 is a widelyused choice of special-purpose alloy as a matrix material compared to several other types of aluminium alloys, due to its favorable mechanical properties combined with efficient formability and corrosion resistance used for marine applications. Aluminium alloys have a meager resistance to wear compared with other metallic materials. To increase toughness and strength, the aluminium alloy must be reinforced. A variety of materials such as silicon carbide (SiC), titanium carbide (TiC), boron carbide (B$_4$C), aluminium oxide (Al$_2$O$_3$), silicon nitride (Si$_3$N$_4$), zirconium dioxide (ZrO$_2$), zirconium silicate (ZrSiO$_4$), boron nitride (BN), and sometimes even softer materials such as graphite and mica,are also used as reinforcements. The materials that stand outare ceramics rather than ferrous alloys. A good composite can be fabricated with ceramic reinforcement, and it exhibits superior qualities that are comparable to, or even greater than, other ferrous alloys. The liquid metallurgy techniques are easier to tackle during manufacturing and have a low-cost manufacturing method, particularly with intermittent reinforcements, compared to various processing techniques. It is observed that these MMCs result in isotropic properties [7]. A decent number of studies were conducted individually on Al/ZrO_2 and Al/graphite. The effect of graphite on the mechanical properties of aluminium composites was explored by Pai et al. [8].

The use of multiple reinforcements improved tribological properties for aluminum matrix hybrid composites more than the use of a single reinforcement. The constituents can interact synergistically, giving rise to better properties. Aluminium matrix composites with zirconia (ZrO_2) reinforcementswith high fracture toughness were produced by squeeze casting to fabricate Al-9Zn-6Mg-3Si composites with additions of 2.5, 5, and 7.5 vol.% ZrO_2. The results reveal that the higher the porosity, the higher the hardness, and the higher the impact values, both in the as-cast condition and after 1 h of ageing at 200 °C [9]. Aluminium metal matrix replaces traditional materials with high melting points and high densities, reducing energy consumption and helping the environment. With the help of AA 6061 and ZrO_2, they produced a low-weight, high-strength composite material using a stir casting technique combined with a squeeze casting configuration [10].The mechanical properties are enhanced, and there is a slight increase in density due to the high density of ZrO_2, up to the addition of 6% ZrO_2 [11].

The compo-casting technique successfully creates aluminium Al7075 alloy composites with varied weight percentages of ZrO_2 reinforcement (3, 6, 9, 12). The micrographs and EDAX show a homogeneous reinforcement distribution in a soft aluminium-rich matrix with micro porosities. The test findings show that the composites' hardness and ultimate tensile strength are increasing up to a reinforcement percentage of 6%, and there is a significant improvement in characteristics, but after that, the improvement is small. The gains in characteristics are mostly attributable to greater reinforcement distribution and interfacial bonding [12]. The mechanical and tribologicalbehaviour of aluminium (Al)-based silicon carbide (SiC, micro particles) and zirconium-oxide (ZrO_2, nano particles)-particle-reinforced hybrid composites. Powder metallurgy (PM) technology was used to add ZrO_2 (0, 3%, 6%, and 9%, weight fraction) to Al-5% SiC composites. The hardness and wear resistance of Al + SiC + ZrO_2 hybrid composites are shown to improve as the ZrO_2 content is increased [13].

Metal matrix composites based on aluminium alloys are becoming more popular in industrial applications that demand a high strength-to-weight ratio. AA6061 aluminium alloy matrices with zirconium-dioxide-particle-reinforced composites were fabricated using the stir casting technique.Composite materials reinforced with zirconium dioxide are fabricated with various weight percentages of reinforcement, such as 2%, 4%, 6%, 8%, and 10%. The composite's metallurgical and mechanical properties are investigated. The particles are equally spread in the matrix alloy, as shown by a scanning electron micrograph. The addition of ceramic particles increases the material's hardness by preventing dislocations in the alloy matrix. The addition of zirconium dioxide particles boosts its strength by 6%, according to tensile test results. In 6% of ZrO_2 and 2% of C inclusion of a 92% of AA6061 matrix material, the maximum strengths are 175 MPa in tensile strength, 45HRB hardness, and 4.56×10^{-9} g/mm. The addition of ZrO_2 raises the hardness of the base metal from 6% to 12%, and increases the ultimate tensile strength from 8% to 15%. The characteristics of the composite material are lowered when the reinforcing particles are added to the highest extent possible [14].

Al 2024 composites reinforced with 5%, 10%, 15%, and 20% ZrO_2 are created via vacuum infiltration. Due to an increase in the ZrO_2 reinforcement ratio, the density steadily rises. The density increase is due to the density of the reinforcing element being greater than the density of the matrix material. The rise in density, on the other hand, is not as great as the increase in the ZrO_2 reinforcement ratio. Reinforcement agglomeration in the composite structure is generated by increasing the ZrO$_2$ reinforcement ratio. In general, graphite as an additive to a composite has an eclectic effect on mechanical properties, whereas it leads to a positive effect on tribological properties [15]. Graphite is accessible in large quantities and at a lower cost. It is used to minimize the energy content, material content, cost, component weight, and improve wear resistance in aluminium castings. Strong interfacial connections exist between the matrix and the graphite particles. Hardness reduces as graphite content rises, but wear characteristics improve [16].

AA7075/graphite composites were produced by the stir casting method. The weight % of the graphite reinforcement in the AA7075 matrix phase varies from 5 to 20% in steps of 5%. A decrease in the ultimate tensile strength of the composite compared to the base matrix with an increase in the addition of graphite in the composite is observed. A significant decrease in the tensile strength is noticed at 5% of graphite compared to other weight percentages of graphite. This is due to graphite, which is brittle in nature, augmenting the tendency of crack initiation and propagation at the metal interface [17].

The tensile test results of Al 6061/B$_4$C/graphite with a constant weight percentage of B$_4$C and varying weigh percentage of graphite are demonstrated. There is an increase in the tensile strength of composite with the addition of B$_4$C, whereas a decrease in the tensile strength with the addition of graphite is noticed. They conclude that the tensile strength of MMC mainly depends on reinforcement strength and interfacial strength between the matrix and the reinforcement. ZrO$_2$ was added to the aluminium 2% graphite matrix in four distinct amounts (3%, 6%, 9%, and 12%). Cold pressing with a pressure of 700 MPa produces green compact samples from composite powders that were mechanically alloyed for 60 min. The green compacts were sintered at 600 °C for 2 h. Microstructure, density, and hardness measurements were taken on the aluminium composites formed. Hardness and density values increase as the amount of ZrO$_2$ increases in the matrix, and decrease as graphite is added as a consequence of the research [18].

The inclusion of graphite to the Al6061 matrix demonstrates a worsening trend in mechanical properties. The influence of SiC and graphite on Al6061 alloy is demonstrated by the fact that the composite's hardness and strength increase with the addition of SiC, while the hardness decreases and the strength increases with the inclusion of graphite to Al6061.

Aluminum alloys have excellent mechanical properties. Zirconium dioxide improves strength, rigidity, and resistance to temperature, with a slight increase in composite density. Graphite decreases density and enhances wear properties. Particulates are readily available, and processing is easy, while fabrication costs are low. Stir casting is the optimal and economical processing route for AMCs. In particular, magnesium improves aluminum/zirconium dioxide andgraphite wettability.

Very few works were identified in analyzing the effect of ZrO$_2$ on mechanical and metallurgical properties [19–21]. No work was identified in analyzing the effect of ZrO$_2$ and graphite on LM5 base material while subjected to mechanical and metallurgical properties [12]. Hence, this work mainly concentrated on developing new composite material by taking LM5 as the base material with 6% ZrO$_2$, and by varying the weight percentage of graphite to 2, 3, and 4% to identify the effect of ZrO$_2$ and graphite on the mechanical and metallurgical properties.

2. Materials

2.1. Matrix Metal

Aluminum alloy LM5 was the matrix material. The alloy is used where very high corrosion resistance from seawater or marine atmospheres is needed, for equipment used in the manufacture of foodstuffs, cooking utensils, and chemical plants, and the casting of highly polished surfaces. Accordingly, they are famous for decorative casts and casts used in dairy and food handling equipment, marine and chemical pipe fittings, and architectural/ornamental marine hardware applications. LM5's chemical composition was examined using optical emission spectrometry as per ASTM E 1251-07 standard, presented in Table 1.

Table 1. Chemical composition of LM5 aluminum alloy.

Cu	Mg	Si	Mn	Fe	Pb	Zn	Al
0.032	3.299	0.212	0.022	0.268	0.02	0.01	Balance

2.2. Reinforcement

Zirconium dioxide (ZrO_2), the crystalline oxide of zirconium, also called zirconia, is a widely studied ceramic element. Zirconia (ZrO_2) was chosen as the reinforcement particle because of its easy accessibility and suitability for high-temperature applications. Producing zirconia comprises the collection and elimination of unnecessary ingredients and impurities. Extraction of zirconia has many routes, including plasma disassociation, chlorine and alkali oxide decomposition, and lime fusion. Similar to other ceramic materials, zirconium oxide is a substrate with a high tolerance to the propagation of cracks. Therefore, zirconium oxide ceramics are often thermally developed; they are also the material of choice for joining ceramics and steel. Very low thermal conductivity and high strength are a desirable combinations of properties [22–24].

2.3. Fabrication of LM5/ZrO_2/Gr Hybrid Composites

The stir casting assembly had a C-Type closed furnace with a capacity of 5KVA (optimal temperature ranges from 500 °C to 1100 °C), with a stirrer assembly (Remi RQM-122/R) utilized to fabricate the composite. The stirrer had a stirring shaft of 350 mm length and 6 mm diameter made of SS304, and had a chuck for easy interchangeability of the shaft. It had a pitched fan type impeller of diameter 38 mm with 4 blades. The impeller was made of high chromium steel with high carbon content coated with zirconia, and was mounted on a vertical variable-speed motor with a range between 20 and 1500 rpm. The silicon carbide crucible was kept within the furnace. Figure 1a shows the stir casting set-up used for this research work, Figure 1b shows the pouring of the molten mixture into the mold, and Figure 1c shows the fabricated composite specimen.

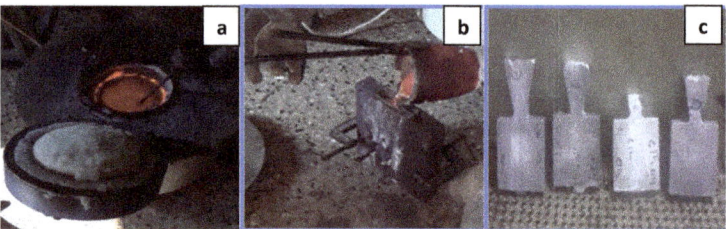

Figure 1. (a) Stir casting set-up; (b) pouring into die; (c) composite specimen.

Initially, the LM5 alloy, as tiny ingots, was charged and heated to about 850 °C in the silicon carbide crucible until the whole alloy was melted in the crucible. Zirconium dioxide and graphite particles were preheated for 20 min to 200 °C in a muffle furnace of 4 KVA, to eliminate the moisture existing in the reinforcement. A vortex was created in the molten metal by the stirrer, which was slowly lowered into the melt. Then, the preheated ZrO_2 of average particles size 70 μm and graphite of 30 μm were slowly mixed into the liquid metal at a steady rate, by maintaining the stirring speed at 600 rpm [25,26]. Magnesium powder 0.5% is added as a wetting agent to enhance the proper mixing of the matrix and the reinforcement [27]. The stirring continued for 7 min even after particle feeding was completed. Hexafluoro ethane tablets were introduced into the mixture before pouring into the mold to minimize the porosity. The temperature for pouring was kept at 750 °C. To achieve uniform solidification, before pouring the mixture into the mold, the mold was also preheated to 650 °C. This process was used to manufacture three sets of novel hybrid composites made of special-purpose aluminum alloy LM5 reinforced with 6% ZrO_2 particles and graphite 2, 3, and 4%. The melt was poured into the preheated mold to fabricate the hybrid composites.

3. Testing of AMCs

3.1. Micro Structural Analysis

3.1.1. Optical Microscopy

An inverted optical microscope was used to examine the MMCs. Abrasive papers were used to make the surfaces of the specimen smooth, and then it was polished through 220 to 1500 mesh with velvet fabric. Before microscopic analysis, the specimens were then etched using HF solution. Metallographic assessments offer strong quality assurance and a practical analysis resource. The specimens obtained from each composite were precisely polished to match the texture of the surface. Figure 2 shows the specimens used for the microstructure analysis.

Figure 2. Composite specimens for microstructural analysis.

The role of a microstructure on a material's physical and mechanical properties is influenced by the numerous flaws that exist or are absent in the structure [28]. These flaws can come in many forms, but the main ones are the pores. These pores play a decisive role in finalizing the characteristics of materials and their formulation. Moreover, for some materials, there can be various phases at the same time. These phases may have various properties, and prevent the material from fracturing if treated correctly.

3.1.2. SEM and EDAX

The crystal structure of materials is a 'fingerprint' of processing. A composite's microstructure is studied to recognize the changes in the structure of the parent metal after the addition of a reinforcement [29,30]. SEM has many advantages such as simple preparation of the specimens, broadest possible magnification scale (commonly between 15 and 50,000 times), and the capacity to observe large regions of the surface of the specimen, including the origin and spread zones. The surface can indeed be placed straight into the microscope, which has excellent field depth for concentrating on largetopographical surfaces.

EDAX is an analytical tool for using a sample's elemental analysis or chemical composition. It focuses on the interactions of a sample by supplying X-ray excitation. The characterization capabilities are primarily due to the underlying theory where each element has a unique atomic structure that causes its electromagnetic emission spectrum to have a unique set of peaks, which is the main principle of spectroscopy. A pulse of X-ray is centered on the examined material to induce the release of characteristic X-rays from a specimen. At resting, an atom inside the sample contains electrons at different energy levels, or electron shells attached to the nucleus in the ground state or unexcited. An energy-dispersive spectrometer can determine the amount and energy of X-rays released from a specimen. Since the X-ray energies indicate the energy difference between the two shells and the emitting element's atomic structure, EDAX makes it possible to measure the elemental composition of the specimen.

3.2. Density

Density is the naturally occurring phenomenon that reveals the characteristics of the composite. Utilizing displacement procedures, the density of a composite is calculated quantitatively, using an electronic weighing machine with a density calculating kit as per the ASTM: D 792-66 test procedure. Theoretical density is the actual density of a material corresponding to the limit that products with total density can achieve without pores. Many materials typically consist of a mixture of structural molecular components, each with their own mass. Archimedes proved very ingeniously that when an object is immersed in water (or any fluid), the force that it experiences is proportional to the mass of the water displaced times gravity (i.e., water weight). Density can be calculated using the standard formula. Porosity is the amount (or volume) of space in a material relative to the total size of the material. It is a mathematical ratio: void volume divided by total volume (vacuum/total); this ratio is usually multiplied by 100 to be compared in percentages rather than decimals [30]. It is calculated using the expression (Equation (1)).

$$\text{Porosity \%} = (\text{Theoretical Density} - \text{Experimental Density}) \times 100 / \text{Theoretical Density} \qquad (1)$$

3.3. Microhardness

Vickers hardness testing tool is often used to evaluate the composite's microhardness. Microhardness tests may be used to provide the data needed to measure discrete microstructures into a broader matrix, to evaluate excellent foils, or to assess a specimen's hardness gradient along a transverse. Microhardness testing refers specifically to the static indentations of 1 kgf or fewer loads. The Vickers hardness test uses a 136° apical angle diamond. The surface to be tested usually needs to be smoothly polished [31,32]. Amicroscope of 500× magnification is required to measure the shaped indents directly. Specifications of the Vickers hardness measurements are ASTM E 92 (for 1 kgf to 120 kgf) and ASTM E 384 (for force inferior to 1 kgf).

3.4. Macrohardness

Macrohardness is the measurement of the hardness of materials tested with high applied loads. The macrohardness measurement of materials is a quick and simple method of obtaining mechanical property data for the bulk material from a small sample. The Rockwell test measures the penetration depth of the indenter under a significant load (large load) in contrast to the penetration made by a minor load (preload). Here, various scales are represented by a solitary letter, which uses various indenters or loads [33–35]. An outcome is a dimensionless number given as HRA, HRB, HRC, HRE, etc., whereas the preceding letter is the Rockwell scale.

3.5. Tensile Strength

Tensile test was performed under atmospheric conditions using a computerized universal testing machine (model FMI F-100), with a cross head speed of 2 mm/min, to assess the manufactured composite materials. The specimens were prepared to ASTM-E8 standards. The findings of the tensile test are used in the selection of engineering materials. The strength of a material is always the prime concern, and can be determined by measuring the stress needed for severe plastic deformation, or the maximum stress tolerated by the material [36,37]. The material's ductility is also of concern, which measures its bend until it breaks. Tensile strength (also known as ultimate tensile strength) is quantified by dividing the maximum force held by the specimen by the initial cross-sectional area of the specimen during the stress test. Both the original gauge length and the percentage increase must be considered when recording the elongation values.

3.6. Compressive Strength

Compression testing was performed under atmospheric conditions using a computerized universal testing machine (model FMI F-100), with a cross head speed of 2 mm/min, to assess the manufactured composite materials. The research specimens (Figure 3) were prepared to ASTM-E8 standards. The findings of the compressive test are used in the selection of engineering materials. The material requirements must provide compressive properties to ensure performance [38,39].

Figure 3. Compressive test specimens.

Compressive testing reveals how the material can react as it is squeezed. Compression testing evaluates the action or reaction of the material against crushing loads, and assesses a material's plastic flow behaviour and ductile fracture limits. A compression test is a procedure for evaluating the behaviour under a compressive load of materials. Compression experiments are carried out by loading the test sample between two parallel plates and then bringing the crossheads together, adding force to the sample [40].

3.7. Impact Strength

Impact resistance of any material is the capability of that material to withstand a force applied or a sudden load [41,42]. Usually, it is distributed as the amount of mechanical energy consumed under the impact load imposed throughout the deformation process, and is presented as energy lost per unit of J/m^3. The Izod impact test is a standard ASTM tool for determining resistance to material impacts. A swivel arm (constant energy potential) is lifted to a particular height, and afterward, it is lowered. The arm falls to a notched plate and breaks it. The energy consumed by the plate is measured by the height at which the arm swings after it hits the plate. In the Izod impact test, the sample (ASTM A-370 standards) is mounted in a cantilever beam configuration, in contrast to a three-point bending configuration. Figure 4 shows the photograph of specimens used for impact strength testing.

Figure 4. Impact test specimen.

4. Results and Discussions

4.1. Microstructural Analysis

4.1.1. Optical Micrograph Analysis

Microstructural analysis's primary purpose is to examine the stratified dispersal of reinforcement particles into the matrix. The optical imaging micrographs demonstrate the homogeneous spread of the zirconia and graphite particles into the matrix. When the weight percentage of the strengthening material (reinforcement) increases, the particles in the particle distribution begin coagulating and disrupting uniformity [43]. Uniform reinforcement spread provides the matrix strength; the same would be the fundamental cause for accelerated mechanical properties. Figure 5 reveals the microstructures of base metal aluminum B.S.1490 grade LM5 aluminum–magnesium alloy and LM5/ZrO_2/Gr composites at 200× magnification.

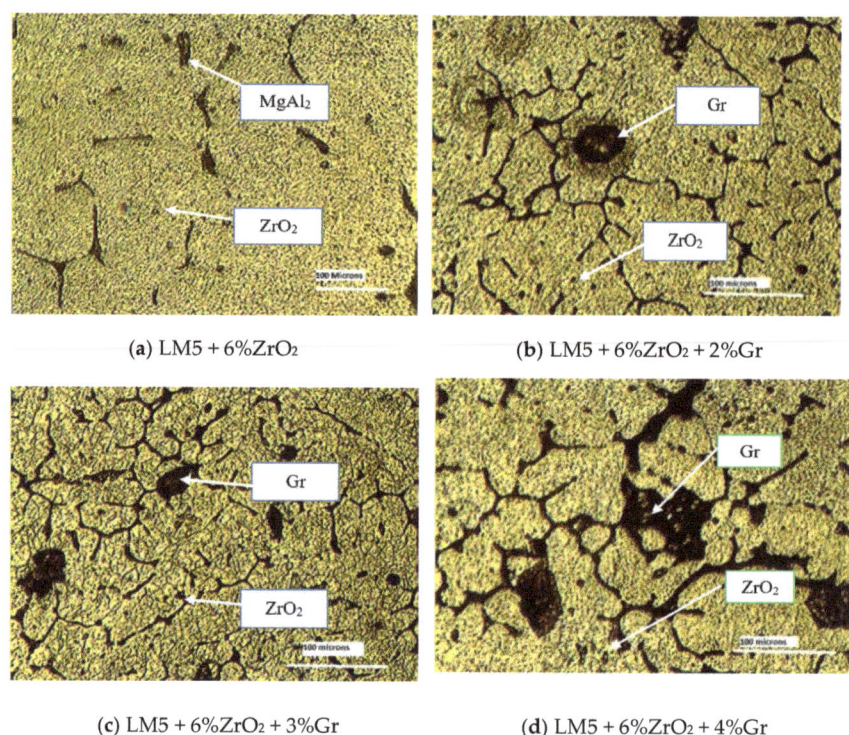

(a) LM5 + 6%ZrO_2
(b) LM5 + 6%ZrO_2 + 2%Gr
(c) LM5 + 6%ZrO_2 + 3%Gr
(d) LM5 + 6%ZrO_2 + 4%Gr

Figure 5. Micrographs of hybrid composites.

The microstructure of LM5 shows the interdendritic pattern of primary aluminum grains. The grain boundaries are precipitated with $MgAl_2$ eutectic particles, which have not dissolved during the solidification. The primary aluminum phase grain size is measured as 40 to 50 microns. The microstructure of the hybrid metal matrix composites with 6% ZrO_2 and 2, 3, and 4% of graphite shows the distribution of particles of ZrO_2 and graphite. The particles are inside the primary aluminum grains. The micrograph shows the resolved particles of the composite particles [44,45]. However, the composite with 6% ZrO_2/Gr particles distribution is observed, and they present as clusters along the grain boundaries.

4.1.2. SEM Analysis

Figure 6 shows the SEM images of LM5 and LM5/ZrO_2/Gr composites. Scanning electron micrographs display the unvarying dispersal of ZrO_2 ceramic particles and graphite in the aluminum MMCs at lower magnifications, and the findings of the SEM display the matrix–particle interfaces at higher magnifications [46,47]. These figures reveal the relatively homogenous distributions of reinforced ZrO_2 particles and graphite with aluminum alloy. In comparison, these statistics demonstrate the uniformity of the composite materials. The properties of MMCs depend on the metal matrix, the weight percentage, the arrangement of particle reinforcement, and the binding of the interface seen among particles and the matrix. No pores are found in either case, suggesting improved wettability between the matrix and the reinforcement particles [48,49].

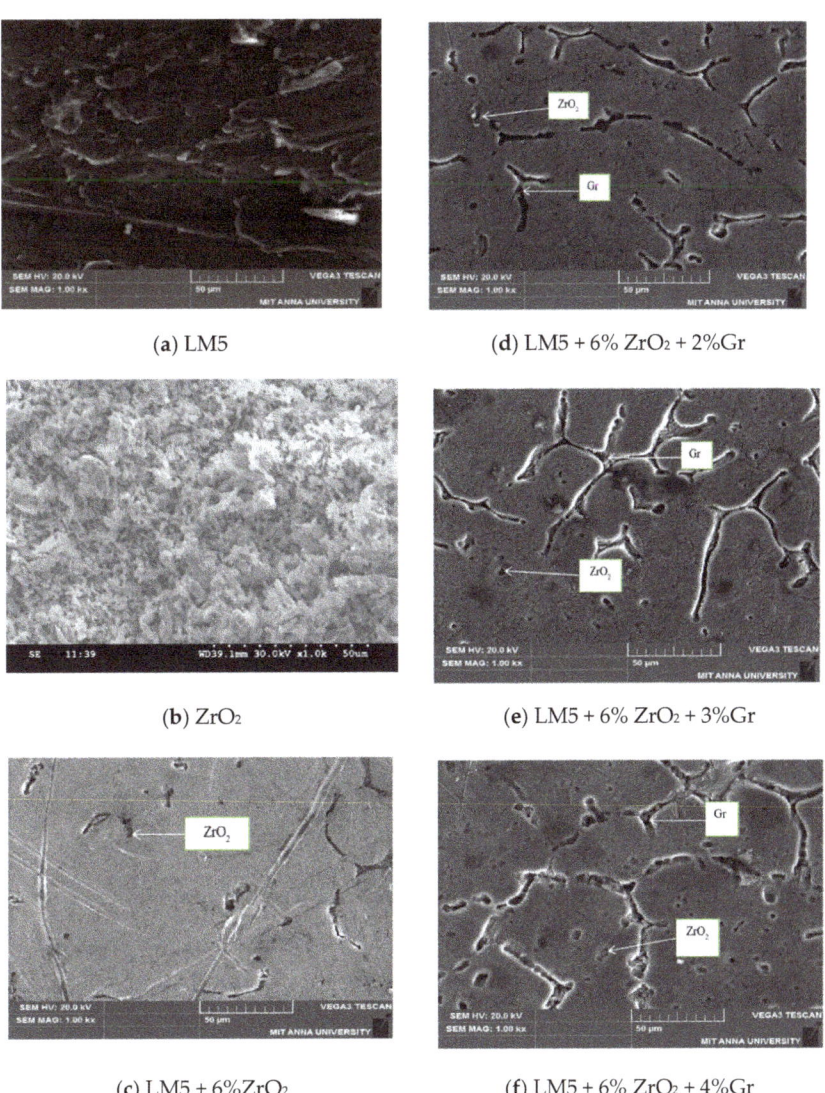

Figure 6. SEM images of hybrid composites.

The interfacial bonding is accomplished in this case due to fast cooling. It is also noticed that the area fraction rises as the weight percentage of ZrO_2 reinforcement rises, seen in the micrographs as a white field, and graphite particles are seen as a black field.

The average grain size of the aluminium LM5 matrix reduces as the ZrO_2 reinforcement weight fraction increases. It is also suspected that mechanical properties are rising due to the rise in the interfacial bonding of the reinforcements with the aluminium matrix alloy. This is due to the gravity of ZrO_2 and graphite, and is consistent with the effective selection of stirring parameters and substantial wetting of preheated ZrO_2 particles before being applied to the alloy of the matrix.

4.1.3. EDAX Analysis

EDAX develops the best solutions for micro-and nano-characterization, where primary and structural information is required, making analysis more accessible and accurate.

The existence of reinforcement and unit percentage of the composites is confirmed by EDAX analysis (Figure 7). EDAX demonstrates the pattern of particles of ZrO_2 and graphite particles scattered in composites of the aluminum matrix.

Element	Weight %	Atomic %	Error %
Mg	4.46	4.95	3.98
Al	93.68	93.86	2.33
Si	0.73	0.70	26.33
Pb	0.13	0.02	66.09
Mn	0.17	0.08	67.48
Fe	0.46	0.22	61.10
Cu	0.29	0.12	67.58
Zn	0.09	0.04	72.18

(a) LM5

Element	Weight %	Atomic %	Error %
C	0.36	0.80	99.99
O	1.72	2.86	18.79
Mg	4.41	4.83	4.11
Al	91.92	90.67	2.40
Si	0.39	0.37	38.16
Zr	0.13	0.04	88.13
Pb	0.16	0.02	74.17
Mn	0.20	0.10	66.93
Fe	0.30	0.14	62.12
Cu	0.20	0.08	71.27
Zn	0.20	0.08	67.32

(b) LM5 + 6%ZrO_2 + 2%Gr

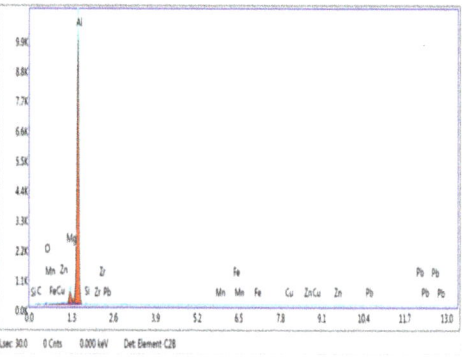

Figure 7. Cont.

Element	Weight %	Atomic %	Error %
C	1.75	3.83	75.93
O	1.37	2.24	25.13
Mg	4.76	5.14	4.04
Al	90.12	87.68	2.48
Si	0.64	0.60	24.16
Zr	0.19	0.06	80.08
Pb	0.18	0.02	76.76
Mn	0.18	0.09	67.98
Fe	0.32	0.15	61.44
Cu	0.23	0.10	69.36
Zn	0.26	0.10	65.44

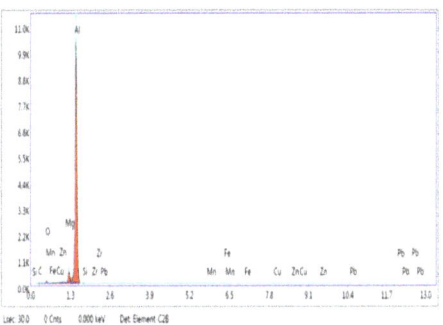

(**c**) LM5 + 6%ZrO$_2$ + 3%Gr

Element	Weight %	Atomic %	Error %
C	0.84	1.83	84.68
O	2.75	4.51	17.01
Mg	4.62	4.98	4.02
Al	90.22	87.76	2.43
Si	0.59	0.55	28.17
Zr	0.22	0.06	78.72
Pb	0.10	0.01	77.12
Mn	0.17	0.08	68.08
Fe	0.22	0.10	63.90
Cu	0.19	0.08	68.23
Zn	0.09	0.04	73.91

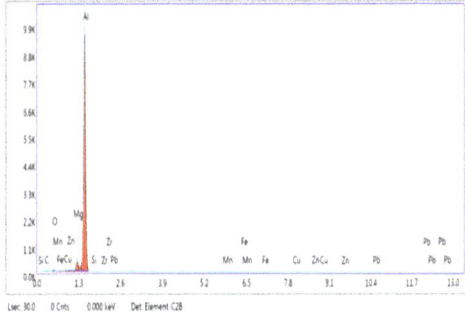

(**d**) LM5 + 6%ZrO$_2$ + 4%Gr

Figure 7. EDAX images of hybrid composites.

4.2. Density

The experimental and theoretical density of the fabricated LM5/6% ZrO$_2$ composite is found to be 2.73 and 2.832 g/cm^3, respectively, and the density of all the graphite composites is less when compared with the density of the ZrO$_2$, because of graphite's low density (2.26 g/cm^3).

The porosity of the produced composites slowly increases, since eutectic alloys have a high tendency to form large pores by increasing the weight percentage of the reinforcement [50]. Figures 8 and 9 shows the effect of graphite on the density and the porosity of AMCs, respectively.

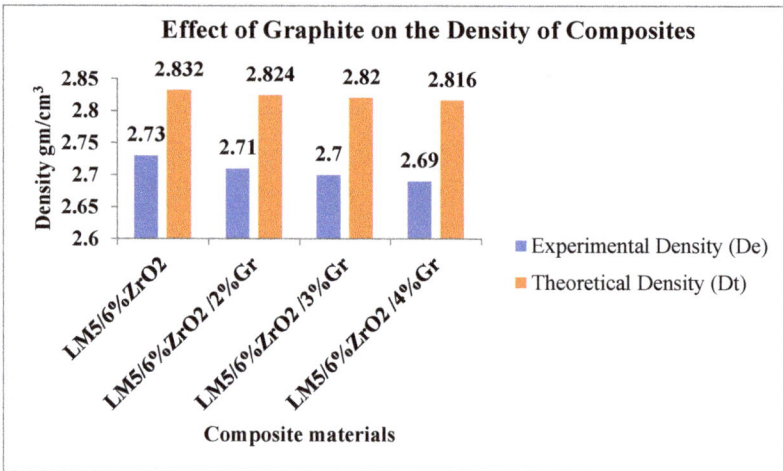

Figure 8. Effect of graphite on the density of composites.

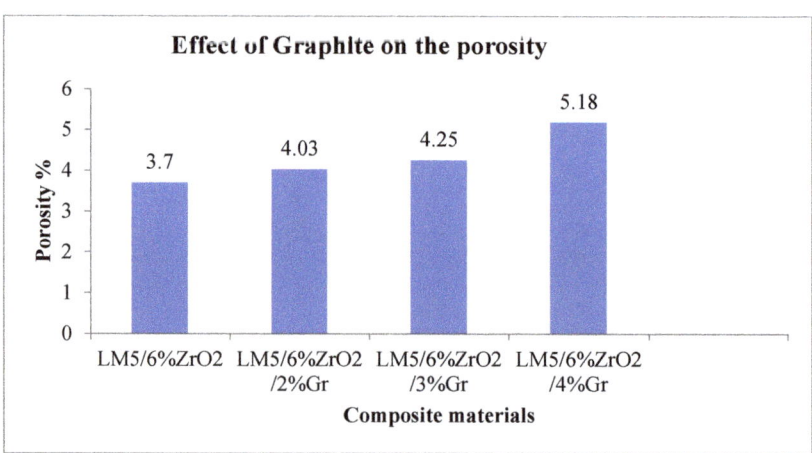

Figure 9. Effect of graphite on the porosity of composites.

4.3. Microhardness

Microhardness tests were performed at room temperature in the Vickers hardness measurement device by introducing a load of 0.5 kgf for 10 s of dwelling time. Figure 10 shows the effect of graphite on the microhardness of AMCs. The highest value is 74 VHN for the 6% ZrO_2 composite. It indicates higher hardness. The introduction of graphite into the metal matrix decreases the hardness value. From Figure 10, it can be observed that the hardness is decreased linearly with an addition of the graphite reinforcement [51,52]. When adding the graphite particles to the composites, the surface area of the reinforcements is increased, and the particle dimensions of the matrix are reduced.

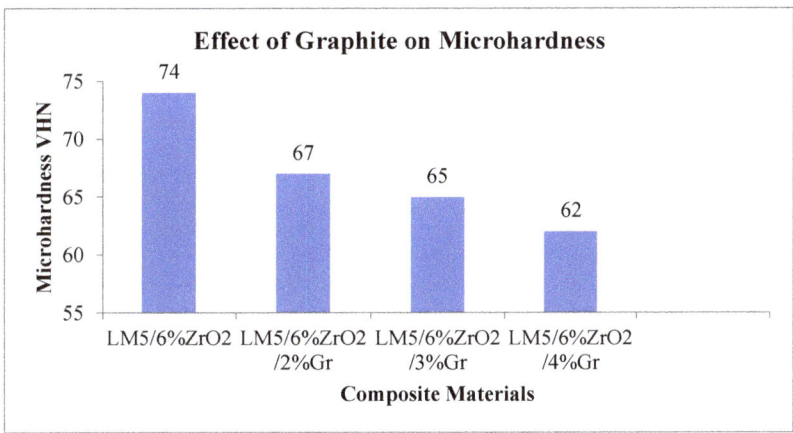

Figure 10. Effect of graphite on the microhardness of composites.

4.4. Macrohardness

The effect of graphite on the macrohardness is shown in Figure 11. The Rockwell hardness value of LM5/6% ZrO_2 alloy is 67HRE. The hardness value decreases dramatically to 60HRE, 59HRE, and 57HRE for 2, 3, and 4 graphite weight percentages, respectively. The appearance of very soft particles of graphite makes plastic distortion more opposed, contributing to decreased material hardness.

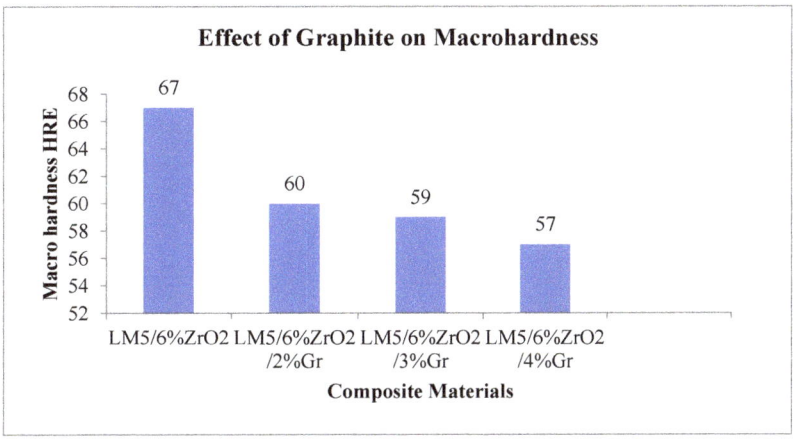

Figure 11. Effect of graphite on the macrohardness of composites.

4.5. Tensile Strength

The tensile test shows that the strength of the composites rises as the weight % of the ZrO_2 particles increases. Sample stress–strain curves of tensile test specimens are shown in Figure 12. Fractured tensile test specimens are shown in Figure 13.

Figure 14 shows the effect of zirconium dioxide and graphite weight percentage on the hybrid aluminum matrix composite's tensile strength. Zirconia has a monoclinic composition, while the composition of aluminum crystallizes in FCC. Their interface strength is due to the different crystalline structures of zirconia and aluminum that sounds incoherent [53,54].

This incoherence, therefore, increases the strength of the composite materials. The high hardness of the Al matrix composites is probably due to the high stiffness rate of the composites during their strain. Improvement of the hardening function may be related to the elastic properties of ZrO_2 particles and their prevention of deformation of the matrix. Thus, in the presence of a suitable interface, the ZrO_2 particles prevent the deformation of the matrix and increase work hardening. In addition, the specific thermal expansion coefficients of zirconia (10×10^{-6} K^{-1}) and aluminum (16×10^{-6} K^{-1}) produce stress that can boost the number of dislocations and, as a result, the strength of the composite. Increasing dislocation density and the piling up behind ZrO_2 particles serve as obstacles in dislocation movement. The greater the sum of ZrO_2, the greater the number of dislocations that are formed [55]. The tensile strength of the reinforced composite of aluminum alloy LM5 with 6% ZrO_2 is 220 MPa, and this value decreases to 216 Mpa for LM5/6% ZrO_2/2% Gr, and decreases again to 215 MPa for the composition of LM5/6% ZrO_2/3% Gr. From the experimental results, it is clear that the tensile strength of the composites slightly decrease compared to the 6% ZrO_2, due to the addition of graphite particles.

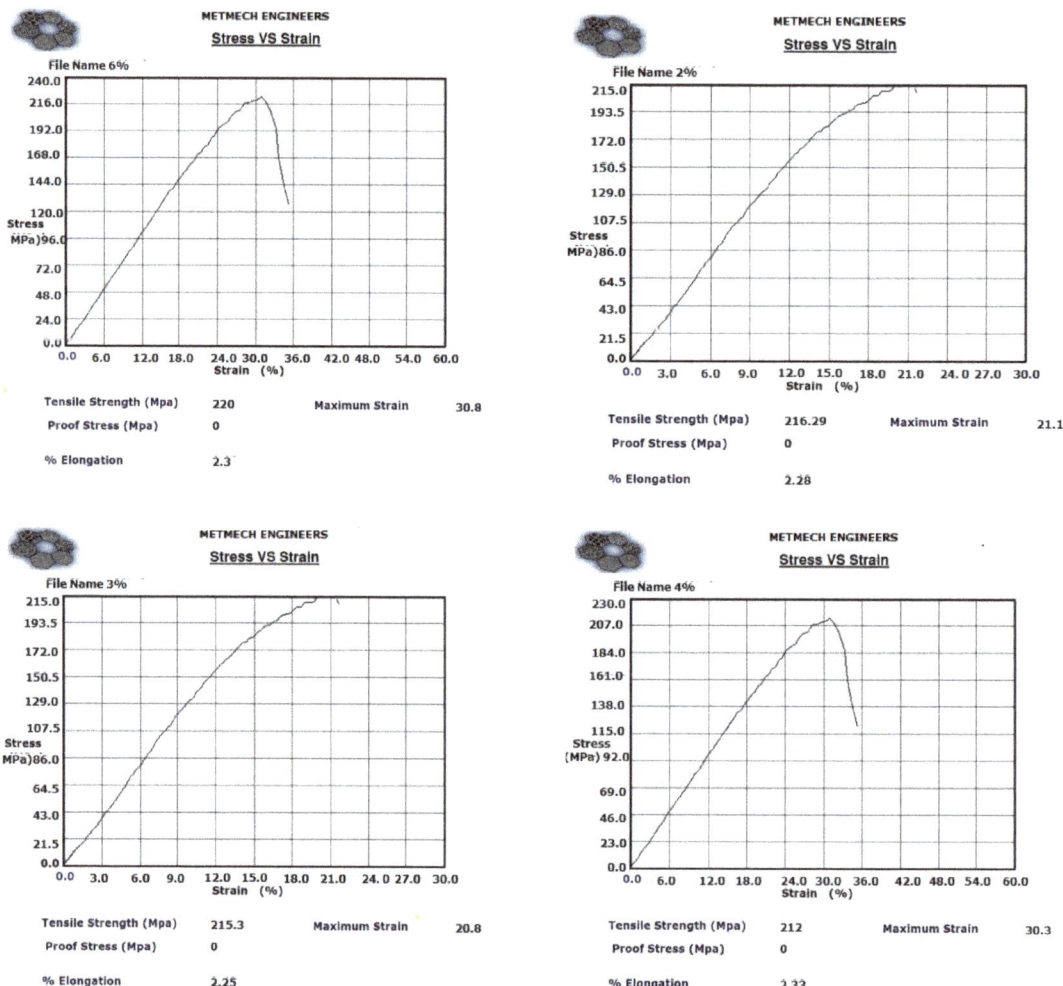

Figure 12. Sample stress–strain curves of tensile test specimens.

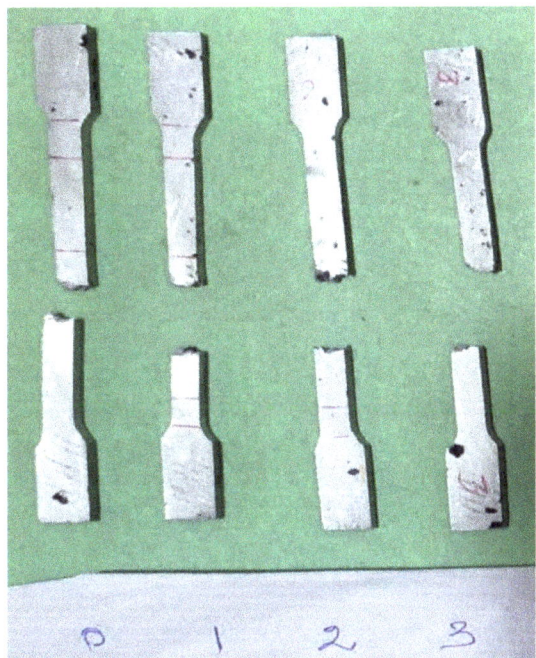

Figure 13. Fractured tensile test specimens.

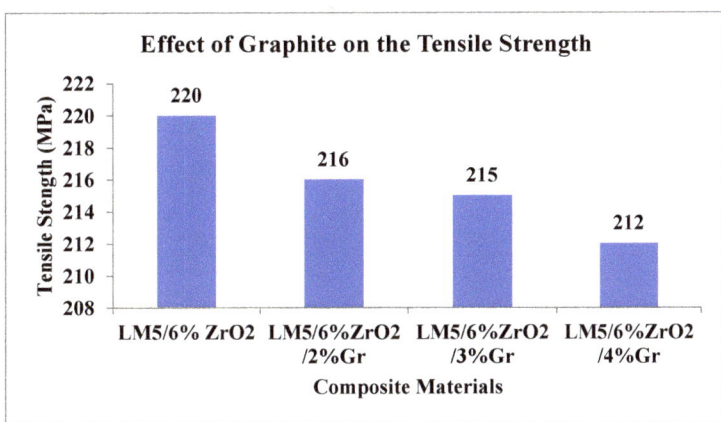

Figure 14. Effect of graphite on the tensile strength of composites.

Figure 15 shows the effect of zirconium dioxide and graphite on the elongation percentage. Results show that by adding the weight percentages of zirconium dioxide, the elongation of the material decreases. The LM5 loses its ductility, and transitions from ductile to brittle by adding the zirconium dioxide. The elongation of LM5 + 6% ZrO_2 is 2.3%; this value is reduced to 2.28%, 2.25%, and 2.23% for the graphite composites. Due to the brittleness of the fabricated composite tensile strength, elongation and strain are less for graphite composites compared to 6% ZrO_2. The maximum ultimate tensile strength is observed at 6% ZrO_2, but graphite may enhance wear properties. Few cluster creation are noticed in the 4% graphite reinforcement, which is projected to cause decreased mechanical properties, particularly tensile strength.

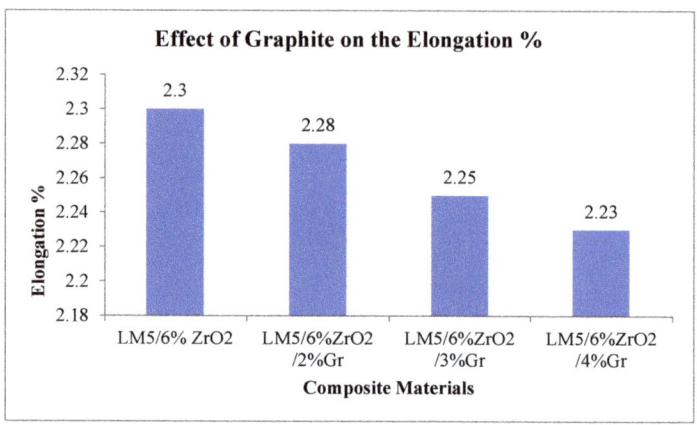

Figure 15. Effect of graphite on the elongation %.

4.6. Compressive Strength

The compressive test exposes that the strength of the LM5/6% ZrO_2/Gr composites rises as the weight % of the particles in Gr increases. Sample stress–strain curves of the compression test specimens are shown in Figure 16. Distorted compression test specimens are shown in Figure 17.

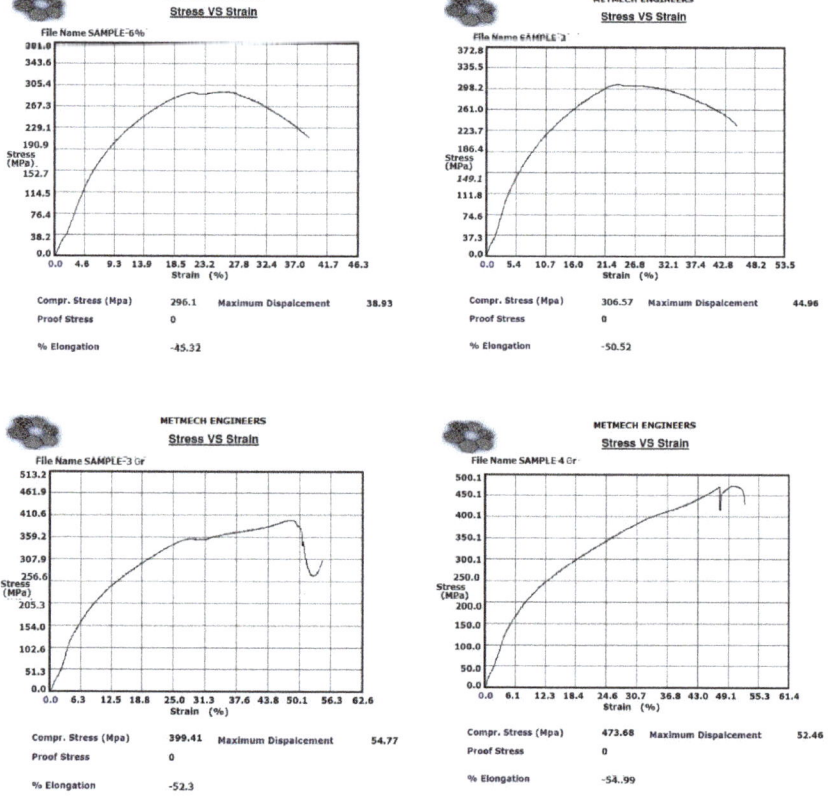

Figure 16. Sample stress-strain curves of compression test specimens.

Figure 17. Distorted compression test specimens.

Figure 18 shows the effect of zirconium dioxide and graphite weight percentage on the LM5/ZrO$_2$/Gr hybrid composites compressive strength. Zirconia has a monoclinic composition, while the composition of aluminum crystallizes in FCC. Their interface is due to the different crystalline structures of zirconia and aluminum that sounds incoherent [56]. This incoherence, therefore, increases the strength of the composite materials.

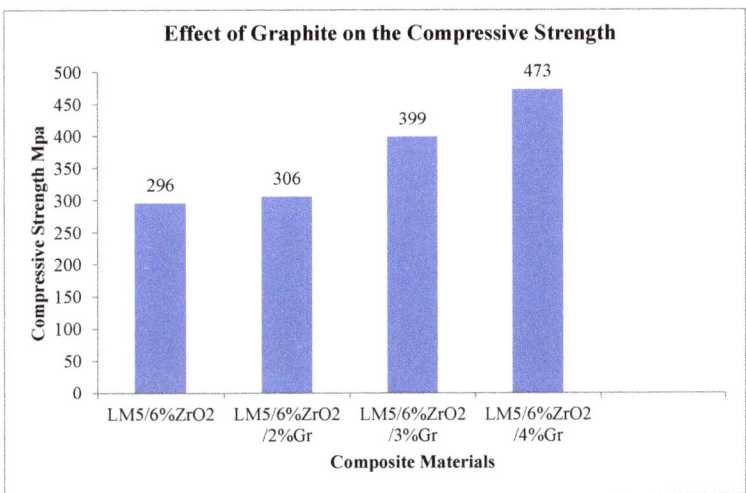

Figure 18. Effect of graphite on the compressive strength of composites.

The high hardness of the Al matrix composites is probably due to the high stiffness rate of the composites during their strain. Improvement of the hardening function may be related to the elastic properties of ZrO$_2$ particles, and their prevention of deformation of the matrix. Thus, in the presence of a suitable interface, the ZrO$_2$ particles prevent deformation of the matrix and increase work hardening. In addition, the specific thermal expansion coefficients of zirconia and aluminum produce stress that can boost the number of dislocations and, as a result, the strength of the composite. Increasing dislocation density and the piling up behind ZrO$_2$ particles serve as obstacles in the dislocation movement. The greater the sum of ZrO$_2$, the greater the number of dislocations that are formed. The compressive strength of the LM5/6% ZrO$_2$ alloy is 296 MPa. Results show that by adding the ZrO$_2$/Gr particles, the compressive strength of the aluminum alloy composite is dramatically increased [57]. The compressive strength of the reinforced composite of aluminum alloy LM5 with 6% ZrO_2/2% Gr is 306 MPa, and this value increases to 399 MPa for LM5 with 6% ZrO_2/3% Gr, and further increases to 473 MPa for the composition of LM5/6% ZrO_2/4% Gr. From the experimental results, it is clear that the compressive strength of the hybrid composites are improved [58].

Figure 19 shows the effect of zirconia and graphite on the compression percentage. By adding 6% zirconium dioxide, the compressive percentage is 45.32%, loses its ductility, and the transition occurs from ductile to brittle. The compression of LM5 + 6% ZrO_2 + 2% graphite is 50.52%, this value is increased to 52.3% for the composition of LM5 + 6% of ZrO_2/3% graphite, and then increases again to 54.99% for the composition of LM5 + 6% of ZrO_2 + 4% graphite. The maximum compressive strength is observed at 6% ZrO_2/4% graphite.

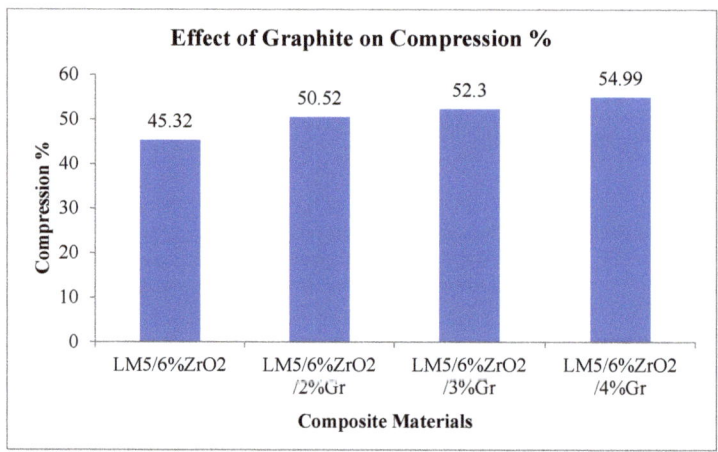

Figure 19. Effect of graphite on the compression of composites.

4.7. Impact Strength

The impact strength of the fabricated composites is determined by conducting an Izod impact test. Fractured impact test specimens are shown in Figure 20.

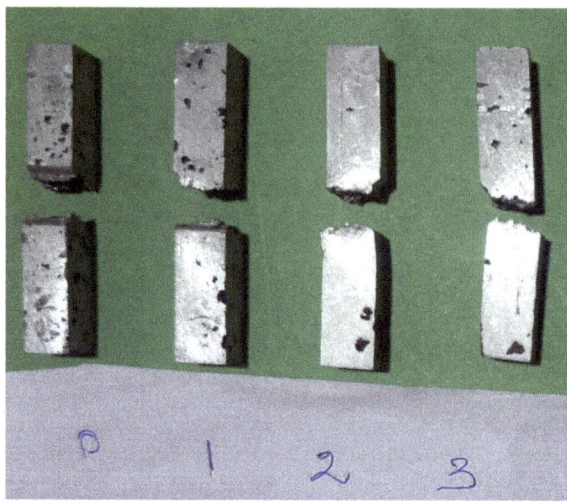

Figure 20. Fractured impact test specimens.

Figure 21 shows that the impact energy of the fabricated composites escalates with the increase in graphite percentage. The impact strength of LM5 reinforced with 6% Gr is 12 joules, and it is decreased to 9.15 joules in 6% ZrO_2/2% graphite, and then gradually increases to 9.25 and 9.35 in the 3% and 4% graphite composites, respectively. The impact energy increase with the strengthening could be due to the tough bond forming among the matrix and the reinforcing ZrO_2 and graphite. It is also noted that the impact strength of all the fabricated composites is relatively greater compared to the impact strength of the LM5 aluminum alloy (7.9 joules), due to the combined effect of zirconia and graphite [59].

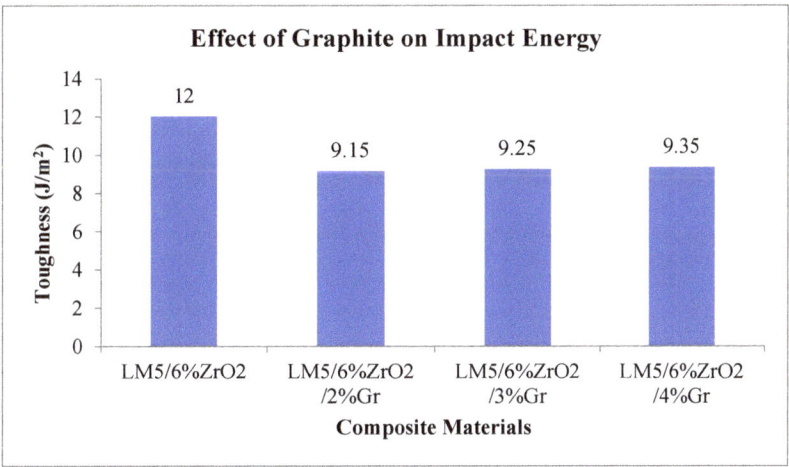

Figure 21. Effect of graphite on the toughness of composites.

5. Conclusions

This research examines the mechanical characterization and microstructural analysis of hybrid composites. The stir casting technique efficaciously fabricates aluminum-based hybrid composites with an even distribution of ZrO_2 and graphite particles. Density is increased slightly, but micro- and macro-hardness are improved magnificently by increasing the fraction of weight of graphite in 6% ZrO_2. The impact, tensile, and compressive strength of MMCs are improved magnificently by increasing the fraction of weight of graphite in 6% ZrO_2. Elongation of the composites decreases due to the transformation of materials from ductile to brittle. An LM5 aluminum alloy reinforced with 6% ZrO_2/4% graphite can be used further for many structural applications, as it has improved mechanical properties such as compressive strength, hardness, and impact strength.

Author Contributions: Conceptualization, J.U.P. and S.J.J.; methodology, S.S.; formal analysis, S.J.J. and J.U.P.; investigation, H.M.A.H.; resources, S.J.J.; data curation, S.S. and S.R.G.; writing—original draft preparation, S.J.J.; writing—review and editing, J.U.P.; visualization, S.S. and S.R.G.; supervision, J.U.P.; project administration, S.S.; funding acquisition, H.M.A.H. All authors have read and agreed to the published version of the manuscript.

Funding: This research received no external funding.

Institutional Review Board Statement: Not applicable.

Informed Consent Statement: Not applicable.

Data Availability Statement: The data presented in this study are available through email upon request to the corresponding author.

Conflicts of Interest: The authors declare that they have no conflict of interest.

References

1. Surappa, M.K. Aluminium matrix composites: Challenges and Opportunities. *Sadhana* **2003**, *28*, 319–334. [CrossRef]
2. Srivatsan, T.S.; Ibrahim, I.A.; Mohamed, F.A.; Lavernia, E.J. Processing techniques for particulate-reinforced metal aluminum matrix composites. *J. Mater. Sci.* **1991**, *26*, 596–5978. [CrossRef]
3. Clyne, T.W.; Withers, P.J. *An Introduction to Metal Matrix Composites*; Cambridge University Press: Cambridge, UK, 1993.
4. Surappa, K.M.; Rohatgi, P.K. Preparation and properties of cast aluminum-ceramic particle composites. *J. Mater. Sci.* **1981**, *16*, 983–993. [CrossRef]
5. Lanker, M.V. *Metallurgy of Aluminum Alloys*; Chapman & Hall Ltd.: Boca Raton, FL, USA, 1967.
6. Hashim, J.; Looney, L.; Hashmi, M.S.J. Metal matrix composites: Production by the stir casting method. *J. Mater. Process. Technol.* **1999**, *92*, 1–7. [CrossRef]
7. Torralba, J.D.; Da Costa, C.E.; Velasco, F. P/M Aluminum matrix composites: An overview. *J. Mater. Process. Technol.* **2003**, *1–2*, 203–206. [CrossRef]
8. Pai, B.C.; Pillai, R.M.; Satyanarayana, K.G. Prospects for graphite aluminium composites in Engineering industries. *Ind. J. Eng. Mater. Sci.* **1994**, *1*, 279–285.
9. Syarifudin, M.; Hale, E.N.; Sofyan, B.T. Effect of ZrO_2 addition on mechanical properties and microstructure of Al-9Zn-6Mg-3Si matrix composites manufactured by squeeze casting. *Mater. Sci. Eng.* **2019**, *517*, 12001. [CrossRef]
10. James, S.J.; Annamalai, A.R. Machinability study of developed composite AA6061-ZrO_2 and analysis of influence of MQL. *Metals* **2018**, *8*, 472. [CrossRef]
11. Babu, L.G.; Ramesh, M.; Ravichandran, M. Mechanical and tribological characteristics of ZrO_2 reinforced Al2014 matrix composites produced via stir casting route. *Mater. Res. Express* **2019**, *6*, 115542.
12. Gunasekaran, T.; Vijayan, S.N.; Prakash, P.; Satishkumar, P. Mechanical properties and characterization of Al7075 aluminum alloy based ZrO_2 particle reinforced metal-matrix composites. *Mater. Today Proc.* **2020**, in press. [CrossRef]
13. Arif, M.; Asif, M.; Ahmed, I. Advanced composite material for aerospace application—A review. *Int. J. Eng. Manuf. Sci.* **2017**, *7*, 393–409.
14. Pandiyarajan, R.; Maran, P.; Murugan, N.; Marimuthu, S.; Sornakumar, T. Friction stir welding of hybrid AA 6061-ZrO_2-C composites FSW process optimization using desirability approach. *Mater. Res. Express* **2019**, *6*, 066553. [CrossRef]
15. Pul, M. Effect of ZrO_2 quantity on mechanical properties of ZrO_2-reinforced aluminum composites produced by the vacuum infiltration technique. *Rev. De Metal.* **2021**, *57*, e195. [CrossRef]
16. Moghanlou, F.S.; Nekahi, S.; Vajdi, M.; Ahmadi, Z.; Motallebzadeh, A.; Shokouhimehr, A.; Shokouhimehr, M.; Jafargholinejad, S.; Asl, M.S. Effects of graphite nano-flakes on thermal and microstructural properties of TiB2–SiC composites. *Ceram. Int.* **2020**, *46*, 11622–11630. [CrossRef]
17. Baradeswaran, A.; ElayaPerumal, A. Effect of graphite on tribological and mechanical properties of AA7075 composites. *Tribol. Trans.* **2015**, *58*, 1–6. [CrossRef]
18. Simsek, I.; Şimsek, D.; Ozyurek, D. The effect of different sliding speeds on wear behavior of ZrO_2 reinforcement aluminium matrix composite materials. *Int. Adv. Res. Eng. J.* **2020**, *4*, 1–7.
19. Rino, J.J.; Sivalingappa, D.; Koti, H.; Jebin, V.D. Properties of Al6063 MMC Reinforced with Zircon Sand and Alumina. *IOSR J. Mech. Civ. Eng.* **2013**, *5*, 72–77. [CrossRef]
20. HimaGireesh, C.; Durga Prasad, K.G.; Ramji, K. Experimental investigation on mechanical properties of an Al6061 hybrid metal matrix composite. *J. Compos. Sci.* **2018**, *2*, 49. [CrossRef]
21. Aruna, K.; Diwakar, K.; Kumar, K.B. Development and Characterization of Al 6061-ZrO_2 Reinforced Metal Matrix Composites. *Int. J. Adv. Res. Comput. Sci. Softw. Eng.* **2018**, *8*, 270–275.
22. Kareem, A.; Qudeiri, J.A.; Abdudeen, A.; Ahammed, T.; Ziout, A. A review on AA 6061 metal matrix composites produced by stir casting. *Materials* **2021**, *14*, 175. [CrossRef]
23. Karthikeyan, G.; Jinu, G.R. Dry sliding wear behaviour of stir cast LM25/ZrO_2 metal matrix composites. *Trans. FAMENA* **2015**, *39*, 89–98.
24. JebaroseJuliyana, S.; UdayaPrakash, J. Drilling parameter optimization of metal matrix composites (LM5/ZrO_2) using Taguchi Technique. *Mater. Today Proc.* **2020**, *33*, 3046–3050. [CrossRef]
25. Madhusudhan, M.; Naveen, G.J.; Mahesha, K. Mechanical characterization of AA7068-ZrO_2 reinforced metal matrix composites. *Mater. Today Proc.* **2017**, *4*, 3122–3130. [CrossRef]
26. Sozhamannan, G.G.; BalasivanandhaPrabu, S.; Venkatajalapathy, V.S.K. Effect of processing parameters on metal matrix composites: Stir casting process. *J. Surf. Eng. Mater. Adv. Technol.* **2012**, *2*, 11–15.
27. Malaki, M.; FadaeiTehrani, A.; Niroumand, B.; Gupta, M. Wettability in metal matrix composites. *Metals* **2021**, *11*, 1034. [CrossRef]
28. Liao, Z.; Standke, Y.; Gluch, J.; Balázsi, K.; Pathak, O.; Höhn, S.; Herrmann, M.; Werner, S.; Dusza, J.; Balázsi, C.; et al. Microstructure and fracture mechanism investigation of porous silicon Nitride–Zirconia–Graphene composite using multi-Scale and In-Situ microscopy. *Nanomaterials* **2021**, *11*, 285. [CrossRef] [PubMed]
29. Alajmi, M.; Shalwan, A. Correlation between mechanical properties with specific wear rate and the coefficient of friction of graphite/epoxy composites. *Materials* **2015**, *8*, 4162–4175. [CrossRef]
30. Molina, J.M.; Rodríguez-Guerrero, A.; Louis, E.; Rodríguez-Reinoso, F.; Narciso, J. Porosity effect on thermal properties of Al-12 wt% Si/graphite composites. *Materials* **2017**, *10*, 177. [CrossRef] [PubMed]

31. Smeulders, R.J.; Mischgofsky, F.H.; Frankena, H.J. Direct microscopy of alloy nucleation, solidification and ageing (coarsening) during stir casting. *J. Cryst. Growth* **1986**, *76*, 151–169. [CrossRef]
32. Zhou, W.; Xu, Z.M. Casting of SiC reinforced metal matrix composites. *J. Mater. Process. Technol.* **1997**, *63*, 358–363. [CrossRef]
33. Callister, W.D., Jr. *Material Science and Engineering—An Introduction*, 2nd ed.; John Wiley: New York, NY, USA, 1991.
34. Ravi, B.; BaluNaik, B.; UdayaPrakash, J. Characterization of Aluminum Matrix Composites (AA6061/B4C) Fabricated by Stir Casting Technique. *Mater. Today Proc.* **2015**, *2*, 2984–2990. [CrossRef]
35. Lindroos, V.K.; Talvitie, M.J. Recent advances in metal matrix composites. *J. Mater. Process. Technol.* **1995**, *53*, 273–284. [CrossRef]
36. Razzaq, A.M.; Majid, D.L.; Basheer, U.M.; Aljibori, H.S.S. Research Summary on the Processing, Mechanical and Tribological Properties of Aluminium Matrix Composites as Effected by Fly Ash Reinforcement. *Crystals* **2021**, *11*, 1212. [CrossRef]
37. Chan, K.F.; Zaid, M.H.M.; Mamat, M.S.; Liza, S.; Tanemura, M.; Yaakob, Y. Recent developments in carbon nanotubes-reinforced ceramic matrix composites: A review on dispersion and densification techniques. *Crystals* **2021**, *11*, 457. [CrossRef]
38. Lokesh, T.; Mallikarjun, U.S. Mechanical and morphological studies of Al6061-Gr-SiC hybrid metal matrix composites. *Appl. Mech. Mater.* **2015**, *813*, 195–202. [CrossRef]
39. Gowrishankar, T.P.; Manjunatha, L.H.; Sangmesh, B. Mechanical and Wear behaviour of Al6061 reinforced with Graphite and TiC Hybrid MMC's. *Mater. Res. Innov.* **2019**, *24*, 179–185.
40. Girisha, K.B.; Chittappa, H.C. Preparation, Characterization and Wear Study of Aluminum Alloy (Al 356.1) Reinforced with Zirconium Nano Particles. *Int. J. Innov. Res. Sci. Eng. Technol.* **2013**, *2*, 3627–3637.
41. Malhotra, S.; Narayan, R.; Gupta, R.D. Synthesis and Characterization of Aluminum 6061 Alloy-Fly ash& Zirconia Metal Matrix Composite. *Int. J. Curr. Eng. Technol.* **2013**, *3*, 1716–1719.
42. Baghchesara, M.A.; Abdizadeh, H.; Baharvandi, H.R. Microstructure and Mechanical Poperties of Aluminum Alloy Matrix Composite Reinforced with ZrO_2 Particles. *Asian J. Chem.* **2010**, *22*, 3824–3834.
43. Yadav, P.; Ranjan, A.; Kumar, H.; Mishra, A.; Yoon, J. A contemporary review of aluminium MMC developed through stir-casting route. *Materials* **2021**, *14*, 6386. [CrossRef]
44. Barabás, R.; Fort, C.I.; Turdean, G.L.; Bizo, L. Influence of HAP on the morpho-structural properties and corrosion resistance of ZrO_2-based composites for biomedical applications. *Crystals* **2021**, *11*, 202. [CrossRef]
45. Nakonieczny, D.S.; Slíva, A.; Paszenda, Z.; Hundáková, M.; Kratošová, G.; Holešová, S.; Majewska, J.; Kałużyński, P.; Sathish, S.K.; SimhaMartynková, G. Simple approach to medical grade alumina and zirconia ceramics surface alteration via acid etching treatment. *Crystals* **2021**, *11*, 1232. [CrossRef]
46. Liu, J.; Sun, K.; Zeng, L.; Wang, J.; Xiao, X.; Liu, J.; Guo, C.; Ding, Y. Microstructure and Properties of Copper–Graphite Composites Fabricated by Spark Plasma Sintering Based on Two-Step Mixing. *Metals* **2020**, *10*, 1506.
47. Malaki, M.; Xu, W.; Kasar, A.K.; Menezes, P.L.; Dieringa, H.; Varma, R.S.; Gupta, M. Advanced metal matrix nanocomposites. *Metals* **2019**, *9*, 330. [CrossRef]
48. Singh, M.; Garg, H.K.; Maharana, S.; Yadav, A.; Singh, R.; Maharana, P.; Nguyen, T.V.; Yadav, S.; Loganathan, M.K. An experimental investigation on the material removal rate and surface roughness of a hybrid aluminum metal matrix composite (Al6061/sic/gr). *Metals* **2021**, *11*, 1449. [CrossRef]
49. Khan, A.; Abdelrazeq, M.W.; Mattli, M.R.; Yusuf, M.M.; Alashraf, A.; Matli, P.R.; Shakoor, R.A. Structural and mechanical properties of Al-SiC-ZrO_2 nanocomposites fabricated by microwave sintering technique. *Crystals* **2020**, *10*, 904. [CrossRef]
50. Díaz, M.; Smirnov, A.; Gutiérrez-González, C.F.; Estrada, D.; Bartolomé, J.F. Microstructure and mechanical properties of zirconia (3Y-TZP)/Zr composites prepared by wet processing and subsequent spark plasma sintering. *Ceramics* **2020**, *3*, 53–64. [CrossRef]
51. Kvashnin, V.I.; Dudina, D.V.; Ukhina, A.V.; Koga, G.Y.; Georgarakis, K. The Benefit of the Glassy State of Reinforcing Particles for the Densification of Aluminum Matrix Composites. *J. Compos. Sci.* **2022**, *6*, 135. [CrossRef]
52. Simoncini, A.; Tagliaferri, V.; Ucciardello, N. High thermal conductivity of copper matrix composite coatings with highly-aligned graphite nanoplatelets. *Materials* **2017**, *10*, 1226. [CrossRef]
53. Sadhana, A.D.; Prakash, J.U.; Sivaprakasam, P.; Ananth, S. Wear behaviour of aluminum matrix composites (LM25/Fly Ash)-A Taguchi approach. *Mater. Today Proc.* **2020**, *33*, 3093–3096. [CrossRef]
54. Prakash, J.U.; Juliyana, S.J.; Pallavi, P.; Moorthy, T.V. Optimization of Wire EDM Process Parameters for Machining Hybrid Composites (356/B4C/Fly Ash) using Taguchi Technique. *Mater. Today Proc.* **2018**, *5*, 7275–7283.
55. Rubi, C.S.; Prakash, J.U. Drilling of Hybrid Aluminum Matrix Composites using Grey-Taguchi Method. *INCAS Bull.* **2020**, *12*, 167–174.
56. Seon, G.; Makeev, A.; Schaefer, J.D.; Justusson, B. Measurement of interlaminar tensile strength and elastic properties of composites using open-hole compression testing and digital image correlation. *Appl. Sci.* **2019**, *9*, 2647.
57. Deng, C.; Li, R.; Yuan, T.; Niu, P.; Wang, Y. Microstructure and Mechanical Properties of a Combination Interface between Direct Energy Deposition and Selective Laser Melted Al-Mg-Sc-Zr Alloy. *Metals* **2021**, *11*, 801.
58. Chandel, R.; Sharma, N.; Bansal, S.A. A review on recent developments of aluminum-based hybrid composites for automotive applications. *Emergent Mater.* **2021**, *4*, 1243–1257.
59. Jadhav, P.R.; Sridhar, B.R.; Nagaral, M.; Harti, J.I. Mechanical behavior and fractography of graphite and boron carbide particulates reinforced A356 alloy hybrid metal matrix composites. *Adv. Compos. Hybrid Mater.* **2020**, *3*, 114–119.

Article

Effect of Laser Energy Density on the Microstructure and Microhardness of Inconel 718 Alloy Fabricated by Selective Laser Melting

Jing Xu [1,†], Zichun Wu [1,†], Jianpeng Niu [1], Yufeng Song [1], Chaoping Liang [2], Kai Yang [3,4], Yuqiang Chen [1] and Yang Liu [1,2,4,*]

[1] Hunan Engineering Research Center of Forming Technology and Damage Resistance Evaluation for High Efficiency Light Alloy Components, Hunan University of Science and Technology, Xiangtan 411201, China
[2] National Key Laboratory of Science and Technology on High-Strength Structural Materials, Central South University, Changsha 410083, China
[3] College of Engineering, Peking University, Beijing 100871, China
[4] Hunan Vanguard Group Co., Ltd., Changsha 410100, China
* Correspondence: liuyang7740038@163.com
† These authors contributed equally to this work.

Abstract: This work focused on the effects of laser energy density on the relative density, microstructure, and microhardness of Inconel 718 alloy manufactured by selective laser melting (SLM). The microstructural architectures, element segregation behavior in the interdendritic region and the evolution of laves phases of the as-SLMed IN718 samples were analyzed by optical metallography (OM), scanning electron microscopy (SEM), energy dispersive spectrometer (EDS), and electron probe microanalysis (EPMA). The results show that with an increase in the laser volume energy density, the relative density and the microhardness firstly increased and then decreased slightly. It also facilitates the precipitation of Laves phase. The variation of mechanical properties of the alloy can be related to the densification degree, microstructure uniformity, and precipitation phase content of Inconel 718 alloy.

Keywords: selective laser melting; Inconel 718 alloy; laser energy density; microstructure; microhardness

Citation: Xu, J.; Wu, Z.; Niu, J.; Song, Y.; Liang, C.; Yang, K.; Chen, Y.; Liu, Y. Effect of Laser Energy Density on the Microstructure and Microhardness of Inconel 718 Alloy Fabricated by Selective Laser Melting. *Crystals* **2022**, *12*, 1243. https://doi.org/10.3390/cryst12091243

Academic Editor: Cyril Cayron

Received: 25 July 2022
Accepted: 30 August 2022
Published: 2 September 2022

Publisher's Note: MDPI stays neutral with regard to jurisdictional claims in published maps and institutional affiliations.

Copyright: © 2022 by the authors. Licensee MDPI, Basel, Switzerland. This article is an open access article distributed under the terms and conditions of the Creative Commons Attribution (CC BY) license (https://creativecommons.org/licenses/by/4.0/).

1. Introduction

Inconel 718 (IN718), a typical precipitation-strengthened nickel-based superalloy, possesses excellent mechanical properties such as good oxidation resistance, good weldability, corrosion resistance, high creep resistance, and high fatigue strength at high temperatures [1–4]. It has been widely used in high-temperature components such as turbine disks, aerospace gas turbine blades, and combustion chambers [5–8]. However, forging and casting, as the conventional manufacturing processes for IN718 alloy, are difficult for producing complex integral structures, owing to high density, high melting point, and severe work hardening [9–11]. Moreover, the higher cost of the traditional machining methods, which are laborious and time-consuming, leads to a sharp increase in production cost [5,12]. The emergence of additive manufacturing provides a feasible way to overcome these disadvantages, which can help reduce material waste to a bare minimum and to reduce production time for complex parts [13].

Selective laser melting (SLM) has advantages such as high forming precision [14] and integrated forming of high mechanical properties complex components [15], so it has become one of the most promising additive manufacturing technologies for metallic materials [4,16–18]. Due to the rapid cooling rate of the molten pool in the SLM process [19], near net shape products with fine microstructure and excellent metallurgical strength can be manufactured layer by layer, such as complex parts manufactured with titanium alloy [20–22], steel [23,24], superalloys [25,26], and other materials [27–29].

Currently, many researchers have carried out research on the microstructure and properties of IN718 by means of SLM, hoping to regulate the microstructure and properties by adjusting the processing parameters, such as laser power, scanning speed, and scanning strategies, hatch spacing, and so on. For example, Wang et al. [30] found that the grain refinement caused by an increase in scanning speed further weakens the anisotropy of mechanical properties, which is conducive to obtaining a more homogenous microstructure and a higher production rate for SLM manufacturing. Pan et al. [31] investigated the independent effects of laser power and scanning speed of SLM on the precipitation and mechanical properties of identically heat-treated IN718 alloys. Liu et al. [32] studied the effects of two scanning strategies (single-directional scanning and cross-directional scanning) on the solidification structure and crystallographic texture of SLM-manufactured IN718 alloy. Ravichander et al. [33] identified the effect of scan strategy on residual stress and studied the metallurgical interactions between the mechanical and microstructural properties within the IN718 superalloy. Wang et al. [34] optimized the process parameters for selective laser melting of Inconel 718 and further established the relationship equation between the relative density of SLMed samples and the energy density coupled by laser parameters. Zhu et al. [35] reported that fine and coarse columnar dendritic microstructure of as-deposited IN718 superalloy were, respectively, obtained at different laser power and laser beam diameter and found that the Nb- and Mo-rich Laves phase is formed in the interdendritic regions, which weakens the mechanical property of the as-deposited IN718 superalloy [8,36–38]. Ma [39] pointed out that with increasing energy input, there was a dendrite-to-cell transition in the dendritic morphology evolution of the as-DLFed IN718 samples; in addition, the size and volume fraction of the Laves phase at the boundary between dendrites increased. However, although the laser parameters have been found to affect the precipitation of IN718 alloy, and the laser energy density was confirmed to control the microstructure and mechanical properties of SLM-manufactured Inconel 718 alloy, the relation between the laser energy density, microstructure, and properties has yet been thoroughly investigated.

Therefore, in this paper, Inconel 718 samples were fabricated by selective laser melting, and the effect of laser energy density on relative density was discussed. The microstructural architectures, the element segregation behavior in the interdendritic region, the evolution of the Laves phases with different volume energy density SLMed IN718 samples were analyzed. The microhardness of the samples was also measured to further clarify the mechanical response to the microstructure evolution of the samples at different volume energy densities. The aim of this study is to provide an insight into the effects of laser energy density on the microstructure and microhardness of the as-SLMed IN718 alloys.

2. Materials and Methods

2.1. Materials and Experimental Equipment

The gas-atomized commercially available Inconel 718 powder (Avimetal Powder Metallurgy Technology Co., Ltd., Beijing, China) was used for the present study. The morphology of Inconel 718 powder was characterized by scanning electron microscope (Nova Nano SEM230, FEI Co., Ltd., Hillsboro, OR, USA). The size distribution of Inconel 718 powders was measured using a laser particle size analyzer (Mastersizer 3000) (Malvery Instruments Ltd., Malvern, UK). The chemical composition of Inconel 718 powder was determined by inductively coupled plasma atomic emission spectrometer (ICP-AES) (Thermo Fisher Scientific Co., Ltd., Waltham, MA, USA). The morphologies and size distributions of the Inconel 718 powders are presented in Figure 1. The powder particles are spherical shape with the size of 15–53 μm. The D10, D50, and D90 of Inconel 718 powders were 19.86 μm, 29.56 μm, and 39.42 μm, respectively. The chemical composition of Inconel 718 powder is listed in Table 1. The SLM experiment was carried out by an DiMetal-100 3D printing machine equipped with a 200 W Yb-fiber laser beam (Guangzhou Lejia Additive Technology Co., Ltd., Guangzhou, China). Prior to the selective laser melting process, the powder was dried in vacuum-drying oven at the temperature of 60 °C for 4 h, to reduce the residual oxygen content and the humidity. During the manufacturing process, argon gas (99.99% purity) was filled in the working chamber with oxygen maintained below 50 ppm, which is conducive to prevent the powder and melt pool from oxidizing during SLM processing.

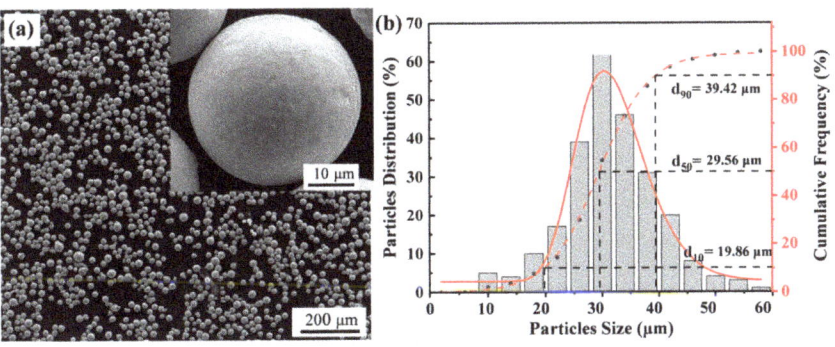

Figure 1. Morphology (**a**) and size distribution (**b**) of Inconel 718 powder.

Table 1. Chemical composition (wt. %) of the Inconel 718 powder.

Elements	Ni	Cr	Nb	Mo	Ti	Al	Co	Fe
Inconel 718	51.13	19.34	5.14	3.04	0.93	0.58	0.03	Balanced

2.2. Sample Fabrications and Characterization

The Taguchi method was used to optimize the parameters for the SLM Inconel 718 alloy in this study. The experiments were conducted with three controllable five-level process parameters: laser power, scanning speed, and scanning interval. The process parameters and their levels are given in Table 2. The design of experiments based on L25 orthogonal array was obtained, as shown in Table 3, using MINITAB statistical software (version 16). A series of samples with 20 mm × 20 mm × 20 mm cubes were prepared at above-mentioned different laser energy density by varying laser power, scanning speed, and scanning interval. The experimental results and associated processing parameter of SLM-fabricated Inconel 718 alloy process, according to L25 orthogonal array, are shown in Table 3.

After SLM fabrication, the as-SLMed Inconel 718 alloy samples were cut parallel and perpendicular to the building direction of the sample using wire electric discharge machining. The specimens were polished by 2000 grit SiC paper and etched in a solution of 10 mL HCl + 3 mL H_2O_2 for 20 s. The microstructure of the as-SLMed Inconel 718 alloy samples was characterized by optical microscope (Olympus BX51 M) (Olympus Co., Ltd., Tokyo, Japan) and scanning electron microscope (Nova Nano SEM230) (FEI Co., Ltd., Hillsboro, OR, USA) using back-scattered mode (SEM-BSE) (FEI Co., Ltd., Hillsboro, OR, USA). The element segregation and Laves phase elemental distribution were analyzed by electron probe microanalysis (EPMA, JXA8530F) (Japan Electronics Co., Ltd., Tokyo, Japan).

Table 2. The process parameters and their levers used in this study.

Parameters	Level 1	Level 2	Level 3	Level 4	Level 5
A: Laser power (W)	160	170	180	190	200
B: Scanning speed (mm/s)	800	900	1000	1100	1200
C: Scanning interval (mm)	0.06	0.07	0.08	0.09	0.1

Table 3. Experimental design of L25 orthogonal array and experimental results for relative density.

Runs	Laser Power (W)	Scanning Speed (mm/s)	Scanning Interval (mm)	Volume Energy Density (J/mm³)	Relative Density (%)
1	150	800	0.06	104.17	99.12
2	150	900	0.07	79.37	99.26
3	150	1000	0.08	62.50	96.13
4	150	1100	0.09	50.51	95.43
5	**150**	**1200**	**0.10**	**41.67**	**92.28**
6	160	800	0.07	95.24	99.68
7	160	900	0.08	74.07	98.33
8	160	1000	0.09	59.26	97.55
9	160	1100	0.10	48.48	94.48
10	160	1200	0.06	74.07	98.62
11	170	800	0.08	88.54	99.57
12	170	900	0.09	69.96	98.63
13	170	1000	0.10	56.67	97.43
14	**170**	**1100**	**0.06**	**85.86**	**99.23**
15	170	1200	0.07	67.46	97.55
16	180	800	0.09	83.33	99.15
17	180	900	0.10	66.67	98.02
18	**180**	**1000**	**0.06**	**100**	**99.53**
19	180	1100	0.07	77.92	98.92
20	180	1200	0.08	62.50	98.21
21	190	800	0.10	79.17	98.5
22	**190**	**900**	**0.06**	**117.28**	**98.92**
23	190	1000	0.07	90.48	99.43
24	190	1100	0.08	71.97	97.88
25	190	1200	0.09	58.64	96.84

The relative density of the SLM-fabricated samples was measured using Archimedes principle, and the results were presented as a percentage of the Inconel 718 alloy density (8.24 g/cm³). In order to reduce the randomness of the measurements, five tests were performed for each sample, and the average of our measurement was used as the value of the relative density. The Vickers microhardness tests were performed on the polished cross section using a digital microhardness instrument with a load of 100 g and a dwell time of 10 s. At least three tests were done for each sample, and the averaged measurement was used as the indicator of the Vickers microhardness.

3. Results and Discussion

3.1. Effect of Laser Energy Density on Relative Density

It is desirable to achieve fully densification in the final part in SLM process, since the retention of a small amount of porosity will severely degrades the mechanical properties [3]. As a result, the output parameter for this Taguchi experiment is chosen to be the relative density in this study. The experimental results and associated processing parameter of LPD process according to L25 orthogonal array were listed in Table 3. For SLM, the scanning speed and laser power are key factors for part forming [21,40]. The response surface graph and contour plot, as shown in Figure 2, were obtained to confirm the interaction effects of laser power and scanning speed on relative density. With a decrease in the scanning speed or an increase in the laser power, the approximate change of relative density first increased and then slightly decreased. The interplay between laser power and scanning speed could be deduced as the determinant factors for the densification process of SLM-fabricated Inconel 718 alloy. This discovery is similar to the features found in other material in the previous studies [41–43].

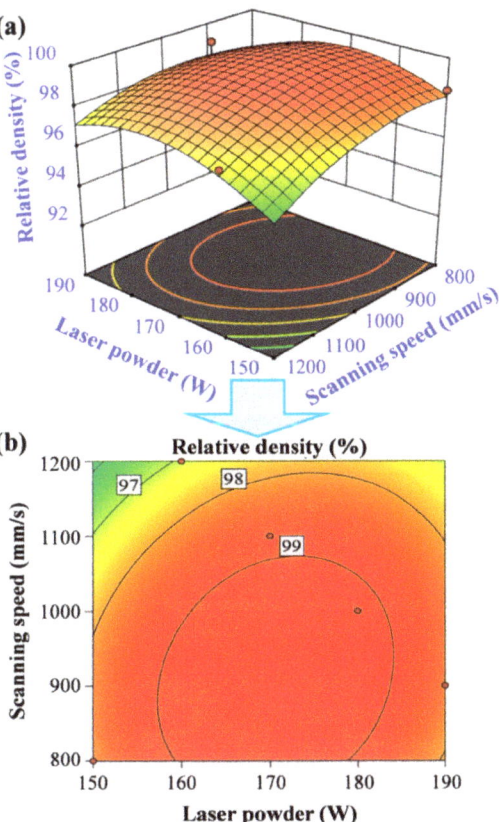

Figure 2. Response surface graph (**a**) and contour plot (**b**) of the effects of laser power and scanning speed on the relative density at scanning interval 0.06 mm.

Volume energy density (E_v) is also a key factor in the densification and significantly affects the microstructure and mechanical properties of SLM-fabricated Inconel 718 alloy [44,45]. The following equation was used to determine the volume energy density (E_v):

$$E_v = \frac{P}{Vht} \tag{1}$$

where P refers to the laser power (W), V is the scanning speed (mm/s), h represents the scanning interval (mm), and t is the powder-bed layer thickness, which kept constant as 0.03 mm. Based on Equation (1), the E_v of each sample are listed in Table 3. In addition, Figure 3 shows the relationship between relative density and volume energy density for the SLM-fabricated Inconel 718 alloy. As can be seen in Figure 3, low energy density (<80 J/mm^3) leads to a low relative density due to the lack of consolidation, while higher energy density is beneficial to the density of SLM-fabricated Inconel 718 alloy. However, when the energy density exceeds approximately 100 J/mm^3, the relative density decreases slightly. This is because the high absorption of heat leads to evaporation of the material. Consequently, gas pores form in the SLMed sample, which can be observed in Figures 4 and 5. Moreover, there is a certain threshold energy density that gives maximum material density in Figure 3. It is approximately 80–110 J/mm^3 for Inconel 718 alloy fabricated by the SLM process, and the relative density can reach more than 99%.

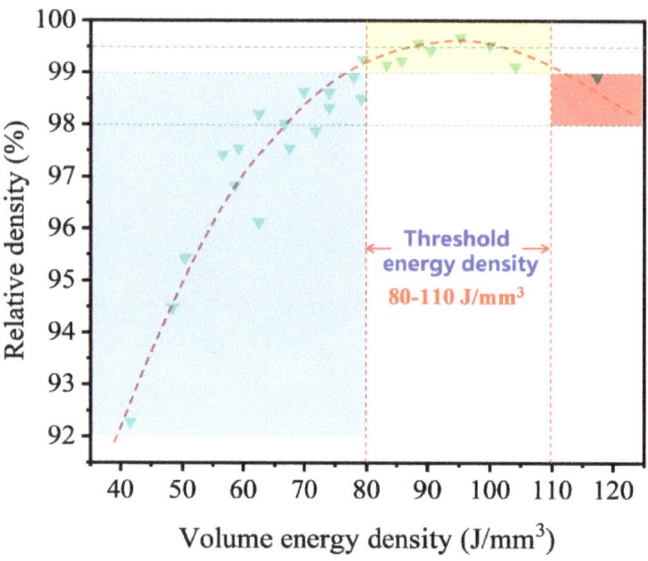

Figure 3. Relationship between relative density and volume energy density for SLM-fabricated Inconel 718 alloy. The points are the results of Table 3.

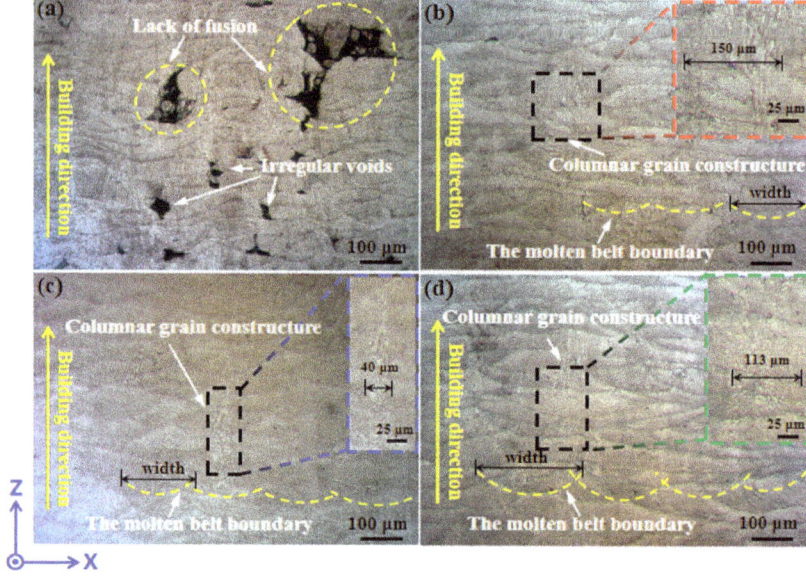

Figure 4. Optical micrographs of the as-SLMed IN718 samples on vertical section (X–Z plane) at the E_vs of (**a**) 41.67 J/mm^3, (**b**) 85.86 J/mm^3, (**c**) 100 J/mm^3, and (**d**) 117.17 J/mm^3.

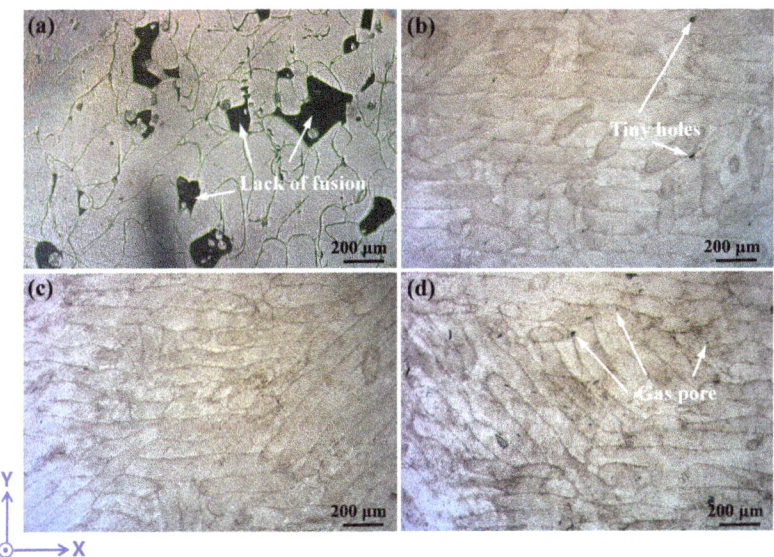

Figure 5. Optical micrographs of the as-SLMed IN718 samples on vertical section (X–Y plane) at the E_v of (**a**) 41.67 J/mm³, (**b**) 85.86 J/mm3, (**c**) 100 J/mm³, and (**d**) 117.17 J/mm³.

3.2. Microstructure Analysis

In this research, four unique samples of #5, #14, #18, and #22, representing different E_v levels, were selected from the above-mentioned 25 samples, for the purpose of clearly explaining the microstructural evolution of as-SLMed Inconel 718 in the following parts. The laser parameters of these four samples selected were marked in bold, as shown in Table 3.

To reveal well the microstructural evolution of all the samples fabricated in this research, the microstructures of these selected four unique samples with different levels of E_v were studied. Figure 4 shows optical micrographs of these four samples on the vertical section (X–Z plane), at E_vs of (a) 41.67 J/mm³, (b) 85.86 J/mm³, (c) 100 J/mm³, and (d) 117.17 J/mm³, respectively, which could reflect the microstructural architectures of all conditions in this study. The morphologies of the molten pools with the shape of the fish scale and a certain number of pores were observed clearly. It can be seen from Figure 4b–d that increasing energy density from 85.86 to 117.17 J/mm³ simultaneously increases the width of the molten pool. Figure 4 also shows the change of the characteristics of pores at different E_vs. The shape of the pores changed from irregular shapes with rough boundaries to round shapes with smooth edges as the laser energy density increased from 45.46 to 85.86 J/mm³. The amount and size of the pores decreased as well. Obviously, the interface was clear and free of pores when the E_v reached 100 J/mm³. With further increase in E_v, gas pores appeared at the E_v of 117.17 J/mm³, as shown in Figure 4d.

Figure 4 also indicates the microstructure of the as-SLMed IN718 samples exists with a certain number of columnar grains and dendrites, as circled by the black dashed rectangle. Continuous dendrites, up to even several millimeters in length, grow along the deposition direction and through several melt pools, as shown in the black dashed rectangle. The growth direction of the dendrites is not exactly parallel to the building direction, declining at an angle to the building direction.

The width of the columnar grains ranges from 40 μm to 150 μm. The columnar grains become smooth and fine at the E_v of 100 J/mm³, while they become coarse and large at E_vs of 85.86 and 117.17 J/mm³. The interface between two adjacent cladding layers is clear without any change in the columnar grains. For the sample at the E_v of 117.17 J/mm³, shown in (Figure 4d), the columnar grains are discontinuous. This is believed to be from the different microstructure between the bottom region of the melt pool (planar interface growth) and the other regions (dendrite growth) [46,47].

The microstructures on the X–Y plane of the SLMed samples at different E_vs are shown in Figure 5. A close connection between the densification of the sample and the laser energy density has been found. At the E_v of 41.67 J/mm^3, the insufficient energy input resulted in poor melt pool fluidity and insufficient filling of the inter-particle voids. This causes the appearance of many discrete melt pools. Additionally, a lack of fusion among powders resulted in the formation of large cavities, which were barely recognizable, and some unmelted particles were trapped in the cavities [48]. The large cavities disappeared. Meanwhile, as the laser energy density increased from 45.46 J/mm^3 to 85.86 J/mm^3. Tiny holes were observed in Figure 5b, and unmelted particles were formed. This is due to the fact that some gases were trapped in the melt pool when solidification of the pool occurred, forming small pores [21,49]. When the laser energy density reaches 100 J/mm^3, the melt pool was built up in an orderly manner with good lap from track to layer. As a result, the as-SLMed Inconel 718 alloy was free of pores. Gas pores still occurred as E_v up to 117.17 J/mm^3, as marked with arrows in Figure 5d. Among the three kinds of pores, gas pores are the most in quantity and the minimum in size [46,50]. It should be stressed here that gas pores appear at the laser energy density of 117.17 J/mm^3, which will damage the mechanical properties of sample. We could conclude that increasing properly the E_v benefits the improvement of densification, but an excessive value is undesirable. At a higher E_v, strong interactions between layers may result in high thermal stress and elements evaporation. The appearance of gas pores will eventually damage the mechanical properties of sample.

The observed microstructural differences indicate a tradeoff between the regions processed with low energy density and high energy density, which would improve the mechanical performance.

To further characterize the aforementioned dendritic growth, Figure 6 shows the SEM micrographs that were taken on the X–Z planes of as-SLMed IN718. As exhibited with the yellow dashed lines in Figure 6a, the boundary of molten pool can be visibly indicated after etching. Figure 6b implies that cellular dendrites grown at the boundary of the molten pool opposite the heat flow inclined away from the building direction at a certain angle. This leads to dendrites' growth parallel to the building direction at the center of the molten pool. The columnar dendrites grow in the direction of the heat flow rather than along the build direction. The solidification shifts from the cellular crystals at the bottom of the melt pool to the dendrites above the side of the melt pool. The subcellular tissue in the high energy input region grows into rough cellular crystals, with the appearance of the remelting zone. Another part of the grains near the remelting zone grows into columnar crystals, along the rough interface under the negative temperature gradient. The cellular crystal–dendrite growth is a typical feature of the solidification mode of IN718 during SLM [51].

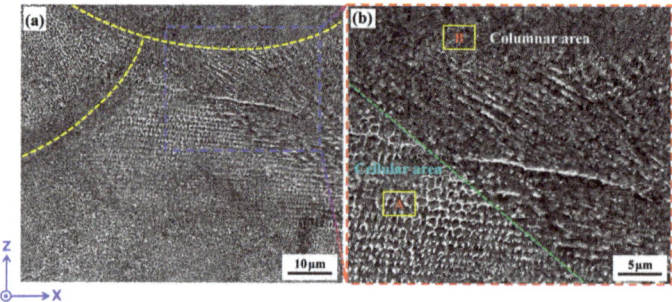

Figure 6. SEM micrographs of the as-built sample in the X–Z plane; (**a**) track–track structures consisting of overlapping area and central fusion area, and (**b**) dendrites and cellular sub-structures contained in the overlapping region.

The element content of the two areas (black matrix and white segregation phase) marked in Figure 6b was analyzed. The dark area (B) was rich in Fe, Cr, and Ni, while the white area (A) was rich in Nb, Mo, and Ti, as indicated in Figure 7 and Table 4. The Laves phase located in the interdendritic regions of the as-SLMed samples was identified [52], This may be a result of segregation during the fast solidification process of the sample via SLM. Recalling the microstructures shown in Figure 6, the microsegregations of Nb and Ti result in the formation of the brittle intermetallic Laves phase. The appearance of the Laves phase will cause a big change in the mechanical properties. For instance, it would affect the mechanical strength of the Inconel 718, making the alloy more brittle and easier to fracture under external force load [37]. The characteristics of the Laves phase are discrepant at different laser energy densities, as shown in Figure 8.

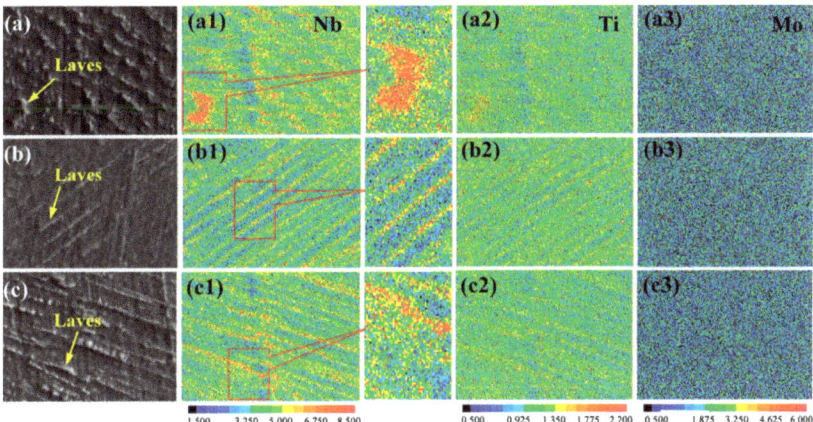

Figure 7. Cross-section EPMA maps of the SLMed Inconel 718 under different volume energy densities, including 85.86 J/mm^3 (**a–a3**), 100 J/mm^3 (**b–b3**), and 117.17 J/mm^3 (**c–c3**).

Table 4. EPMA quantitative analysis results for the position P1 (A) and P2 (B) of Figure 6b.

Element (Wt. %)	Ni	Cr	Fe	Nb	Mo	Ti	Al
White area (A)	48.528	17.681	16.974	7.898	3.175	1.259	0.463
Dark area (B)	49.490	18.714	18.213	5.164	2.882	1.041	0.445

Figure 7 also shows the magnification of the dendrites and the electron probe microanalysis at different E_vs. It shows the appearance of the Laves phase and the segregation of Nb, Ti, and Mo in the interdendritic region. The Laves phase is lighter yellow, as shown in Figure 7a–c. As the E_v increases from 85.86 J/mm^3 to 100 J/mm^3, the Laves phase becomes less and less obvious. With increasing energy density, the Nb element is less segregated, especially in some interdendritic regions (as shown in the red box). The even distribution of Mo, Nb, and Ti in the crystal and grain boundaries is observed at an energy density of 100 J/mm^3. This indicates that a higher energy density leads to the uniform diffusion of the elements, causing the decrease in the Laves phase. Moreover, the volume fraction of the Laves phase and Nb segregation strongly depend on the solidification processes [35]. The faster cooling rate at a higher energy density greatly improves the dendrite growth rate and solute trapping, so that Nb does not have enough time to diffuse from the solid phase to the liquid phase. This allows more Nb elements to be trapped in the solid phase (matrix), while fewer Nb elements are segregated to structure the Laves phase. However, when the energy density exceeds a critical value, the resulting Nb segregation increases again. The increase in energy density in turn facilitates the precipitation of the Laves phase in Inconel 718 prepared by SLM.

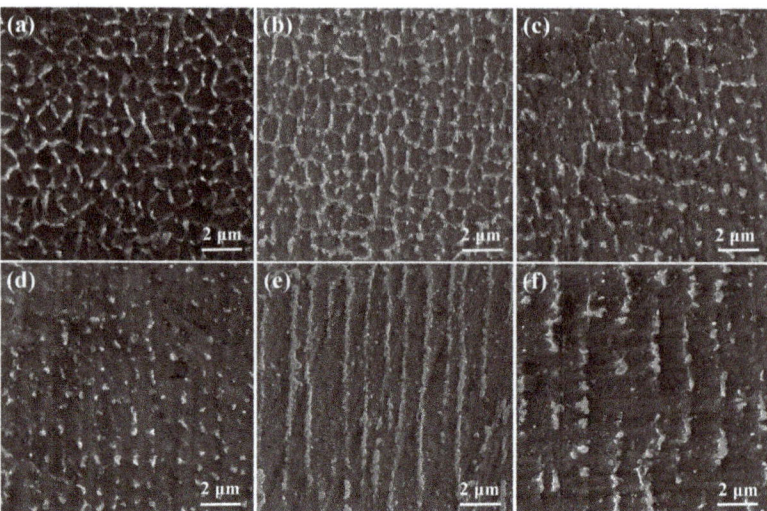

Figure 8. High-magnification SEM graphs of cellular dendrites and columnar dendrites in the X–Z plane of SLMed IN718 samples with different laser volume energy densities; (**a,d**) 85.86 J/mm^3, (**b,e**) 100 J/mm^3, and (**c,f**) 117.17 J/mm^3.

It is known that segregation is a phenomenon largely dependent on time. As a result, it has a tight connection with the cooling rate [35]. Due to the non-equilibrium fast solidification conditions that prevail during SLM, the appearance of the Laves phase easily occurs in as-SLMed IN718 alloy, through the segregation of high atomic diameter elements such as Nb, Mo, and Ti. Accordingly, the Laves phase is easier to appear in the interdendritic boundaries at a low energy density or high energy density. In summary, the proper increase in the volume energy density helps to reduce the precipitation of the Laves phase, while excessive energy will facilitate the formation of the Laves phase in the SLM-fabricated Inconel 718.

To further reveal the influence of volume energy density (E_v) on the microstructures of SLM-processed Inconel 718, Figure 8 shows even higher magnifications on the specific spots in the structures. Two kinds of dendrites of the as-SLMed IN718 alloys on vertical section (X–Z plane) were observed at the E_vs of (a) 85.86 J/mm^3, (b) 100 J/mm^3, and (c) 117.17 J/mm^3. At a lower E_v, cellular dendrites with clear boundaries are formed, and the grains are distributed at a certain angle, as evidenced in Figure 8a,d. At the E_v level of 100 J/mm^3, the dendrites' arrays are refined (Figure 8b), and the directional solidified slender columnar constructures with visible boundaries are observed, as shown in Figure 8e. At a high E_v, there is no obvious cellular dendrite formed, but large-area particle agglomeration occurs. The coarsened and large columnar dendrites are distributed directionally with a fragmentized characteristic. Disconnected dendrite constructures with incomplete precipitants at the interdendritic region were exhibited. It is difficult to distinguish a single dendrite due to its clustering during the fast solidification process in the molten pool, as shown in Figure 8c,f. The primary columnar dendrites are dispersed at an increased volume energy density, despite the columnar dendrites being irregularly arranged.

3.3. Microhardness Distribution

Figure 9 shows the microhardness distribution and average microhardness of different energy densities measured on the polished sections of as-SLMed parts from bottom to top. Upon increasing the E_v from 85 J/mm^3 to 100 J/mm^3, the average microhardness increased from 266.13 HV$_{0.1}$ to 300.13 HV$_{0.1}$. This result was mainly due to the appearance of tiny pores at lower energy densities, which tend to expand the size and number of pores under load during the Vickers hardness test. Moreover, the samples with different energy densities have different standard deviations on measurement, and microhardness measured by the different positions of the same sample generates a large change during the hardness test. This large distribution in hardness is caused by the inhomogeneous distribution of the microstructure [53]. The appearance of fine microstructure for the sample at the E_v of 100 J/mm^3 allows the γ matrix to acquire more Nb elements. It is worth noting that the morphology and concentration of the Laves phase were found to be the most critical factors in the microstructure of Inconel 718 alloy [54,55]. The Laves phase is the Nb-rich phase in the γ matrix (see Figures 7 and 8). Moreover, the addition of Nb promotes the formation of supersaturated solid solution and enhances the solid-phase strengthening effect. According to the above analysis, the microhardness of the present Inconel 718 parts was enhanced by densification behavior, grain refinement, precipitation strengthening, and solid-solution strengthening.

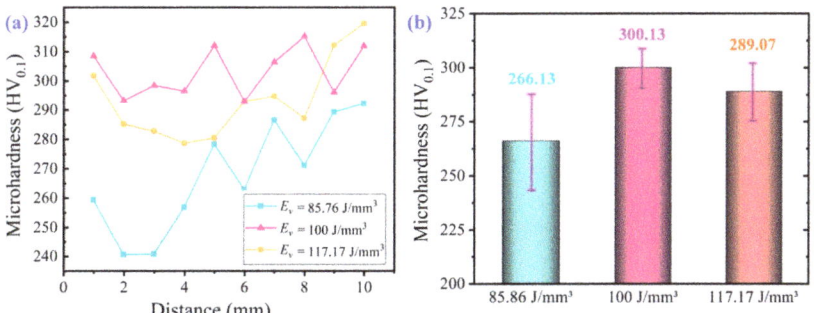

Figure 9. Vickers hardness in the X–Z plane of SLMed IN718 samples with different laser volume energy densities: (**a**) hardness distribution and (**b**) average hardness value.

4. Conclusions

Inconel 718 samples were fabricated by SLM with different process parameters. The effects of laser energy density on the relative density, microstructure, and microhardness of the as-SLMed Inconel 718 samples were analyzed. The following conclusions can be drawn:

1. The relative density firstly increased and then decreased slightly with the increase in the laser volume energy density. When the laser volume energy density was 100 J/mm^3, the material density reached a peak value of 99.53%.
2. When the laser energy density was 41.67 J/mm^3, the insufficient energy input resulted in poor melt pool fluidity and insufficient filling of the inter-particle voids, which resulted in the appearance of many discrete melt pools. The proper increase in E_v benefited the improvement of densification. However, an excessive value of laser energy density (117.17 J/mm^3) resulted in high thermal stress and elements' evaporation, causing the appearance of gas pores, which will damage the mechanical properties of the sample.
3. When the E_v rose from 85.86 J/mm^3 to 100 J/mm^3, the microhardness of the Inconel 718 alloy fabricated by the SLM process firstly increased from 266.13 HV$_{0.1}$ to 300.13 HV$_{0.1}$. When the E_v further increased to 117.17 J/mm^3, the microhardness showed a slight decrease to 289.07 HV$_{0.1}$. The fluctuation of the microhardness was related to the densification degree, microstructure uniformity, and precipitation phase content of Inconel 718 alloy.

Author Contributions: Conceptualization, J.X. and Z.W.; methodology, J.N. and Y.S.; validation, Y.C. and C.L.; formal analysis, K.Y. and Y.C; investigation, J.X., Z.W. and Y.S.; resources, Y.C and Y.L.; data curation, J.X. and Z.W.; writing—original draft, J.X. and Z.W.; writing—review and editing, J.N., C.L. and Y.L.; supervision, K.Y. and Y.C; project administration, Y.L.; funding acquisition, Y.S., C.L. and Y.L. All authors have read and agreed to the published version of the manuscript.

Funding: The work was supported by the Key Research and Development Program of Hunan Province of China (2022GK2043), the National Natural Science Foundation of China (52105334), the Natural Science Foundation of Hunan Province of China (2021JJ40206, 2022JJ20025), and the Educational Commission of Hunan Province of China (21B0472).

Institutional Review Board Statement: Not applicable.

Informed Consent Statement: Not applicable.

Data Availability Statement: The data presented in this study are available on request from the corresponding author.

Conflicts of Interest: The authors declare no conflict of interest.

References

1. Zhang, Y.; Yu, J.; Lin, X.; Guo, P.; Yu, X.; Zhang, S.; Liu, J.; Huang, W. Passive Behavior of Laser Directed Energy Deposited Inconel 718 after Homogenization and Aging Heat Treatment. *Corros. Sci.* **2022**, *205*, 110439. [CrossRef]
2. Rai, A.K.; Paul, C.P.; Mishra, G.K.; Singh, R.; Rai, S.K.; Bindra, K.S. Study of Microstructure and Wear Properties of Laser Borided Inconel 718. *J. Mater. Process. Technol.* **2021**, *298*, 117298. [CrossRef]
3. Ji, H.; Song, Q.; Wang, R.; Cai, W.; Liu, Z. Evaluation and Prediction of Pore Effects on Single-Crystal Mechanical and Damage Properties of Selective Laser Melted Inconel-718. *Mater. Des.* **2022**, *219*, 110807. [CrossRef]
4. Yoo, J.; Kim, S.; Jo, M.C.; Park, H.; Jung, J.E.; Do, J.; Yun, D.W.; Kim, I.S.; Choi, B.G. Investigation of Hydrogen Embrittlement Properties of Ni-Based Alloy 718 Fabricated via Laser Powder Bed Fusion. *Int. J. Hydrogen Energy* **2022**, *47*, 18892–18910. [CrossRef]
5. Blakey-Milner, B.; Gradl, P.; Snedden, G.; Brooks, M.; Pitot, J.; Lopez, E.; Leary, M.; Berto, F.; du Plessis, A. Metal Additive Manufacturing in Aerospace. A Review. *Mater. Des.* **2021**, *209*, 110008. [CrossRef]
6. Shao, S.; Khonsari, M.M.; Guo, S.; Meng, W.J.; Li, N. Overview: Additive Manufacturing Enabled Accelerated Design of Ni-Based Alloys for Improved Fatigue Life. *Addit. Manuf.* **2019**, *29*, 100779. [CrossRef]
7. Sreeramagiri, P.; Bhagavatam, A.; Ramakrishnan, A.; Alrehaili, H.; Dinda, G.P. Design and Development of a High-Performance Ni-Based Superalloy WSU 150 for Additive Manufacturing. *J. Mater. Sci. Technol.* **2020**, *47*, 20–28. [CrossRef]
8. Zhang, S.; Wang, L.; Lin, X.; Yang, H.; Huang, W. The Formation and Dissolution Mechanisms of Laves Phase in Inconel 718 Fabricated by Selective Laser Melting Compared to Directed Energy Deposition and Cast. *Compos. Part B Eng.* **2022**, *239*, 109994. [CrossRef]
9. De Bartolomeis, A.; Newman, S.T.; Jawahir, I.S.; Biermann, D.; Shokrani, A. Future Research Directions in the Machining of Inconel 718. *J. Mater. Process. Technol.* **2021**, *297*, 117260. [CrossRef]
10. Sugihara, T.; Enomoto, T. High Speed Machining of Inconel 718 Focusing on Tool Surface Topography of CBN Tool. *Procedia Manuf.* **2015**, *1*, 675–682. [CrossRef]
11. Umbrello, D. Investigation of Surface Integrity in Dry Machining of Inconel 718. *Int. J. Adv. Manuf. Technol.* **2013**, *69*, 2183–2190. [CrossRef]
12. Zheng, Y.; Liu, F.; Zhang, W.; Liu, F.; Huang, C.; Gao, J.; Li, Q. The Microstructure Evolution and Precipitation Behavior of TiB2/Inconel 718 Composites Manufactured by Selective Laser Melting. *J. Manuf. Process.* **2022**, *79*, 510–519. [CrossRef]
13. Vrancken, B.; Thijs, L.; Kruth, J.P.; Van Humbeeck, J. Heat Treatment of Ti6Al4V Produced by Selective Laser Melting: Microstructure and Mechanical Properties. *J. Alloys Compd.* **2012**, *541*, 177–185. [CrossRef]
14. Chen, L.Y.; Liang, S.X.; Liu, Y.; Zhang, L.C. Additive Manufacturing of Metallic Lattice Structures: Unconstrained Design, Accurate Fabrication, Fascinated Performances, and Challenges. *Mater. Sci. Eng. R Rep.* **2021**, *146*, 100648. [CrossRef]
15. Zhu, G.; Pan, W.; Wang, R.; Wang, D.; Shu, D.; Zhang, L.; Dong, A.; Sun, B. Microstructures and Mechanical Properties of GTD222 Superalloy Fabricated by Selective Laser Melting. *Mater. Sci. Eng. A* **2021**, *807*, 140668. [CrossRef]
16. Hosseini, E.; Popovich, V.A. A Review of Mechanical Properties of Additively Manufactured Inconel 718. *Addit. Manuf.* **2019**, *30*, 100877. [CrossRef]
17. Zhang, S.; Lin, X.; Wang, L.; Yu, X.; Hu, Y.; Yang, H.; Lei, L.; Huang, W. Strengthening Mechanisms in Selective Laser-Melted Inconel718 Superalloy. *Mater. Sci. Eng. A* **2021**, *812*, 141145. [CrossRef]
18. Sui, S.; Tan, H.; Chen, J.; Zhong, C.; Li, Z.; Fan, W.; Gasser, A.; Huang, W. The Influence of Laves Phases on the Room Temperature Tensile Properties of Inconel 718 Fabricated by Powder Feeding Laser Additive Manufacturing. *Acta Mater.* **2019**, *164*, 413–427. [CrossRef]
19. Amato, K.N.; Gaytan, S.M.; Murr, L.E.; Martinez, E.; Shindo, P.W.; Hernandez, J.; Collins, S.; Medina, F. Microstructures and Mechanical Behavior of Inconel 718 Fabricated by Selective Laser Melting. *Acta Mater.* **2012**, *60*, 2229–2239. [CrossRef]
20. Lv, J.; Luo, K.; Lu, H.; Wang, Z.; Liu, J.; Lu, J. Achieving High Strength and Ductility in Selective Laser Melting Ti-6Al-4V Alloy by Laser Shock Peening. *J. Alloys Compd.* **2022**, *899*, 163335. [CrossRef]

21. Yi, J.H.; Kang, J.W.; Wang, T.J.; Wang, X.; Hu, Y.Y.; Feng, T.; Feng, Y.L.; Wu, P.Y. Effect of Laser Energy Density on the Microstructure, Mechanical Properties, and Deformation of Inconel 718 Samples Fabricated by Selective Laser Melting. *J. Alloys Compd.* **2019**, *786*, 481–488. [CrossRef]
22. Pede, D.; Li, M.; Virovac, L.; Poleske, T.; Balle, F.; Mu, C. Microstructure and Corrosion Resistance of Novel β-Type Titanium Alloys Manufactured by Selective Laser Melting. *J. Mater. Res. Technol.* **2022**, *19*, 4598–4612. [CrossRef]
23. Riemer, A.; Leuders, S.; Thöne, M.; Richard, H.A.; Tröster, T.; Niendorf, T. On the Fatigue Crack Growth Behavior in 316L Stainless Steel Manufactured by Selective Laser Melting. *Eng. Fract. Mech.* **2014**, *120*, 15–25. [CrossRef]
24. Cruz, V.; Chao, Q.; Birbilis, N.; Fabijanic, D.; Hodgson, P.D.; Thomas, S. Electrochemical Studies on the Effect of Residual Stress on the Corrosion of 316L Manufactured by Selective Laser Melting. *Corros. Sci.* **2020**, *164*, 108314. [CrossRef]
25. Lu, Y.; Wu, S.; Gan, Y.; Huang, T.; Yang, C.; Junjie, L.; Lin, J. Study on the Microstructure, Mechanical Property and Residual Stress of SLM Inconel-718 Alloy Manufactured by Differing Island Scanning Strategy. *Opt. Laser Technol.* **2015**, *75*, 197–206. [CrossRef]
26. Ivanov, D.; Travyanov, A.; Petrovskiy, P.; Cheverikin, V.; Alekseeva, E.; Khvan, A.; Logachev, I. Evolution of Structure and Properties of the Nickel-Based Alloy EP718 after the SLM Growth and after Different Types of Heat and Mechanical Treatment. *Addit. Manuf.* **2017**, *18*, 269–275. [CrossRef]
27. Wang, L.-Z.; Wang, S.; Wu, J.-J. Experimental Investigation on Densification Behavior and Surface Roughness of AlSi10Mg Powders Produced by Selective Laser Melting. *Opt. Laser Technol.* **2017**, *96*, 88–96. [CrossRef]
28. Wen, Z.; Song, X.; Chen, D.; Fan, T.; Liu, Y.; Cai, Q. Electrospinning Preparation and Microstructure Characterization of Homogeneous Diphasic Mullite Ceramic Nanofibers. *Ceram. Int.* **2020**, *46*, 12172–12179. [CrossRef]
29. Song, X.; Zhang, K.; Song, Y.; Duan, Z.; Liu, Q.; Liu, Y. Morphology, Microstructure and Mechanical Properties of Electrospun Alumina Nanofibers Prepared Using Different Polymer Templates: A Comparative Study. *J. Alloys Compd.* **2020**, *829*, 154502. [CrossRef]
30. Wang, R.; Chen, C.; Liu, M.; Zhao, R.; Xu, S.; Hu, T.; Shuai, S.; Liao, H.; Ke, L.; Vanmeensel, K.; et al. Effects of Laser Scanning Speed and Building Direction on the Microstructure and Mechanical Properties of Selective Laser Melted Inconel 718 Superalloy. *Mater. Today Commun.* **2022**, *30*, 103095. [CrossRef]
31. Pan, H.; Dahmen, T.; Bayat, M.; Lin, K.; Zhang, X. Independent Effects of Laser Power and Scanning Speed on IN718's Precipitation and Mechanical Properties Produced by LBPF plus Heat Treatment. *Mater. Sci. Eng. A* **2022**, *849*, 143530. [CrossRef]
32. Liu, X.B.; Xiao, H.; Xiao, W.J.; Song, L.J. Microstructure and Crystallographic Texture of Laser Additive Manufactured Nickel-Based Superalloys with Different Scanning Strategies. *Crystals* **2021**, *11*, 591. [CrossRef]
33. Ravichander, B.B.; Mamidi, K.; Rajendran, V.; Farhang, B.; Ganesh-Ram, A.; Hanumantha, M.; Shayesteh Moghaddam, N.; Amerinatanzi, A. Experimental Investigation of Laser Scan Strategy on the Microstructure and Properties of Inconel 718 Parts Fabricated by Laser Powder Bed Fusion. *Mater. Charact.* **2022**, *186*, 111765. [CrossRef]
34. Wang, W.; Wang, S.; Zhang, X.; Chen, F.; Xu, Y.; Tian, Y. Process Parameter Optimization for Selective Laser Melting of Inconel 718 Superalloy and the Effects of Subsequent Heat Treatment on the Microstructural Evolution and Mechanical Properties. *J. Manuf. Process.* **2021**, *64*, 530–543. [CrossRef]
35. Zhu, L.; Xu, Z.F.; Liu, P.; Gu, Y.F. Effect of Processing Parameters on Microstructure of Laser Solid Forming Inconel 718 Superalloy. *Opt. Laser Technol.* **2018**, *98*, 409–415. [CrossRef]
36. Liu, H.; Guo, K.; Sun, J.; Shi, H. Effect of Nb Addition on the Microstructure and Mechanical Properties of Inconel 718 Fabricated by Laser Directed Energy Deposition. *Mater. Charact.* **2022**, *183*, 111601. [CrossRef]
37. Chen, Y.; Guo, Y.; Xu, M.; Ma, C.; Zhang, Q.; Wang, L.; Yao, J.; Li, Z. Study on the Element Segregation and Laves Phase Formation in the Laser Metal Deposited IN718 Superalloy by Flat Top Laser and Gaussian Distribution Laser. *Mater. Sci. Eng. A* **2019**, *754*, 339–347. [CrossRef]
38. Kim, H.; Cong, W.; Zhang, H.C.; Liu, Z. Laser Engineered Net Shaping of Nickel-Based Superalloy Inconel 718 Powders onto Aisi 4140 Alloy Steel Substrates: Interface Bond and Fracture Failure Mechanism. *Materials* **2017**, *10*, 341. [CrossRef]
39. Ma, M.; Wang, Z.; Zeng, X. Effect of Energy Input on Microstructural Evolution of Direct Laser Fabricated IN718 Alloy. *Mater. Charact.* **2015**, *106*, 420–427. [CrossRef]
40. McLouth, T.D.; Witkin, D.B.; Bean, G.E.; Sitzman, S.D.; Adams, P.M.; Lohser, J.R.; Yang, J.M.; Zaldivar, R.J. Variations in Ambient and Elevated Temperature Mechanical Behavior of IN718 Manufactured by Selective Laser Melting via Process Parameter Control. *Mater. Sci. Eng. A* **2020**, *780*, 139184. [CrossRef]
41. Moussaoui, K.; Rubio, W.; Mousseigne, M.; Sultan, T.; Rezai, F. Effects of Selective Laser Melting Additive Manufacturing Parameters of Inconel 718 on Porosity, Microstructure and Mechanical Properties. *Mater. Sci. Eng. A* **2018**, *735*, 182–190. [CrossRef]
42. Liu, Y.; Liu, C.; Liu, W.; Ma, Y.; Tang, S.; Liang, C.; Cai, Q.; Zhang, C. Optimization of Parameters in Laser Powder Deposition AlSi10Mg Alloy Using Taguchi Method. *Opt. Laser Technol.* **2019**, *111*, 470–480. [CrossRef]
43. Liu, Y.; Liu, C.; Liu, W.; Ma, Y.; Zhang, C.; Liang, C.; Cai, Q. Laser Powder Deposition Parametric Optimization and Property Development for Ti-6Al-4V Alloy. *J. Mater. Eng. Perform.* **2018**, *27*, 5613–5621. [CrossRef]
44. Kladovasilakis, N.; Charalampous, P.; Tsongas, K.; Kostavelis, I.; Tzovaras, D.; Tzetzis, D. Influence of Selective Laser Melting Additive Manufacturing Parameters in Inconel 718 Superalloy. *Materials* **2022**, *15*, 1362. [CrossRef] [PubMed]
45. Su, C.H.; Rodgers, K.; Chen, P.; Rabenberg, E.; Gorti, S. Design, Processing, and Assessment of Additive Manufacturing by Laser Powder Bed Fusion: A Case Study on INCONEL 718 Alloy. *J. Alloys Compd.* **2022**, *902*, 163735. [CrossRef]
46. Liu, F.; Lin, X.; Leng, H.; Cao, J.; Liu, Q.; Huang, C.; Huang, W. Microstructural Changes in a Laser Solid Forming Inconel 718 Superalloy Thin Wall in the Deposition Direction. *Opt. Laser Technol.* **2013**, *45*, 330–335. [CrossRef]

47. Wang, H.; Wang, L.; Cui, R.; Wang, B.; Luo, L.; Su, Y. Differences in Microstructure and Nano-Hardness of Selective Laser Melted Inconel 718 Single Tracks under Various Melting Modes of Molten Pool. *J. Mater. Res. Technol.* **2020**, *9*, 10401–10410. [CrossRef]
48. King, W.E.; Barth, H.D.; Castillo, V.M.; Gallegos, G.F.; Gibbs, J.W.; Hahn, D.E.; Kamath, C.; Rubenchik, A.M. Observation of Keyhole-Mode Laser Melting in Laser Powder-Bed Fusion Additive Manufacturing. *J. Mater. Process. Technol.* **2014**, *214*, 2915–2925. [CrossRef]
49. Wang, L.; Cui, R.; Li, B.-Q.; Jia, X.; Yao, L.-H.; Su, Y.-Q.; Guo, J.-J.; Liu, T. Influence of Laser Parameters on Segregation of Nb during Selective Laser Melting of Inconel 718. *China Foundry* **2021**, *18*, 379–388. [CrossRef]
50. Wan, H.Y.; Zhou, Z.J.; Li, C.P.; Chen, G.F.; Zhang, G.P. Effect of Scanning Strategy on Grain Structure and Crystallographic Texture of Inconel 718 Processed by Selective Laser Melting. *J. Mater. Sci. Technol.* **2018**, *34*, 1799–1804. [CrossRef]
51. Deng, D.; Peng, R.L.; Brodin, H.; Moverare, J. Microstructure and Mechanical Properties of Inconel 718 Produced by Selective Laser Melting: Sample Orientation Dependence and Effects of Post Heat Treatments. *Mater. Sci. Eng. A* **2018**, *713*, 294–306. [CrossRef]
52. Parimi, L.L.; Ravi, G.; Clark, D.; Attallah, M.M. Microstructural and Texture Development in Direct Laser Fabricated IN718. *Mater. Charact.* **2014**, *89*, 102–111. [CrossRef]
53. Choi, J.P.; Shin, G.H.; Yang, S.; Yang, D.Y.; Lee, J.S.; Brochu, M.; Yu, J.H. Densification and Microstructural Investigation of Inconel 718 Parts Fabricated by Selective Laser Melting. *Powder Technol.* **2017**, *310*, 60–66. [CrossRef]
54. Olakanmi, E.O. Selective Laser Sintering/Melting (SLS/SLM) of Pure Al, Al-Mg, and Al-Si Powders: Effect of Processing Conditions and Powder Properties. *J. Mater. Process. Technol.* **2013**, *213*, 1387–1405. [CrossRef]
55. Park, J.H.; Bang, G.B.; Lee, K.A.; Son, Y.; Kim, W.R.; Kim, H.G. Effect on Microstructural and Mechanical Properties of Inconel 718 Superalloy Fabricated by Selective Laser Melting with Rescanning by Low Energy Density. *J. Mater. Res. Technol.* **2021**, *10*, 785–796. [CrossRef]

Article

Optimization of Parameters in Laser Powder Bed Fusion TA15 Titanium Alloy Using Taguchi Method

Yang Liu [1,2,*], Zichun Wu [1], Qing Wang [3], Lizhong Zhao [3], Xichen Zhang [1], Wei Gao [1], Jing Xu [1], Yufeng Song [1,*], Xiaolei Song [4,*] and Xuefeng Zhang [3]

[1] Hunan Engineering Research Center of Forming Technology and Damage Resistance Evaluation for High Efficiency Light Alloy Components, School of Mechanical Engineering, Hunan University of Science and Technology, Xiangtan 411201, China
[2] National Key Laboratory of Science and Technology on High-Strength Structural Materials, Central South University, Changsha 410083, China
[3] Institute of Advanced Magnetic Materials, College of Materials and Environmental Engineering, Hangzhou Dianzi University, Hangzhou 310018, China
[4] Key Laboratory of Advanced Structural Materials, Ministry of Education, College of Materials Science and Engineering, Changchun University of Technology, Changchun 130012, China
* Correspondence: liuyang7740038@163.com (Y.L.); federer.song@163.com (Y.S.); songxiaolei@ccut.edu.cn (X.S.)

Abstract: In this work, laser powder bed fusion (LPBF) was explored to fabricate TA15 (Ti-6Al-2Zr-1Mo-1V) titanium alloy based on the experimental design obtained by using the Taguchi method. The impact of processing parameters (including laser power, scanning speed, and scanning interval) on the density and microhardness of the as-LPBFed TA15 titanium alloy was analyzed using the Taguchi method and analysis of variance (ANOVA). The interaction among parameters on the density of the as-LPBFed TA15 titanium alloy was indicated by a response surface graph (RSR). When the laser energy density was adjusted to 100 J/mm^3, the highest relative density could reach 99.7%. The further increase in the energy input led to the reduction in relative density, due to the formation of tiny holes caused by the vaporization of material at a high absorption of heat. Furthermore, in order to better reveal the correlation between relative density and processing parameters, the regression analysis was carried out for relative density. The results showed that the experimental and predicted values obtained by the regression equation were nearly the same.

Keywords: laser-based powder bed fusion; titanium alloy; Taguchi; process optimization; density

1. Introduction

TA15 (Ti-6Al-2Zr-1Mo-1V) titanium alloy is a high-aluminum-equivalent α titanium alloy, which has good specific strength, high-temperature creep resistance, thermal stability, and corrosion resistance. This alloy has also been widely used in the key load-bearing components and engine structure parts of aerospace applications [1–3]. However, titanium alloy has high activity, low thermal conductivity, and high deformation resistance [4–7], which makes it very difficult to manufacture by traditional manufacturing methods, such as casting [8,9], forging [10,11], and welding [12–14]. In addition, aerospace parts tend to be functional, lightweight, and have a structural integrated design [15–23]. Meanwhile, it is increasingly difficult for traditional manufacturing technology to meet the manufacturing needs of aerospace titanium alloy parts with complex structures. Hence, new types of manufacturing techniques should be explored to fabricate the complex structures.

Metal additive manufacturing (AM) techniques, including laser powder bed fusion (LPBF) [24–26] and laser powder deposition (LPD) [27–29], have been confirmed to successfully fabricate complex structures of titanium alloy parts. Compared with others, research has mainly concentrated on the directed-energy-deposition-fabricated TA15 titanium alloy [30,31], which can directly fabricate the large-scale complex structural components.

However, in order to obtain the TA15 structural parts with higher precision, some studies have conducted research on LPBF-processed TA15 titanium alloy. Laser powder bed fusion is a promising metal AM technique that utilizes a focused laser beam to selectively melt metal powder layer by layer under the guidance of a computer-aided-design (CAD) model [3,6]. It has the advantages of high material utilization rate, high precision, and directly manufacturing near-full-density complex-shaped metal parts [32], which indicates that the application of LPBF to manufacture TA15 titanium alloy parts has the potential to extend its application in aerospace [33]. However, the characteristics of high-temperature melting and rapid solidification in the LPBF process may result in some issues [34,35]. For example, the evaporation of elements in the heating process and the volume shrinkage in the cooling process both easily lead to pore generation [36]. There is even a lack of fusion defects existing due to improper adjustment of process parameters [37]. These pores and lack of fusion will reduce the relative density of LPBF-fabricated metal parts, resulting in seriously degrading the mechanical properties [38]. Therefore, achieving a high density of LPBF-fabricated metal parts has become a target for researchers.

As for LPBF Ti-Al-Zr-Mo-V titanium alloy, Li et al. [3] investigated the effect of LPBF process parameters on the relative density of the TA15 titanium alloy sample, and found that the volume energy density (E_v) range of 125–167 J/mm^3 is the optimal LPBF process window for manufacturing high-density TA15 titanium alloy. At the same time, Li et al. [39] also reported that the effect of the different levels of scanning strategies on the relative density of LPBF-fabricated TA15 titanium alloy caused a significant difference. Thijs et al. [40] systematically investigated the correlation between the density of Ti6Al4V and various parameters via experiments and simulation, but did not consider the effect of parameter–parameter interactions on the density.

The Taguchi method has been proven to be a powerful means to optimize multiple parameters with the consideration of the interaction among process parameters [41–43]. Kumar et al. [44] investigated the effect of wire electrical discharge machining (WEDM) process parameters on the rate of material removal (MRR), surface roughness (SR), and corrosion rate (CR) of ZE41A magnesium alloy using the Taguchi method, and determined the optimized parameter combinations of each of them based on considering the interaction among process parameters. Pandel et al. [45] reported the influence of input parameters (including cross-sectional area and TE leg length) on the output parameter (power output of Mg2(Si–Sn) thermoelectric generators) using the Taguchi method, and the contributions of each parameter were obtained by ANOVA analysis with the results of 35.22% and 27.62%, respectively. In our study, adopting the Taguchi method can minimize the number of experiments required to achieve a fuller understanding of the effects of processing parameters on the relative density of LPBF-fabricated TA15 titanium alloy. In addition, according to the Taguchi experimental design, the optimized parameters can be obtained by the approach that compares the mean of the signal-to-noise (S/N) ratio. The important parameters and percentage contribution of the single process parameter on density were obtained by ANOVA.

In this work, the Taguchi method was utilized to optimize LPBF process parameters targeting high-density TA15 titanium alloy samples. The effect of laser energy density on the relative density of the as-LPBFed TA15 titanium alloy was discussed. The optimized parameter results were validated by confirmation analysis. This work aimed to provide the database and guidance for LPBF fabrication of TA15 titanium alloy.

2. Materials and Methods
2.1. Materials

The gas-atomized TA15 titanium alloy powder used in this work was purchased from Avimetal Powder Metallurgy Technology Co., Ltd., China. The chemical composition of TA15 titanium alloy powder is listed in Table 1. The main alloying elements of TA15 titanium alloy are Al, V, Zr, Mo, and Ti, as indicated in Table 1. The morphology and size distribution of TA15 titanium alloy powder are presented in Figure 1. The powder particles were spherically shaped with an equivalent spherical diameter of 15–53 μm (D10 = 21.46 μm, D50 = 33.73 μm, and D90 = 48.50 μm).

Table 1. Chemical composition (Wt. %) of the TA15 titanium alloy powder.

Elements	Al	V	Zr	Mo	Si	Fe	Ti
TA15	6.42	1.94	1.93	1.43	0.02	0.03	Balance

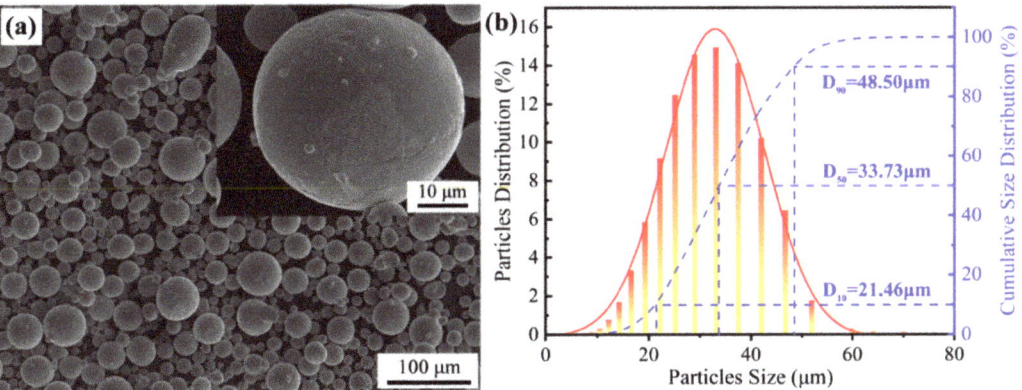

Figure 1. Morphology (a) and size distribution (b) of TA15 titanium alloy powder.

2.2. Sample Fabrications and Optimization of Parameters Using Taguchi Method

The LPBF experiments were carried out by a DiMetal-100 3D printing machine (Guangzhou Leijia Additive Technology Co., Ltd., Guangzhou, China) with an oxygen concentration below 100 ppm. High density is the premise for the sample to have excellent mechanical properties, and the density of the LPBF-fabricated TA15 titanium alloy was largely affected via process parameters, such as laser power, laser scanning speed, scanning interval, and powder-bed layer thickness.

In this investigation, the Taguchi method was utilized to optimize the parameters for the density of the LPBF-fabricated TA15 titanium alloy. Based on a previous study on the effect of laser energy density on the densification of titanium alloy [46] and the fact that the low power was expected to obtain a satisfactory surface quality, the regions of the process parameters were determined. The three controllable five-level process parameters are listed in Table 2. Considering the interaction among parameters, the experimental parameters combinations were determined by the orthogonal test method using the Taguchi method. An L_{25} orthogonal array was obtained as shown in Table 3 using MINITAB statistical software (MINITAB 16, Pennsylvania State University, Pennsylvania, USA). The 25 parametric combinations listed in Table 3 were then applied to fabricate 10 × 10 × 10 mm³ cubes for the sake of parametric optimization. The experimental results are displayed in the form of S/N ratio, which could be separated from the three types of performance features: nominal-the-better, smaller-the-better, and larger-the-better. In this study, the objective was to obtain maximum density in the LPBF-fabricated TA15 titanium alloy. Thereafter, the larger-the-better feature was chosen. The larger-the-better S/N ratio can be obtained based on the following equation:

$$\frac{S}{N} = -10 \log \left(\frac{1}{n} \sum_{i=1}^{n} \frac{1}{y_i^2} \right) \tag{1}$$

where y_i refers to the value of density for the *i*th experiment, and n represents the total number of experiments.

Table 2. The process parameters and their levels used in this study.

Parameters	Level 1	Level 2	Level 3	Level 4	Level 5
A: Laser power (W)	150	160	170	180	190
B: Scanning speed (mm/s)	800	900	1000	1100	1200
C: Scanning interval (mm)	0.06	0.07	0.08	0.09	0.10

Table 3. Experimental design as L_{25} orthogonal array and experimental results for relative density.

Runs	Laser Power (W)	Scanning Speed (mm/s)	Scanning Interval (mm)	Laser Energy Density (J/mm^3)	Microhardness (HV$_{0.1}$)	Relative Density (%) Experimental Value	Relative Density (%) Predicted Value	Error (%)
1	150	800	0.06	107.64	365.56	99.46	99.26	−0.2007
2	150	900	0.07	79.36	336.00	98.96	99.10	0.1437
3	150	1000	0.08	62.5	350.66	98.78	98.77	−0.0138
4	150	1100	0.09	50.51	321.18	98.32	98.25	−0.0681
5	150	1200	0.10	41.67	319.53	97.74	97.56	−0.1821
6	160	800	0.07	95.24	347.96	99.52	99.30	−0.2206
7	160	900	0.08	74.07	327.26	99.26	99.12	−0.1384
8	160	1000	0.09	59.26	353.04	98.79	98.75	−0.0376
9	160	1100	0.10	48.48	324.46	98.24	98.19	−0.0497
10	160	1200	0.06	74.07	328.90	99.18	99.01	−0.1746
11	170	800	0.08	88.54	327.84	99.43	99.32	−0.1068
12	170	900	0.09	69.96	331.06	99.08	99.11	0.0323
13	170	1000	0.10	56.67	327.00	99.02	98.69	−0.3292
14	170	1100	0.06	85.86	326.94	99.37	99.3	−0.0620
15	170	1200	0.07	67.46	328.96	99.36	98.99	−0.3688
16	180	800	0.09	84.51	322.14	99.62	99.33	−0.2907
17	180	900	0.10	66.67	322.50	99.43	99.07	−0.3617
18	180	1000	0.06	100	317.46	99.7	99.43	−0.2746
19	180	1100	0.07	77.92	329.76	99.19	99.28	0.0872
20	180	1200	0.08	62.5	317.50	99.14	98.91	−0.2359
21	190	800	0.10	79.17	318.70	99.2	99.32	0.1214
22	190	900	0.06	117.28	330.66	99.37	99.36	−0.0099
23	190	1000	0.07	90.48	324.74	99.64	99.39	−0.2509
24	190	1100	0.08	71.97	317.10	99.15	99.18	0.0351
25	190	1200	0.09	58.64	308.12	98.83	98.74	−0.0864

Finally, the percent contribution of each parameter and significant parameters for the density were obtained by the method of analysis of variance (ANOVA).

2.3. Characterizations

Inductively coupled plasma-atomic emission spectrometry (ICP-AES) was employed to determine the chemical component of TA15 titanium alloy powder. Scanning electron microscopy (Nova Nano SEM230) was performed for the morphology of TA15 titanium alloy powder. The size distribution of TA15 titanium alloy powder was counted by a laser particle size analyzer (Mastersizer 3000). The microstructure of the LPBF-fabricated TA15 titanium alloy samples was characterized by a scanning electron microscope (Nova Nano SEM230) using back-scattered mode (SEM-BSE).

The Archimedes principle was used to measure the relative density of LPBF-fabricated TA15 titanium alloy samples, and the results were indicated with a percentage of the TA15 titanium alloy density (4.45 g/cm^3) [47]. To decrease the randomness of the tests, five measurements were carried out for every sample, and the mean of the measurements was represented as the experimental value of the relative density. The vertical section (X–Y plane) of each sample was polished for Vickers microhardness tests and the tests were conducted by a digital microhardness instrument at a load of 100 g and a dwell time of 10 s. The results obtained for each set of samples were the average values of at least three measurements.

3. Results and Discussion

3.1. Effect of Processing Parameters on Density and Microhardness of as-LPBFed TA15 Titanium Alloy

3.1.1. Effects of Laser Power

Figure 2 shows the effect of laser power on the density and microhardness of the as-LPBFed TA15 titanium alloy. Here, the value of each bar represents the average value of the experimental results obtained by five parameter combinations under a certain level of laser power. It can be seen from Figure 2a that the relative density of the as-LPBFed TA15 titanium alloy was below 99% with an average value of 98.65% when the laser power was 150 W, but it was beyond 99% with the increase in laser power from 150 W to 170 W. The reason was that the low energy input corresponded to the low depth, width, and height of the molten pool, which was why some of the powders could not be fully melted in the LPBF process, resulting in the decrease in relative density, and the higher levels of laser power could melt more alloy powders in the molten pool to obtain higher relative density [48]. Subsequently, the laser power increased from 170 W to 180 W, further increasing the relative density. It is worth noticing that the relative density could reach up to 99.5% when the laser power was 180 W. However, the higher laser power of 190 W caused the decrease in relative density, which was owing to the excessive energy input to the elements by burning [41], resulting in a decreasing relative density of samples. Interestingly, as shown in Figure 2b, the samples of the lowest relative density indicated the highest microhardness, and the microhardness of these five levels of laser power from 150 W to 190 W represented a decreasing trend from 338.5 $HV_{0.1}$ to 319.8 $HV_{0.1}$, from which it could be indicated that the mechanical properties of samples were not only determined by relative density but also by many factors. Meanwhile, the values of the relative density exhibited a significant change when the laser power increased from 150 W to 190 W, which indicated the significant contribution of laser power to the relative density of LPBF TA15 alloys.

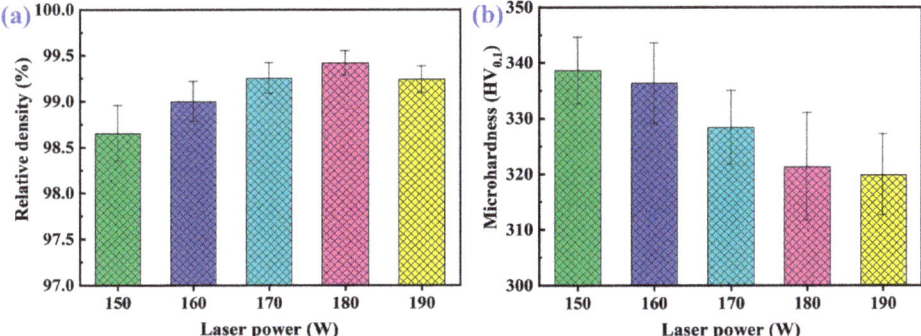

Figure 2. The density (**a**) and microhardness (**b**) of as-LPBFed TA15 titanium alloy under various laser powers.

3.1.2. Effects of Scanning Speed

Figure 3 shows the influence of scanning speed on the density and microhardness of the as-LPBFed TA15 titanium alloy. Here, the value of each bar represents the average value of experimental results obtained by five parameter combinations under a certain level of scanning speed. As the scanning speed increased from 800 mm/s to 1200 mm/s at an interval of 100 mm/s, the relative density of the as-LPBFed TA15 titanium alloy decreased from 99.44% to 98.85%, as shown in Figure 3a, which was due to the laser energy density decreasing as the scanning speed increased. Consequently, the generated molten pool caused by insufficient energy input could not fully catch the alloy powders, leading to the decrease in relative density [49]. Meanwhile, Figure 3b indicates the effect of scanning speed on the microhardness of the as-LPBFed TA15 titanium alloy, from which it could be seen that the trend of microhardness of the as-LPBFed TA15 titanium alloy was decreasing from 336.4 $HV_{0.1}$ to 329.4 $HV_{0.1}$ first upon the increase in scanning speed from 800 mm/s 900 mm/s. Then, the further increase in scanning speed from 900 mm/s to 1000 mm/s simultaneously improved the microhardness, and the microhardness reached 334.5 $HV_{0.1}$ when the scanning speed was 1000 mm/s. Then, the scanning speed increased from

1000 mm/s to 1200 mm/s, giving rise to a steep descent in microhardness from 334.5 $HV_{0.1}$ to 320.6 $HV_{0.1}$.

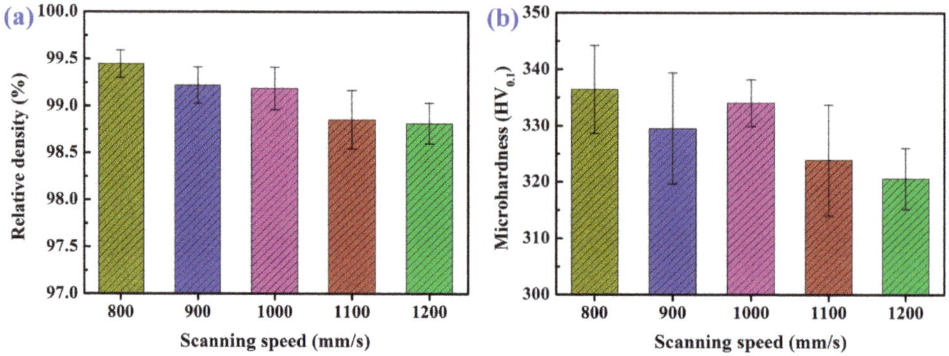

Figure 3. The density (**a**) and microhardness (**b**) of as-LPBFed TA15 titanium alloy under various scanning speeds.

3.1.3. Effects of Scanning Interval

Figure 4 shows the relationships between scanning interval and density and microhardness of the as-LPBFed TA15 titanium alloy. Here, the value of each bar represents the average value of experimental results obtained by five parameter combinations under a certain level of scanning interval. As the scanning interval increased from 0.06 mm to 0.10 mm with an interval of 0.01 mm, the density of the as-LPBFed TA15 titanium alloy decreased correspondingly from 99.41% to 98.72% with a smooth trend. The descent trend of density with increasing scanning interval was owing to the lower energy density per unit volume, leading to the decreasing energy adopted by the TA15 powders [41]. Meanwhile, the microhardness of the as-LPBFed TA15 titanium alloy also represented a reduction trend from 333.9 $HV_{0.1}$ to 322.4 $HV_{0.1}$ with the increase in scanning interval from 0.06 mm to 0.10 mm.

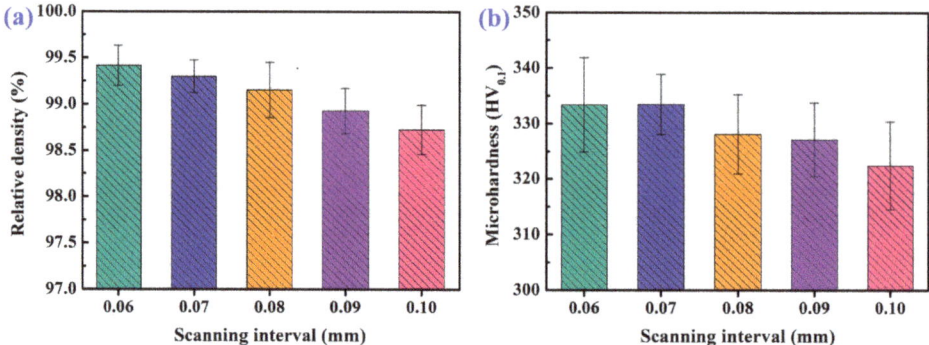

Figure 4. The density (**a**) and microhardness (**b**) of as-LPBFed TA15 titanium alloy under various scanning intervals.

3.2. Optimization of Parameters for Density

3.2.1. Analysis of the Signal-to-Noise (S/N) Ratio

The output parameter (relative density) was utilized to measure the mean and signal-to-noise ratios of every input parameter for the best quality of the as-LPBFed TA15 titanium alloy specimens. The mean value and signal-to-noise ratio (S/N) were obtained to evaluate the effect of every process parameter on the as-LPBFed TA15 titanium alloy specimens.

In this investigation, the larger the criterion, the better the model used to select the mean and S/N ratio to identify the response of the process parameters. Tables 4 and 5 show the response tables for relative density of the mean value and the signal-to-noise ratio, respectively. The larger the distinction between the S/N values, the more significant the process parameters were. Thus, it can be indicated that the laser power had the greatest effect on the relative density. In addition, the primary effect curves of the mean value and S/N ratio on the densities are shown in Figures 5 and 6. It can be seen that the highest relative density of the as-LPBFed TA15 titanium alloy specimens was obtained at the process parameters of laser power of 180 W, scanning speed of 800 mm/s, and scanning interval of 0.06 mm.

Table 4. Mean response table for relative density.

Level	Laser Power	Scanning Speed	Scanning Interval
1	98.65	99.45	99.42
2	99.00	99.22	99.33
3	99.25	99.19	99.15
4	99.42	98.85	98.93
5	99.24	98.85	98.73
Delta	0.76	0.60	0.69
Rank	1	3	2

Table 5. S/N ratio response table for relative density.

Level	Laser Power	Scanning Speed	Scanning Interval
1	39.88	39.95	39.95
2	39.91	39.93	39.94
3	39.93	39.93	39.93
4	39.95	39.90	39.91
5	39.93	39.90	39.89
Delta	0.07	0.05	0.06
Rank	1	3	2

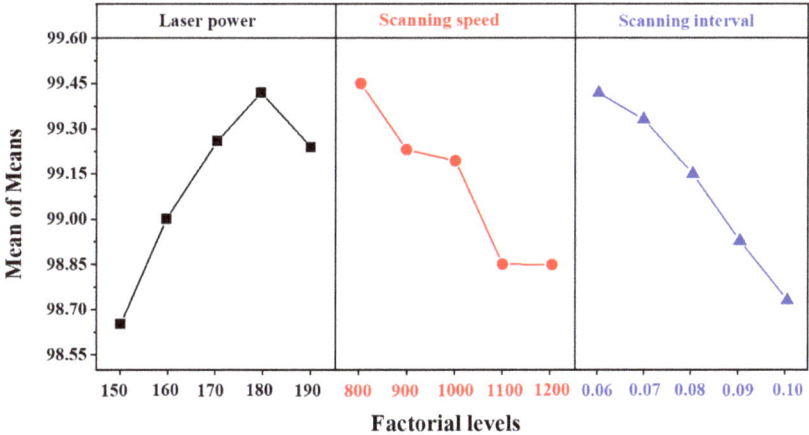

Figure 5. Main effect plots of means—relative density.

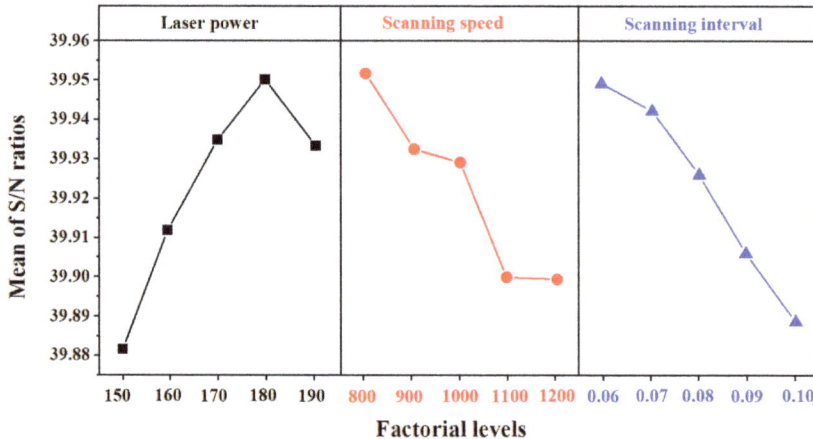

Figure 6. Main effect plots of S/N ratio—relative density.

3.2.2. Analysis of Variance (ANOVA)

The percentage contribution of every parameter was calculated by ANOVA. ANOVA facilitated the formal testing of the results of all the main factors and their relationships by assessing the mean squared deviation of the experimental error approximation at a defined confidence level. In this study, the percentage contribution of each process parameter to obtain the best molding quality of the specimen was performed by MINITAB software. The most significant process parameter obtained by calculating the percentage contribution was the laser power of 33.86%, followed by the scanning interval of 31.33% and finally the scanning speed of 25.35%. Table 6 indicates the results obtained by ANOVA of the densities. The results show that the laser power was the most significant process parameter affecting the relative density of the as-LPBFed TA15 titanium alloy.

Table 6. Results acquired from ANOVA—relative density.

Source	DF	Seq. SS	Adj. SS	Adj. MS	F	P	Percentage of Contribution (%)
Laser power	4	1.7624	1.7624	0.44061	10.72	0.001	33.86
Scanning speed	4	1.3195	1.3195	0.32988	8.03	0.002	25.35
Scanning interval	4	1.6307	1.6307	0.40769	9.92	0.001	31.33
Error	12	0.4930	0.4930	0.04108	—	—	9.47
Total	24	5.2057	—	—	—	—	100

DF, degree of freedom; Seq. SS, sequential sum of squares; Adj. SS, adjusted sum of squares; Adj. MS, adjusted mean squares; F, statistical test; P, statistical value [46].

3.3. Effect of Laser Energy Density on Relative Density of as-LPBFed TA15 Titanium Alloy

In this section, the interaction influences of scanning speed and laser power on the relative density of the alloy are discussed. Figure 7 shows the response surface graph and contour plot for the influence of laser power and scanning speed on the relative density obtained by using Design Expert software, from which it can be directly seen that the relative density of samples first increased obviously and then slightly reduced with the reduction in the scanning speed or the increase in the laser power. The effect of the scanning speed on relative density appeared more remarkable at lower laser powers, and so did laser power at high scanning speeds. As reported, volume energy density (E_v) is the critical element that determines the relative density, and it can be represented as follows:

$$E_v = P/vht \qquad (2)$$

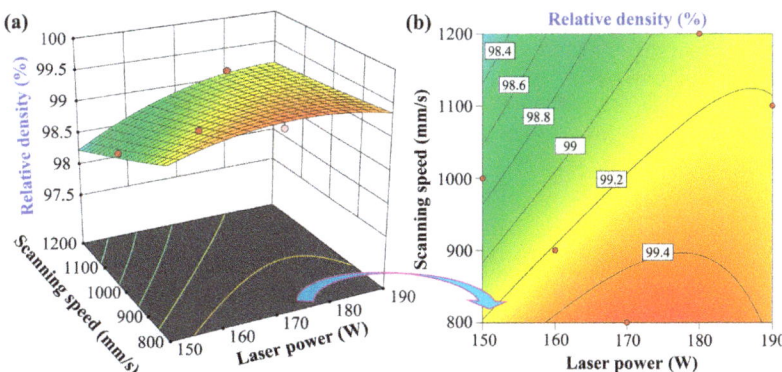

Figure 7. Response surface graph (**a**) and contour plot (**b**) of the effects of laser power and scanning speed on the relative density at scanning interval of 0.08 mm.

Here, P represents the laser power (W), v refers to the laser scanning speed (mm/s), h is the laser scanning interval (mm), and t is the layer thickness (mm), which remains at 0.03 mm throughout the investigation. Based on Equation (2), the E_v of each sample is listed in Table 3.

Figure 8 indicates the correlation between volume energy density and relative density of the TA15 alloy fabricated via LPBF processes. It can be found from Figure 8 that a lower relative density of 98.24% occurred at a lower energy density of 48.48 J/mm³, which was due to the low level of laser energy input not being able to make the powder surface melt completely, resulting in the lack of fusion, as shown in Figure 9a. Thus, a higher energy density could be utilized to realize higher values of relative density. When the laser energy density was adjusted to 100 J/mm³, the highest relative density could reach 99.7%, as shown in Table 3. There are almost no defects such as pores, as indicated in (Figure 9b). Notably, the further increase in the energy input would lead slightly to the reduction in the relative density due to the vaporization of material caused by the high absorption of heat from melt pool turbulence or the interaction zone [36], which would result in the formation of tiny holes, as shown in Figure 9c.

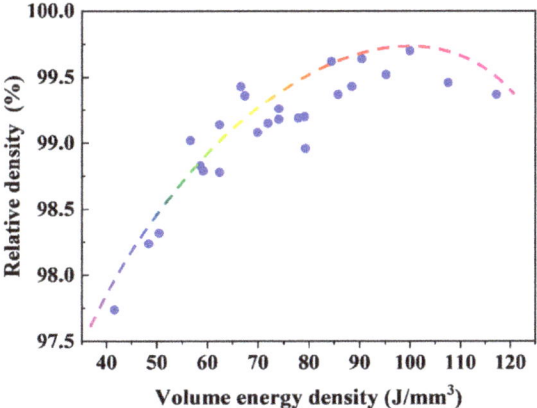

Figure 8. Relationship between relative density and volume energy density for LPBF-fabricated TA15 titanium alloy.

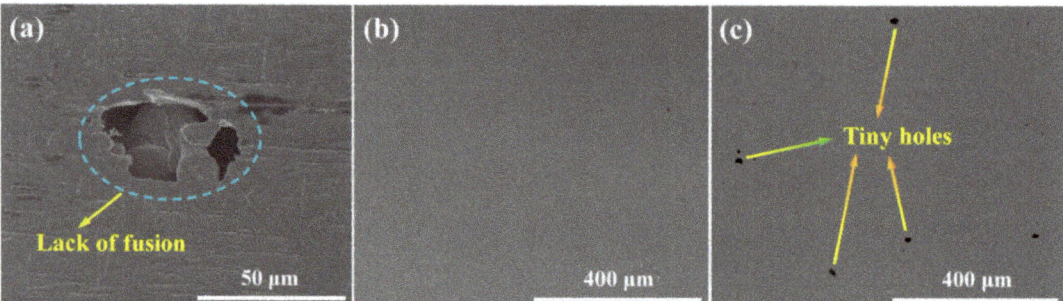

Figure 9. Various types of defects present across the as-LPBFed TA15 titanium alloy at different energy densities: (**a**) 48.48 J/mm^3, (**b**) 100 J/mm^3, and (**c**) 117.28 J/mm^3.

3.4. Confirmation Analysis

In order to better reveal the correlation between relative density and laser power, and scanning speed and scanning interval, the regression analysis was carried out for relative density. With the regression expression given according to the response parameter (relative density) and the three input process parameters (laser power (A), scanning speed (B), and scanning interval (C)) with the expression via a second-order polynomial, the equation is as follows [50]:

$$\text{Relative density} = a_0 + a_1(A) + a_2(B) + a_3(C) + a_4(AB) + a_5(AC) + a_6(BC) + a_7(ABC) \quad (3)$$

Table 7 shows the corresponding value of coefficients from a_0 to a_7 for the relative density. By replacing each value of these three process parameters and corresponding coefficients in the regression expression, the predicted value for relative density could be obtained. Recall above Table 3 that the values of prediction and experiment for relative density were clearly shown, and the values of prediction and experiment for relative density were compared. The result indicates that the difference between their values was not remarkable and their percentage of error was less than 0.5%. The confirmation analysis had been finished to identify that the values of prediction and experiment obtained by means of the regression equation were nearly the same. Directly, the high correlation between experimental values and predicted values for relative density is directly shown in Figure 10.

Table 7. The regression equation coefficients for relative density.

Coefficient	The Corresponding Value
a_0	111.61
a_1	-1.06×10^{-1}
a_2	-1.64×10^{-2}
a_3	5.23
a_4	1.35×10^{-4}
a_5	4.91×10^{-1}
a_6	1.89×10^{-2}
a_7	-7.18×10^{-4}

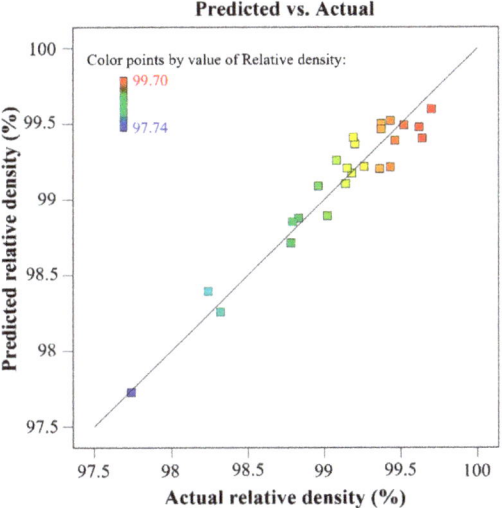

Figure 10. Correlation graph of the experimental values and predicted values of relative density.

4. Conclusions

(1) With the increase in laser power, the relative density first increased and then decreased. When the laser power was 180 W, the relative density could reach the peak value of 99.5%. However, the higher laser power of 190 W caused the decrease in relative density, which was owing to the excessive energy input to the elements by burning, resulting in the decrease in relative density.

(2) As the scanning interval increased from 0.06 mm to 0.10 mm with an interval of 0.01 mm, the density of the as-LPBFed TA15 titanium alloy decreased correspondingly from 99.41% to 98.72% with a smooth trend. The descent trend of density with increasing scanning interval was owing to the lower energy density per unit volume, causing the decreased energy absorbed by the TA15 powders.

(3) The correlation between relative density, laser power, scanning speed, and scanning interval was analyzed by the regression expression (A: laser power, B: scanning speed, and C: scanning interval) as follows:

Relative density = $111.61 - 1.06 \times 10^{-1}$ (A) $- 1.64 \times 10^{-2}$ (B) $+ 5.23$ (C) $+ 1.35 \times 10^{-4}$ (AB) $+ 4.91 \times 10^{-1}$ (AC) $+ 1.89 \times 10^{-2}$ (BC) $- 7.18 \times 10^{-4}$ (ABC)

Author Contributions: Conceptualization, Z.W. and Q.W.; methodology, Q.W. and L.Z.; validation, Y.S., X.S. and X.Z. (Xuefeng Zhang); formal analysis, L.Z. and Q.W.; investigation, Q.W., X.Z. (Xichen Zhang) and W.G.; resources, Y.L., X.S. and X.Z. (Xuefeng Zhang); data curation, Z.W. and J.X.; writing—original draft, Z.W. and Q.W.; writing—review and editing, Y.L., Y.S. and X.S.; supervision, L.Z. and X.Z. (Xuefeng Zhang); project administration, Y.L.; funding acquisition, Y.L. and X.Z. (Xuefeng Zhang). All authors have read and agreed to the published version of the manuscript.

Funding: This research was funded by the Natural Science Foundation of Zhejiang Province (2021C01023), Natural Science Foundation of China (52105334), Natural Science Foundation of Hunan Province of China (2021JJ40206, 2022JJ20025), Key Research and Development Program of Hunan Province of China (2022GK2043), and Education Department of Jilin Province (JJKH20210733KJ).

Institutional Review Board Statement: Not applicable.

Informed Consent Statement: Not applicable.

Data Availability Statement: The data presented in this study are available on request from the corresponding author.

Conflicts of Interest: The authors declare no conflict of interest.

References

1. Wang, C.S.; Li, C.L.; Chen, R.; Qin, H.Z.; Ma, L.; Mei, Q.S.; Zhang, G.D. Multistep Low-to-High-Temperature Heating as a Suitable Alternative to Hot Isostatic Pressing for Improving Laser Powder-Bed Fusion-Fabricated Ti-6Al-2Zr-1Mo-1V Microstructural and Mechanical Properties. *Mater. Sci. Eng. A* **2022**, *841*, 143022. [CrossRef]
2. Huang, S.; Sun, B.; Guo, S. Microstructure and Property Evaluation of TA15 Titanium Alloy Fabricated by Selective Laser Melting after Heat Treatment. *Opt. Laser Technol.* **2021**, *144*, 107422. [CrossRef]
3. Li, S.; Lan, X.; Wang, Z.; Mei, S. Microstructure and Mechanical Properties of Ti-6.5Al-2Zr-Mo-V Alloy Processed by Laser Powder Bed Fusion and Subsequent Heat Treatments. *Addit. Manuf.* **2021**, *48*, 102382. [CrossRef]
4. Takase, A.; Ishimoto, T.; Morita, N.; Ikeo, N.; Nakano, T. Comparison of Phase Characteristics and Residual Stresses in Ti-6al-4v Alloy Manufactured by Laser Powder Bed Fusion (L-Pbf) and Electron Beam Powder Bed Fusion (Eb-Pbf) Techniques. *Crystals* **2021**, *11*, 796. [CrossRef]
5. Zhang, Y.; Zhang, S.; Zou, Z.; Shi, Y. Achieving an Ideal Combination of Strength and Plasticity in Additive Manufactured Ti–6.5Al–2Zr–1Mo–1V Alloy through the Development of Tri-Modal Microstructure. *Mater. Sci. Eng. A* **2022**, *840*, 142944. [CrossRef]
6. Zhang, S.; Zhang, Y.; Zou, Z.; Shi, Y.; Zang, Y. The Microstructure and Tensile Properties of Additively Manufactured Ti–6Al–2Zr–1Mo–1V with a Trimodal Microstructure Obtained by Multiple Annealing Heat Treatment. *Mater. Sci. Eng. A* **2022**, *831*, 142241. [CrossRef]
7. Hu, M.; Wang, L.; Li, G.; Huang, Q.; Liu, Y.; He, J.; Wu, H.; Song, M. Investigations on Microstructure and Properties of Ti-Nb-Zr Medium-Entropy Alloys for Metallic Biomaterials. *Intermetallics* **2022**, *145*, 107568. [CrossRef]
8. Tao, P.; Shao, H.; Ji, Z.; Nan, H.; Xu, Q. Numerical Simulation for the Investment Casting Process of a Large-Size Titanium Alloy Thin-Wall Casing. *Prog. Nat. Sci. Mater. Int.* **2018**, *28*, 520–528. [CrossRef]
9. Uwanyuze, R.S.; Kanyo, J.E.; Myrick, S.F.; Schafföner, S. A Review on Alpha Case Formation and Modeling of Mass Transfer during Investment Casting of Titanium Alloys. *J. Alloys Compd.* **2021**, *865*, 158558. [CrossRef]
10. Sun, Z.C.; Zhang, J.; Yang, H.; Wu, H.L. Effect of Workpiece Size on Microstructure Evolution of Different Regions for TA15 Ti-Alloy Isothermal near-β Forging by Local Loading. *J. Mater. Process. Technol.* **2015**, *222*, 234–243. [CrossRef]
11. Zhang, R.; Wang, D.J.; Yuan, S.J. Effect of Multi-Directional Forging on the Microstructure and Mechanical Properties of TiBw/TA15 Composite with Network Architecture. *Mater. Des.* **2017**, *134*, 250–258. [CrossRef]
12. Liu, W.; Sheng, Q.; Ma, Y.; Cai, Q.; Wang, J.; Liu, Y. Interfacial Microstructures, Residual Stress and Mechanical Analysis of Hot Isostatic Pressing Diffusion Bonded Joint of 93W–4.9Ni–2.1Fe Alloy and 30CrMnSiNi2A Steel. *Fusion Eng. Des.* **2020**, *156*, 111602. [CrossRef]
13. Li, J.; Shen, J.; Hu, S.; Zhang, H.; Bu, X. Microstructure and Mechanical Properties of Ti-22Al-25Nb/TA15 Dissimilar Joint Fabricated by Dual-Beam Laser Welding. *Opt. Laser Technol.* **2019**, *109*, 123–130. [CrossRef]
14. Wang, C.; Guo, Q.; Shao, M.; Zhang, H.; Wang, F.; Song, B.; Ji, Y.; Li, H. Microstructure and Corrosion Behavior of Linear Friction Welded TA15 and TC17 Dissimilar Joint. *Mater. Charact.* **2022**, *187*, 111871. [CrossRef]
15. Samal, S.K.; Vishwanatha, H.M.; Saxena, K.K.; Behera, A.; Nguyen, T.A.; Behera, A.; Prakash, C.; Dixit, S.; Mohammed, K.A. 3D-Printed Satellite Brackets: Materials, Manufacturing and Applications. *Crystals* **2022**, *12*, 1148. [CrossRef]
16. Liu, C.; Liu, Y.; Wang, T.; Liu, W.; Ma, Y. Effects of Heat Treatment on the Microstructure and Properties of Graded-Density Powder Aluminum Alloys. *Met. Sci. Heat Treat.* **2022**, *63*, 590–598. [CrossRef]
17. Song, X.; Zhang, K.; Song, Y.; Duan, Z.; Liu, Q.; Liu, Y. Morphology, Microstructure and Mechanical Properties of Electrospun Alumina Nanofibers Prepared Using Different Polymer Templates: A Comparative Study. *J. Alloys Compd.* **2020**, *829*, 154502. [CrossRef]
18. Wen, Z.; Song, X.; Chen, D.; Fan, T.; Liu, Y.; Cai, Q. Electrospinning Preparation and Microstructure Characterization of Homogeneous Diphasic Mullite Ceramic Nanofibers. *Ceram. Int.* **2020**, *46*, 12172–12179. [CrossRef]
19. Song, Y.; Ding, X.; Xiao, L.; Liu, W.; Chen, Y.; Zhao, X. The Effect of Ni Plating on the Residual Stress and Micro-Yield Strength in an Al-Cu-Mg Alloy Under Different Diffusion Treatments. *JOM* **2019**, *71*, 4370–4377. [CrossRef]
20. Xiao, L.R.; Tu, X.X.; Zhao, X.J.; Cai, Z.Y.; Song, Y.F. Microstructural Evolution and Dimensional Stability of TiC-Reinforced Steel Matrix Composite during Tempering. *Mater. Lett.* **2020**, *259*, 8–12. [CrossRef]
21. Song, Y.F.; Ding, X.F.; Zhao, X.J.; Xiao, L.R.; Yu, C.X. The Effect of SiC Addition on the Dimensional Stability of Al-Cu-Mg Alloy. *J. Alloys Compd.* **2018**, *750*, 111–116. [CrossRef]
22. Tang, X.; Din, C.; Yu, S.; Liu, Y.; Luo, H.; Zhang, D.; Chen, S. Synthesis of Dielectric Polystyrene via One-Step Nitration Reaction for Large-Scale Energy Storage. *Chem. Eng. J.* **2022**, *446*, 137281. [CrossRef]
23. Yu, S.; Ding, C.; Liu, Y.; Liu, Y.; Zhang, Y.; Luo, H.; Zhang, D.; Chen, S. Enhanced Breakdown Strength and Energy Density over a Broad Temperature Range in Polyimide Dielectrics Using Oxidized MXenes Filler. *J. Power Sources* **2022**, *535*, 231415. [CrossRef]
24. Cai, C.; Wu, X.; Liu, W.; Zhu, W.; Chen, H.; Qiu, J.C.D.; Sun, C.N.; Liu, J.; Wei, Q.; Shi, Y. Selective Laser Melting of Near-α Titanium Alloy Ti-6Al-2Zr-1Mo-1V: Parameter Optimization, Heat Treatment and Mechanical Performance. *J. Mater. Sci. Technol.* **2020**, *57*, 51–64. [CrossRef]
25. Xu, J.; Wu, Z.; Niu, J.; Song, Y.; Liang, C.; Yang, K.; Chen, Y.; Liu, Y. Effect of Laser Energy Density on the Microstructure and Microhardness of Inconel 718 Alloy Fabricated by Selective Laser Melting. *Crystals* **2022**, *12*, 1243. [CrossRef]

26. Yang, H.; Liu, B.; Niu, P.; Fan, Z.; Yuan, T.; Wang, Y.; Liu, Y.; Li, R. Effect of Laser Scanning Angle on Shear Slip Behavior along Melt Track of Selective Laser Melted 316l Stainless Steel during Tensile Failure. *Mater. Charact.* **2022**, *193*, 112297. [CrossRef]
27. Liu, Y.; Liu, W.; Ma, Y.; Liang, C.; Liu, C.; Zhang, C.; Cai, Q. Microstructure and Wear Resistance of Compositionally Graded Ti–Al Intermetallic Coating on Ti6Al4V Alloy Fabricated by Laser Powder Deposition. *Surf. Coat. Technol.* **2018**, *353*, 32–40. [CrossRef]
28. Liu, Y.; Liu, C.; Liu, W.; Ma, Y.; Zhang, C.; Cai, Q.; Liu, B. Microstructure and Properties of Ti/Al Lightweight Graded Material by Direct Laser Deposition. *Mater. Sci. Technol.* **2018**, *34*, 945–951. [CrossRef]
29. Liu, Y.; Wu, Z.; Liu, W.; Ma, Y. Microstructure Evolution and Reaction Mechanism of Continuously Compositionally Ti/Al Intermetallic Graded Material Fabricated by Laser Powder Deposition. *J. Mater. Res. Technol.* **2022**, *20*, 4173–4185. [CrossRef]
30. Li, R.; Wang, H.; He, B.; Li, Z.; Zhu, Y.; Zheng, D.; Tian, X.; Zhang, S. Effect of α Texture on the Anisotropy of Yield Strength in Ti–6Al–2Zr–1Mo–1V Alloy Fabricated by Laser Directed Energy Deposition Technique. *Mater. Sci. Eng. A* **2021**, *824*, 141771. [CrossRef]
31. Li, R.; Wang, H.; Zheng, D.; Gao, X.; Zhang, S. Texture Evolution during Sub-Critical Annealing and Its Effect on Yield Strength Anisotropy of Laser Directed Energy Deposited Ti-6Al-2Zr-1Mo-1V Alloy. *Mater. Sci. Eng. A* **2022**, *850*, 143556. [CrossRef]
32. Abd-elaziem, W.; Elkatatny, S.; Abd-elaziem, A.; Khedr, M.; El-baky, M.A.A.; Ali Hassan, M.; Abu-Okail, M.; Mohammed, M. On the Current Research Progress of Metallic Materials Fabricated by Laser Powder Bed Fusion Process: A Review. *J. Mater. Res. Technol.* **2022**, *20*, 681–707. [CrossRef]
33. Wu, X.; Zhang, D.; Guo, Y.; Zhang, T.; Liu, Z. Microstructure and Mechanical Evolution Behavior of LPBF (Laser Powder Bed Fusion)-Fabricated TA15 Alloy. *J. Alloys Compd.* **2021**, *873*, 159639. [CrossRef]
34. Yao, Z.; Yang, T.; Yang, M.; Jia, X.; Wang, C.; Yu, J.; Li, Z.; Han, H.; Liu, W.; Xie, G.; et al. Martensite Colony Engineering: A Novel Solution to Realize the High Ductility in Full Martensitic 3D-Printed Ti Alloys. *Mater. Des.* **2022**, *215*, 110445. [CrossRef]
35. Liu, Y.; Liang, C.; Liu, W.; Ma, Y.; Liu, C.; Zhang, C. Dilution of Al and V through Laser Powder Deposition Enables a Continuously Compositionally Ti/Ti6Al4V Graded Structure. *J. Alloys Compd.* **2018**, *763*, 376–383. [CrossRef]
36. King, W.E.; Barth, H.D.; Castillo, V.M.; Gallegos, G.F.; Gibbs, J.W.; Hahn, D.E.; Kamath, C.; Rubenchik, A.M. Observation of Keyhole-Mode Laser Melting in Laser Powder-Bed Fusion Additive Manufacturing. *J. Mater. Process. Technol.* **2014**, *214*, 2915–2925. [CrossRef]
37. Wang, W.; Liang, S.Y. Physics-Based Predictive Model of Lack-of-Fusion Porosity in Laser Powder Bed Fusion Considering Cap Area. *Crystals* **2021**, *11*, 1568. [CrossRef]
38. Chowdhury, S.; Yadaiah, N.; Prakash, C. Laser Powder Bed Fusion: A State-of-the-Art Review of the Technology, Materials, Properties & Defects and Numerical Modelling. *J. Mater. Res. Technol.* **2022**, *20*, 2109–2172. [CrossRef]
39. Li, S.; Yang, J.; Wang, Z. Multi-Laser Powder Bed Fusion of Ti-6.5Al-2Zr-Mo-V Alloy Powder: Defect Formation Mechanism and Microstructural Evolution. *Powder Technol.* **2021**, *384*, 100–111. [CrossRef]
40. Thijs, L.; Verhaeghe, F.; Craeghs, T.; Humbeeck, J.V.; Kruth, J.P. A Study of the Microstructural Evolution during Selective Laser Melting of Ti-6Al-4V. *Acta Mater.* **2010**, *58*, 3303–3312. [CrossRef]
41. Liu, Y.; Liu, C.; Liu, W.; Ma, Y.; Tang, S.; Liang, C.; Cai, Q.; Zhang, C. Optimization of Parameters in Laser Powder Deposition AlSi10Mg Alloy Using Taguchi Method. *Opt. Laser Technol.* **2019**, *111*, 470–480. [CrossRef]
42. Jiang, H.Z.; Li, Z.Y.; Feng, T.; Wu, P.Y.; Chen, Q.S.; Feng, Y.L.; Li, S.W.; Gao, H.; Xu, H.J. Factor Analysis of Selective Laser Melting Process Parameters with Normalised Quantities and Taguchi Method. *Opt. Laser Technol.* **2019**, *119*, 105592. [CrossRef]
43. Canel, T.; Zeren, M.; Sınmazçelik, T. Laser Parameters Optimization of Surface Treating of Al 6082-T6 with Taguchi Method. *Opt. Laser Technol.* **2019**, *120*, 105714. [CrossRef]
44. Kumar, R.; Katyal, P.; Mandhania, S. Grey Relational Analysis Based Multiresponse Optimization for WEDM of ZE41A Magnesium Alloy. *Int. J. Light. Mater. Manuf.* **2022**, *5*, 543–554. [CrossRef]
45. Pandel, D.; Kumar Singh, A.; Kumar Banerjee, M.; Gupta, R. Optimization of Mg2(Si-Sn) Based Thermoelectric Generators Using the Taguchi Method. *Mater. Today Proc.* **2020**, *44*, 4124–4130. [CrossRef]
46. Liu, Y.; Liu, C.; Liu, W.; Ma, Y.; Zhang, C.; Liang, C.; Cai, Q. Laser Powder Deposition Parametric Optimization and Property Development for Ti-6Al-4V Alloy. *J. Mater. Eng. Perform.* **2018**, *27*, 5613–5621. [CrossRef]
47. Wei, M.; Chen, S.; Liang, J.; Liu, C. Effect of Atomization Pressure on the Breakup of TA15 Titanium Alloy Powder Prepared by EIGA Method for Laser 3D Printing. *Vacuum* **2017**, *143*, 185–194. [CrossRef]
48. Li, Y.; Hu, Y.; Cong, W.; Zhi, L.; Guo, Z. Additive Manufacturing of Alumina Using Laser Engineered Net Shaping: Effects of Deposition Variables. *Ceram. Int.* **2017**, *43*, 7768–7775. [CrossRef]
49. Jiang, J.; Chen, J.; Ren, Z.; Mao, Z.; Ma, X.; Zhang, D.Z. The Influence of Process Parameters and Scanning Strategy on Lower Surface Quality of TA15 Parts Fabricated by Selective Laser Melting. *Metals* **2020**, *10*, 1228. [CrossRef]
50. Read, N.; Wang, W.; Essa, K.; Attallah, M.M. Selective Laser Melting of AlSi10Mg Alloy: Process Optimisation and Mechanical Properties Development. *Mater. Des.* **2015**, *65*, 417–424. [CrossRef]

Article

Engulfment of a Particle by a Growing Crystal in Binary Alloys

Qingyou Han [1,*], Yanfei Liu [2], Cheng Peng [3] and Zhiwei Liu [3]

[1] School of Engineering Technology (Retired), Purdue University, West Lafayette, IN 47906, USA
[2] R&D Department, China Academy of Launch Vehicle Technology, Beijing 100048, China
[3] State Key Laboratory for Mechanical Behavior of Materials, School of Materials Science and Engineering, Xi'an Jiaotong University, Xi'an 710049, China
* Correspondence: hanq@purdue.edu

Abstract: Under quasi-steady particle pushing conditions in an alloy, fresh liquid has to flow to the gap separating a particle and an advancing solid–liquid interface of a crystal to feed the volume change associated with the liquid–solid phase transformation. In the meantime, solute rejected by the growing crystal has to diffuse out of the gap against the physical feeding flow. An inequality equation was derived to estimate the pushing-to-engulfment transition (PET) velocity of the crystal under which the particle is pushed by the growing crystal. Experiments were performed in an Al-4.5 wt.%Cu-2 wt.% TiB_2 composite under isothermal coarsening conditions. TiB_2 particles were indeed engulfed by the growing aluminum dendrites as predicted using the inequality equation. Predictions of the inequality equation also agreed reasonably well with literature data from the solidification of distilled water containing particles obtained under minimal convection conditions. The inequality equation suggests that the PET velocity is much smaller in a binary alloy than that in a pure material. Without the influence of fluid flow or other factors that put a particle in motion in the liquid, the particle should be engulfed by the growing crystal in alloys solidified under normal cooling rates associated with convectional casting conditions.

Keywords: A1. dendrite; A1. particle pushing; A1. diffusion; A2. growth from melt; B1. alloys

1. Introduction

A great deal of work has been published concerning the interaction of insoluble particles with an advancing solid–liquid interface [1–7] owing to its importance in various fields, including the processing of composites (metal matrix or polymer matrix) [8,9], preservation of biological materials [10], and freezing of soils [11,12]. It has been generally recognized that certain types of particles are pushed by the solid–liquid interface advancing at small velocities. There is a critical growth velocity, V_C, or pushing-to-engulfment transition (PET) velocity, above which a particle ceases to be pushed but is engulfed in the growing crystal. Engulfment likely leads to a uniform distribution of particles within a solid crystal, while pushing results in particle segregation in the solidified material.

The PET velocity has been defined in various forms [1–7], but generally, it is proportional to the difference in interfacial energies, $\Delta\sigma$, of the particle/material system, the minimum separation between the particle and the advancing solid–liquid interface, h, and the reciprocal of particle radius, R. For discussion purpose, V_C can be written as:

$$V_C \propto \frac{h\Delta\sigma}{R^\alpha} \qquad (1)$$

where α is a constant with a value of 1, 3/2, or 4/3 from various models [1,2,4], and $\Delta\sigma$ is defined as:

$$\Delta\sigma = \sigma_{SP} - (\sigma_{SL} + \sigma_{PL}) \qquad (2)$$

Equation (2) is proposed based on interfacial energy interactions [1,2,4], suggesting that when the solid–particle interfacial energy, σ_{SP}, is greater than the sum of the solid–liquid interfacial energy, σ_{SL}, and the particle–liquid interfacial energy, σ_{PL}, a repulsive force arises that acts on the particle as the solid–liquid interface approaches the particle within h. Analytical models in the form of Equation (1) predict that, for a 2-μm particle satisfying Equation (2) in a pure material, V_C is in the order of a few micrometers per second, which is in general agreement with experimental observation, but the h value has to be altered in order to fit experimental data [13].

In an alloy or in a pure material where impurity has to be considered, the boundary conditions at the advancing solid–liquid interface become too complicated to obtain simple analytical solutions on V_C. Pötschke and Rogge [6] developed a numerical model to describe the influence of solute content on V_C. Kao et al. [6] presented a more rigorous numerical model for the capturing of a foreign particle by an advancing solid–liquid interface in binary alloys. These models predict that V_C is in the form:

$$V_C \propto \frac{h \Delta \sigma G^\beta}{R^\alpha C_0^\varepsilon} \quad (3)$$

where G is the temperature gradient, C_0 is the alloy composition, and β and ε are positive constants in the range of 0–1 [4,6,7]. The values of α, β, and ε are affected by $\Delta \sigma$ [7]. V_C in a binary alloy can be orders of magnitude smaller than that in a pure material [7]. One issue with the numerical models is the lack of accurate data for $\Delta \sigma$, especially when a solute is enriched in the thin gap between the particle and the advancing solid–liquid interface. Interface interactions related to $\Delta \sigma$ define the morphology of the solid–liquid interface beneath the particle and the resultant h, which, in turn, affects the force balance during particle pushing.

A major drawback in the numerical models is associated with the length scales of h and that of interface morphology. The repulsive force is related to the van der Waals forces, which become substantial when h is in the order of a few atomic diameters. In most binary alloys, the first solid crystal precipitating from a liquid usually consists of a non-faceted phase. The solid–liquid interface of such a non-faceted crystal is only a few atoms thick and is zigzag at the nanoscale. This means that the minimum separation, h, is in the same order of interface roughness at the nanoscale level and should vary from atom to atom substantially. However, numerical models so far developed have been focusing on calculating a macroscopically smooth solid–liquid interface and a resultant constant h using rigorous macroscopic boundary conditions. Consequently, these models may not be meaningful in describing the van der Walls forces since the h value cannot be obtained simply by solving the macroscopic diffusion equations. It is noted that numerical models predict that particle pushing occurs when the depth of a depression on the solid–liquid interface is around 0.8–1.2R [7]. Such s deep depression on the advancing solid–liquid interface during steady-state particle pushing has not yet been observed experimentally. Much sophisticated theoretical work involving multi-length scale modeling is required to describe particle pushing.

Experimental validation of model prediction yielded conflicting results. The predicted V_C is in the order of a few micrometers per second in a pure material [1–7] and orders of magnitude smaller than that in alloys [6,7]. Such a V_C is much smaller than the average growth rates of solid solidified under normal casting conditions. As a result, particles should have been engulfed by the growing solid during the normal solidification of composite materials. Experimental data on particle pushing during dendritic solidification in metal matrix composites do not support theoretical predictions [14–18]. Instead of being engulfed, particles are actually pushed by dendrites growing at rates a few orders of magnitude greater than predicted [14–19]. In fact, except for particles that nucleate the growing solid phase, i.e., a particle that exists in the center of a growing grain, there is a lack of convincing experimental observation conforming that particles being actually engulfed within a dendrite arm under cooling conditions, so far tested regardless of $\Delta \sigma$ being positive [14,15] or negative [17–19]. Obviously, convection in liquid plays a significant role in particle pushing [20–22]. However, convection or mechanical disturbance in the melt has been difficult to avoid in most of the experiments performed so far.

Engulfment of insoluble particles by advancing solid–liquid interface is still not fully understood owing to the deficiencies both in modeling and in experiments described above. The purpose of this work was to examine the effect of solute on particle pushing/engulfment in an alloy under convection-less conditions. Simple analytical equations were derived considering only the first-order phenomena governing particle pushing so that the boundary conditions do not need to be as rigorous as that used in numerical modeling [6,7,23,24]. A unique experimental method, which minimized convection and mechanical disturbance, was designed to capture particles within dendrite arms. These simple equations can be used for estimating V_C in binary alloys without using interfacial energies. The method can be used for observing the capturing of particles by an advancing solid–liquid interface under convection-less conditions.

2. Analysis on the Pet Velocity

As shown in Figure 1, a depression should be gradually formed on a growing dendrite arm as the arm approaches an insoluble foreign particle. The existence of the particle in the vicinity of the dendrite arm will certainly affect solute diffusion ahead of the advancing solid–liquid interface of the arm. In the gap between the particle and the solid–liquid interface, solute rejected by the growing dendrite arm is difficult to be transferred out of the gap. The enrichment of solute in the gap decreases the local melting temperature if the partition coefficient of the solute element, k, is smaller than one. As a result, the solid–liquid interface under the gap will be concave, forming a depression on the dendrite arm to allow the particle to dwell. Such type of depression formation has been confirmed by Sen et al. [25] in molten aluminum by means of X-ray imaging.

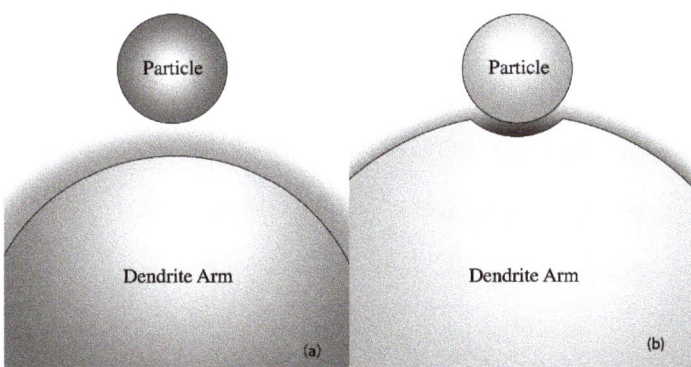

Figure 1. Schematic illustration showing (**a**) as a particle is approached by an advancing solid–liquid interface of a dendrite arm, (**b**) a depression is formed on the growing dendrite arm if $k < 1$. The shade indicates the solute field ahead of the solid–liquid interface of the dendrite arm. The solute concentration is illustrated as proportional to the darkness of the shade.

Consider the condition of quasi-steady particle pushing, i.e., a particle that dwells on an advancing solid–liquid interface advancing at a constant rate is pushed steadily by the solid–liquid interface over a certain distance [1]. In order to push the particle under such conditions, fresh liquid has to constantly flow into the gap and freeze on the growing solid. At the same time, in an alloy, the solute must diffuse against this physical flow at a rate sufficient to prevent being built up in the gap.

Figure 2 depicts the temperature and solute fields in the vicinity of the solid–liquid interface. Following Pötschke and Rogge [6], the balance of mass and solute under steady-state conditions in the gap region can be described. At any point in the gap shown in Figure 2, the mass of liquid, M_L, flowing into the gap per unit of time is:

$$M_L = 2\pi h \left[(R + \frac{1}{2}h) \sin\theta \right] V_L \rho_L \qquad (4)$$

where V_L is the flow rate, ρ_L is the density of the liquid, and θ is the angle shown in Figure 2.

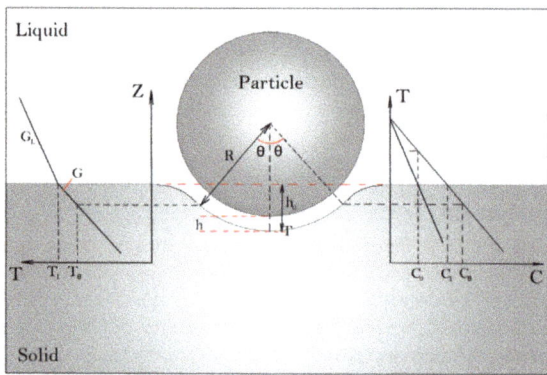

Figure 2. Schematic illustration of the temperature and solute fields in the gap between a particle and an advancing solid–liquid interface.

The mass of liquid, M_L, which is transformed into solid growing at a velocity, V_S, per unit of time in the region defined by any θ is:

$$M_L = \pi[(R+h)\sin\theta]^2 V_S \rho_S \qquad (5)$$

where ρ_S is the density of the solid. Mass balance defined by equating Equations (4) and (5) gives:

$$V_L = \frac{(R+h)^2 \rho_S \sin\theta}{2h(R+\frac{1}{2}h)\rho_L} V_S \qquad (6)$$

Ignoring the solute entering the growing solid, one can write the solute balance equation at any location in the gap as [5]:

$$V_L C_\theta + \frac{D}{R+h/2} \frac{\partial C_\theta}{\partial \theta} = 0 \qquad (7)$$

Under directional solidification conditions, the temperature field is fixed so the temperature distribution can be estimated using:

$$T = T_I + GZ \qquad (8)$$

where G is the temperature gradient in the solid, T_I is the temperature at the planar solid–liquid interface far away from the particle, and Z is the vertical distance from any point in the solid to the planar solid–liquid interface.

The relationship between solute and temperature on a phase diagram is described by the phase diagram of the alloy system. For an alloy with a diluted solute content, the liquidus slope, m, on the phase diagram is usually a constant. The change in temperature is related to the change in composition by:

$$\Delta T = m \Delta C \qquad (9)$$

Under steady-state growth conditions and ignoring the kinetic growth undercooling, the temperature, T_I, at the planar solid–liquid interface far away from the particle is:

$$T_I = T_0 + m\frac{C_0}{k}(k-1) \tag{10}$$

where C_0 is the bulk solute content, T_0 is the liquidus temperature of the alloy, and k is the partition coefficient. The composition at the liquid side of the planar solid–liquid interface is:

$$C_I = \frac{C_0}{k} \tag{11}$$

Assuming that the depression shown in Figure 2 has a constant radius, a, the temperature at any given θ is:

$$T_\theta = T_I + m(C_I - C_\theta) - \Delta T_a - \Delta T_K \tag{12}$$

where ΔT_a is curvature undercooling and ΔT_K is kinetic undercooling. Under the assumption of a constant gap radius, ΔT_a and ΔT_K are constants in the gap region. The temperature difference between θ_0 and θ in the gap region can be calculated in the following equation by either using Equation (9) or Equation (12).

$$\Delta T = m(C_{\theta_0} - C_\theta) \tag{13}$$

Combining Equations (8) and (12), one can write the vertical distance, ΔZ, between θ_0 and θ in the gap as:

$$m(C_\theta - C_{\theta_0}) = G\Delta Z \tag{14}$$

i.e.,

$$\Delta Z = \frac{m}{G}(C_\theta - C_{\theta_0}) \tag{15}$$

Let the depth of the depression be h_0, the vertical distance from the planar solid–liquid interface to the center of the particle is $R-h_0$, as shown in Figure 2, and the edge of the planar interface neighboring the gap is the location corresponding to θ_0. At any given θ, the vertical distance from the point corresponding to θ in the gap to the planar interface is given by:

$$\Delta Z = (R+h)\cos\theta - (R-h_0) \tag{16}$$

Combining Equations (15) and (16) gives:

$$C_\theta = C_{\theta_0} - \frac{G}{m}[(R+h)\cos\theta - (R-h_0)] \tag{17}$$

where

$$C_{\theta_0} \approx C_I \tag{18}$$

Differentiating Equation (17) gives:

$$\frac{\partial C_\theta}{\partial \theta} = \frac{G}{m}(R+h)\sin\theta \tag{19}$$

Substituting Equations (17) and (19) into Equation (7) yields:

$$V_L\left(C_I - \frac{G}{m}[(R+h)\cos\theta - (R-h_0)]\right) + \frac{D}{R+h/2}\left[\frac{G}{m}(R+h)\sin\theta\right] = 0 \tag{20}$$

Combining Equations (6), (11) and (20) gives:

$$\frac{(R+h)^2\rho_S}{2h(R+\frac{1}{2}h)\rho_L}V_S\left(\frac{C_0}{k} - \frac{G}{m}[(R+h)\cos\theta - (R-h_0)]\right) + \frac{DG(R+h)}{m(R+h/2)} = 0 \tag{21}$$

Considering mass balance at the entrance of the gap where:

$$R - h_0 = (R + h) \cos \theta \tag{22}$$

and ignoring h in the $R + h$ and $R + h/2$ terms since h is only a few atomic layers thick [1–6] and is much smaller than R, one can rewrite Equation (21) as:

$$\frac{C_0 R \rho_S}{2hk\rho_L} V_S + \frac{DG}{m} = 0 \tag{23}$$

Rearranging Equation (23) gives:

$$V_S = \frac{2hD\rho_L}{\rho_S} \frac{k}{mC_0} \frac{1}{R} G \tag{24}$$

In deriving Equation (6), solute entering the growing solid is ignored. To account for this amount of solute, the PET velocity, V_C, of the solid has to be smaller than V_S described in Equation (24) in order to maintain the particle being pushed under steady-state, i.e.,

$$V_C \leq \frac{2hD\rho_L}{\rho_S} \frac{k}{mC_0} \frac{1}{R} G \tag{25}$$

Similar to Equation (2), the PET velocity in Equation (25) is proportional to h and the reciprocal of R. The uniqueness of Equation (25) is that the PET velocity is a linear function of composition, diffusion coefficient, and temperature gradient as well. Compared to Equation (3), the values of α, β, and ε in Equation (25) are different, and the term of interfacial energies is not included because of the assumption that the radius of the gap is a constant. Still, the minimum separation between the particle and solid–liquid interface has to be determined by the forces acting on the particle. Furthermore, m and k in Equation (25) should not be zero because of the division operations performed in obtaining these equations. When $m = 0$, no depression will form on the growing solid behind the particle because $T_I = T_0$ and $C_L = C_0$. Thus, the solute should have no effect on particle pushing. This is also held true for $C_0 = 0$ and $k = 1$. Under such conditions, solute diffusion is not a dominant factor during the engulfment of a particle, so particle pushing is governed by surface energy interactions when this is no convection in the liquid.

3. Validation
3.1. G vs. V_C

Equations (24) and (25) predict that the PET velocity, V_C, is zero when the temperature gradient, G, is zero, meaning that particles should be engulfed by a growing crystal/solid at any growth rate when $G = 0$ under no convection conditions. Unfortunately, no data on particle pushing is available in the open literature on conditions of $G = 0$ and no convection in the liquid. Such conditions can only be achieved during isothermal coarsening of an alloy at a semisolid state where (1) the liquid film between dendrites is so thin that convection in the liquid film is inhibited by the friction of the solid–liquid interfaces to fluid flow, (2) the slow growth of larger solid grains at the expense of smaller grains is spontaneous at $G = 0$ due to the Ostwald ripening phenomenon, and (3) the fraction solid does not change, so that macro-scale flow (across a few grains) due to shrinkage feeding is prevented.

Isothermal coarsening experiments on particle pushing by globular grains were performed in Al-4.5 wt. %Cu-2 wt.%TiB$_2$. Globular grains were intentionally used in order to illustrate clearly than TiB2 particles could be entrapped with the grains. The alloy was prepared using pure copper (99.9% purity), pure aluminum (99.7% purity), and an Al-10 wt.% TiB$_2$ master alloy containing TiB$_2$ particles in a size range of 0.1 to 3 µm. Around 400 g of raw materials with a composition of Al-4.5 wt.%Cu-2 wt.%TiB$_2$ were melted in a graphite crucible in an electrical resistance furnace. When the temperature of the melt reached and remained stable at 730 °C, ultrasound was introduced into the melt through an

Nb probe for 5 min in order to (1) disperse TiB$_2$ particles in the matrix and (2) cleanse the surfaces of the particles for improved grain refining efficiency. The ultrasonically treated melt was then poured into a steel mold to form an ingot. Specimens of 25 mm by 6 mm were cut from the ingot, heated up from room temperatures to 635 °C, and held at that temperature for various times up to 4 hrs before being cooled in air to room temperature. The microstructure of the specimens was examined using a scanning electron microscope (SEM). In order to observe α-Al grains evolution, specimens were anodized with a Baker's reagent (10 mL HBF4 (40%)+200 mL distilled water) applying 20V DC for 90–120 s and then were observed by polarized light using ZEISS OM. The grain size of the specimens was measured as a function of isothermal coarsening times to determine the average growth rates of the aluminum phase.

Figure 3 illustrates globular aluminum grains and the distribution of TiB$_2$ particles in Al-4.5 wt.%Cu-2 wt.%TiB$_2$ alloy before and after isothermal heat treatment. The size of the globular aluminum grains increases with increasing isothermal coarsening times, as shown in Figure 4.

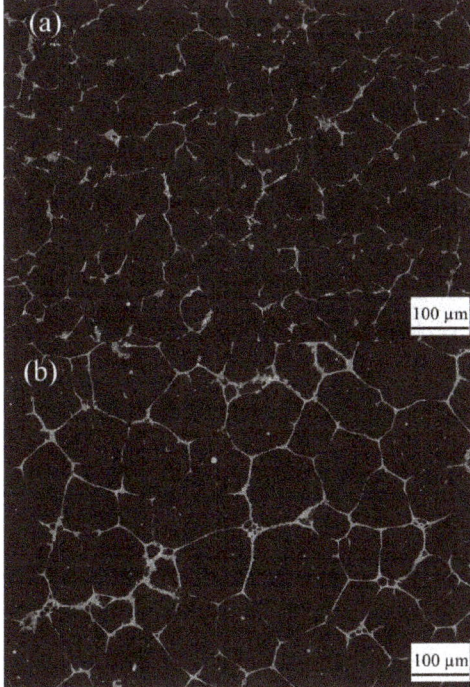

Figure 3. SEM images of the microstructure in Al-4.5 wt.%Cu-2 wt.%TiB$_2$ alloy under (**a**) as-cast conditions and (**b**) after isothermal coarsening for 1 h (3600 s). The dark globular phase is the aluminum grains. The bright regions surrounding the aluminum grains consist of eutectic phases and TiB$_2$ particles.

In the as-cast specimen where convection existed during its solidification, TiB$_2$ particles were all pushed by the aluminum grains/dendrites and were distributed in the eutectic regions where solidification was completed last. As a result, no small TiB$_2$ particles are found within the dark aluminum dendrites shown in Figure 3a. Such results are in agreement with experimental results reported in Refs. [17–19].

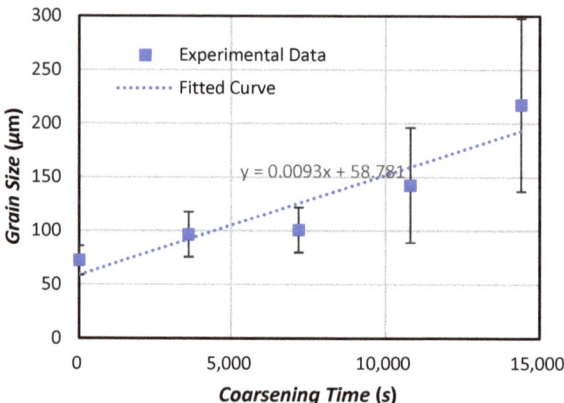

Figure 4. Relationship between dendrite arm size and times during isothermal coarsening of Al-4.5 wt.%Cu-2 wt.%TiB$_2$ alloy.

In the specimen subjected to isothermal coarsening shown in Figure 3b, two types of particles exist within every aluminum dendrite arm: Al$_2$Cu and TiB$_2$. The large and bright ones are eutectic Al$_2$Cu particles, and the small but less bright ones are TiB2 particles. Al$_2$Cu particles existed at the original grain boundaries or interdendritic regions in the as-cast samples. During isothermal coarsening, the eutectic Al$_2$Cu particles remelted, became part of the liquid films or spherical droplets that were entrapped at the grooves of grain boundaries as the globular aluminum grains grew. The size of Al$_2$Cu particles was much greater than the TiB$_2$ particles. Most TiB$_2$ particles also existed in the grain boundaries. Still, some particles that happened to rest on top of aluminum grains could be engulfed. Indeed, there are numerous small particles in the periphery of aluminum grains shown in Figure 3b. Some of them are TiB$_2$ particles that were engulfed when aluminum grains grow during the isothermal coarsening process. We believe that Figure 3b is the first SEM image showing that small TiB$_2$ particles are indeed engulfed within aluminum grains.

The average coarsening rate of the globular aluminum grains during isothermal coarsening was about 0.0093 μm/s, which is the slope of the linear curve that fits with the experimental data shown in Figure 4. TiB$_2$ particles that rested on the coarsening grains were indeed engulfed at such a small average growth rate. It is important to note that although the average coarsening rate is about two orders of magnitude smaller than the V_C predicted by models for pure materials [1,2,4–6,25], it is still not the V_C for this particle/alloy system. The PET velocity, V_C, below which a TiB$_2$ particle is pushed by a growing aluminum crystal/grain in this alloy, could be much smaller than that shown in Figure 4. In fact, during isothermal coarsening, small grains shown in Figure 3b initially grow at the expense of smaller ones. They then start to dissolve after the smaller ones are totally dissolved because larger grains exist nearby. As a result, the small grains shown in Figure 3b could grow at various rates below the average growth rates, including zero. Therefore, particle engulfment in grains of various sizes during isothermal coarsening does suggest that the PET velocity is zero when $G = 0$, as predicted by Equation (25).

When $G > 0$, V_C increases with increasing G linearly, as predicted by Equation (25). Such a linearly relationship was supported by Körber et al. [26], as shown in Figure 5 but was not supported by Cissé and Bolling [27], who performed experiments at two gradient levels but admitted that they did not have good control of the temperature gradient. More experimental data are needed to validate model predictions.

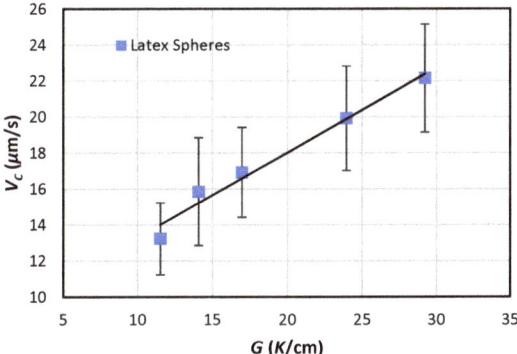

Figure 5. Relationship between PET velocity, V_C, and temperature gradient for latex particles of 5.7 μm mean diameter in distilled water solidified horizontally in thin specimens [26].

3.2. R vs. V_C

Validation of Equations (24) and (25) was performed using experimental data obtained in distilled water and degassed water solidified both vertically [27] and horizontally [26]. The vertical solidification was carried out in a large container freezing from the bottom to the top [27]. Convection in such a large container was unavoidable, so data on heavy particles (silica, copper, and tungsten) were used, and data on hollow carbon spheres were discounted. The horizontal solidification experiments of the latex/water were performed in a thin tube [26]. The density of latex was 1.05 g/cm^3 and was only slightly denser than water. The specimen hosted in a thin tube was withdrawn using a mechanical translation system [26]. It is unclear if the solid–liquid interface was free from mechanical disturbance or vibration from the mechanical translation system.

Figure 6 illustrates the relationship between PET velocity, V_C, and particle radius, R. Experimental data are marked with dots, squares, triangles, and diamonds for various particles. The curves are calculated using Equation (25). Data used in Equation (25) are $D = 10^{-9}$ m^2/s, $G = 10^3$ K/m, and $\rho_L/\rho_S = 1$. Impurities in the distilled and degassed water are not provided in the literature, so we have to assume that $mC_0/k = 0.001$ K. Given the value of mC_0/k, the h values, shown in Figure 6, can be determined by fitting Equation (25) with experimental data.

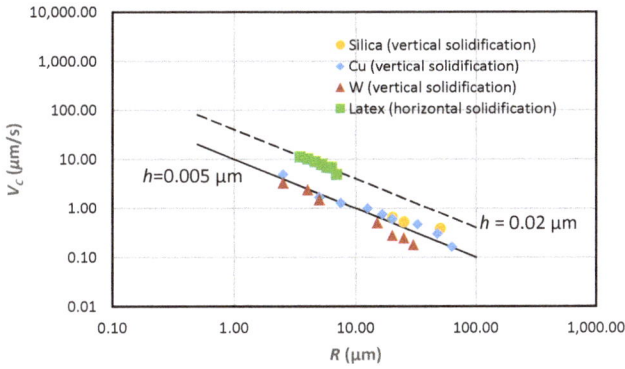

Figure 6. Critical growth rate, V_C, vs. particle radius, R, in distilled water solidified vertically or horizontally. Equation (25) is used to calculate the dash line assuming $h = 0.02$ μm and the solid line assuming $h = 0.005$ μm. The solid markers are experimental data from literature. ● Silica (vertical solidification) [27], ◆ Cu (vertical solidification) [27], ▲ W (vertical solidification) [27], and ■ (horizontal solidification) [26].

Generally, predictions using Equation (25) fit experimental data quite well in terms of the trend of V_C vs. particle radius, R, indicating that V_C is proportional to $1/R$ as described by Equation (25). The good agreement of the predicted V_C with experimental data may not mean much since such a good agreement is achieved by assuming values for mC_0/k and h. Still, it is interesting to note that the h value associated with the denser particles under vertical solidification conditions is much smaller than that associated with the neutrally-buoyant latex particles under horizontal solidification conditions. Such results are reasonable according to the force balance during steady-state particle pushing.

The driving force, F_σ, for particle pushing is given by [2,6]:

$$F_\sigma = 2\pi R \Delta\sigma \left(\frac{a_0}{h}\right)^2 \tag{26}$$

where a_0 is the interatomic distance of the solid. The force resisting to the motion of the particle is a viscous force, F_D, given by [2,6]:

$$F_D = 6\pi\mu V_C \frac{R^2}{h} \tag{27}$$

where μ is the viscosity of the liquid. Under horizontal solidification conditions, the driving force is balanced by the viscous force, giving rise to:

$$h = \frac{\Delta\sigma_0 a_0^2}{3\mu V_C R} \tag{28}$$

Equation (28) indicates that the minimum separation is a function of the viscosity of the liquid and the difference in interfacial energy. The minimum separation decreases with the increasing viscosity of the liquid. The PET velocity described in Equation (25) should decrease with the increasing viscosity of the liquid. Indeed, Uhlmann et al. [1] reported that the critical growth velocities in liquids of higher viscosity were higher than that of lower viscosity.

Under vertical solidification conditions, the gravitational force, F_W, is given by:

$$F_W = \frac{4}{3}\pi R^3 (\rho_P - \rho_L) g \tag{29}$$

where ρ_P is the density of the particle. For a particle denser than the liquid and the solid growing against gravity, force balance gives:

$$F_\sigma = F_D + F_W \tag{30}$$

Substituting Equations (26), (27) and (29) into (30) and rearranging the equation yield:

$$\frac{2}{3} R^2 (\rho_P - \rho_L) g h^2 + 3\mu V_C R h = \Delta\sigma_0 a_0^2 \tag{31}$$

To differentiate Equation (31) against $(\rho_P-\rho_L)$ yields $dh/d(\rho_P-\rho_L) < 0$, indicating that an increase in the density of the particle would lead to a decrease in h. The gravitational force acting on the particle seems to push the particle closer to the solid–liquid interface, resulting in a smaller h and resultant lower V_C under vertical solidification conditions than that under horizontal solidification conditions. The slightly large scatter in data obtained under vertical solidification conditions shown in Figure 6 seems to be related to the densities of the particles tested as well. The densities of the particles are 2250, 8960, and 19,360 kg/m^3 for silica, copper, and tungsten, respectively [2]. The PET velocity for tungsten particles is lower than that of the silica and copper particles, as shown in Figure 6.

4. Discussion

The analytical models of this work are derived under simplified boundary conditions in order to obtain analytical solutions. The macroscopic morphology of the solid–liquid interface is not solved because it cannot represent the morphologies of the solid–liquid interface and the resultant separation, h, at the atomic scale. As a result, the curvature undercooling of the interface has to be estimated, assuming a constant radius for the solid–liquid interface in the gap region. Another simplification in obtaining the analytical models is that the solute composition at the edges of the gap is assumed to be that at the planar solid–liquid interface, i.e., Equation (18). In spite of these oversimplified assumptions and less rigorous boundary conditions, results shown in Figures 3–6 indicate that the predictions made by Equation (25) are generally in agreement with experimental observations in distilled water during directional solidification and in an aluminum alloy during isothermal coarsening.

Data shown in Figure 6 include only those obtained in distilled water. Experimental data in other particle/matrix systems [1,3,28] are not plotted because each system would require its unique h and $m\bar{C}_0/k$ data in order to use Equation (25). Still, for a particle of 1 µm in radius, the PET velocities are in the range of between a few micrometers per second to a few tens of micrometers per second. Such growth rates are comparable to our model predictions since the viscosity and the interfacial energies are quite different from that of the particle/distilled water system. Experimental data in systems containing sub-millimeter-sized particles [29,30] are not considered in this study because the thermal conductivities of the materials in the systems are not included in our study. These physical properties have less effect on particle pushing for small particles [1,2,5] but could have a significant effect on large particles [6,25,29–31].

In comparison with experimental data in particle/distilled water systems, the h value used in Equation (21) is in the range of 0.005 to 0.02 µm, which is one to two orders of magnitude greater than an interatomic spacing [32]. Such a minimum separation between the particle and the solid–liquid interface is in agreement with what is reported in the literature [1–3]. PET velocity increases with increasing h because it would be less difficult for solute elements or impurities to diffuse out of a thicker gap than a thinner one. Solute or impurity elements have to diffuse out of the gap, against the feeding current, to maintain steady-state particle pushing. Otherwise, the particle should be engulfed by the growing solid.

The dependence of the PET velocity on particle size is also related to the diffusion of solutes/impurities out of the gap between the particle and the growing solid crystal. The length of the gap and the resultant diffusion length are proportional to the particle size. Thus, it would be more difficult for the solutes/impurities to diffuse out of a longer gap than a shorter one, given a minimum separation, h. Our diffusion model, i.e., Equation (25), indicates that V_C is proportional to $1/R$. This is quite different from the models based on interface interactions where V_C is proportional to $1/R^\alpha$ and $\alpha > 1$ [6,7]. The dependence of V_C to $1/R$ seems to fit with experimental data better than V_C to $1/R^\alpha$ where $\alpha > 1$.

The solute content has a major effect on V_C. Unfortunately, experiments carried out in alloys on particle pushing under diffusion-only conditions, i.e., no convection, are scarce. The purity of matrix material used is usually of high purity (about 99.999 wt. %). One experiment by Körber et al. [28] indicates that the addition of 0.56% $NaMnO_4$ in water does not alter the PET velocity of water containing latex particles. However, no phase diagram of the water/$NaMnO_4$ system is available, so it would be impossible to determine the values of m and k using Equation (25). A solute element should have no effect on V_C if $m = 0$ or $k = 1$.

The PET velocity for a 2-µm particle is around a few micron meters per second for a particle in the pure matrix, usually of 99.999 wt.% purity. When the impurity or solute content is increased by a few orders of magnitude, the PET velocity should decrease by a few orders of magnitude according to Equation (25). For an alloy containing 1 wt. % of solute, the PET velocity for pushing a 2-µm particle could be in the nanometer per second range, which is a few orders of magnitude smaller than the average growth velocity of solids during normal cooling conditions for normal gravity casting. This would mean that particles should be engulfed during dendritic solidification in a casting. Experimental results in alloys suggest otherwise. Particles are pushed by the growing dendrites under casting conditions so far tested [14,15,18]. Clearly, the fluid flow has a major effect on particle pushing during solidification. A particle traveling in the melt can bounce off the solid–liquid interface and thus is prevented from being engulfed by the growing solid [18,22]. A particle resting on a depression can be dislodged from the depression on the growing solid crystal by a rolling/sliding mechanism [18,20,21]. Vibration or other mechanical disturbance is also capable of dislodging a particle from a depression [18,20]. As long as the particle is dislodged from a depression and starts motion, it should not be engulfed until it settles on a depression again.

Without convection and other mechanical disturbance in the liquid, a small particle (large enough than what is subjected to Brownian motion) should be engulfed by the growing solid in an alloy if it rests on a depression on a growing crystal under normal casting conditions. Convection can be suppressed in an alloy during isothermal coarsening or during the late stage of solidification when the growth rates of the solid are slow, and the fraction solid is relatively large. Indeed, we have found that particles are engulfed during isothermal coarsening of Al-4.5 wt. %Cu alloy in this study.

5. Conclusions

An analytical inequality equation has been derived for describing particle pushing by a planar advancing solid–liquid interface in an alloy. The PET velocity at which the pushing-to-engulfment transition occurs is proportional to the minimum separation between the particle and the solid–liquid interface and the temperature gradient at the interface and is inversely proportional to the particle radius and the solute content of the alloy. The PET velocities in alloys are much smaller than that in pure materials. Without fluid flow in the liquid, a particle that dwells on an advancing solid–liquid interface should be engulfed by the growing solid crystal under normal casting conditions. The fact that particles are pushed by dendrites in casting is due to fluid flow in the mushy zone. Engulfment of small TiB_2 particles by growing grains/dendrites is observed in Al-4.5 wt. %-2 wt.%TiB_2 alloy during isothermal coarsening at 0.5 fraction solid where convection in the remaining liquid in the mushy zone is suppressed.

Author Contributions: Investigation: C.P., Y.L., Z.L. and Q.H.; resources: Z.L. and Q.H.; writing—original draft preparation: Y.L. and Q.H.; writing—review and editing: Z.L., Y.L. and Q.H.; supervision: Q.H. and Z.L. All authors have read and agreed to the published version of the manuscript.

Funding: This research was funded by the Center for Materials Processing Research at Purdue University (W. Lafayette, IN, USA); and the National Natural Science Foundation of China (Nos. 51974224 and 52174372, Beijing, China).

Acknowledgments: The authors would like to thanks Purdue Center for Materials Processing Research for supporting the work of Yanfei Liu while he was a graduate student in the School of Engineering Technology, Purdue University. The authors would also like to thank the National Natural Science Foundation of China (Grant Nos. 51974224 and 52174372) for supporting the work at Xi'an Jiaotong University.

Conflicts of Interest: The authors declare that they have no known competing financial interests of personal relationships that could have appeared to influence the work reported in this paper.

References

1. Uhlmann, D.R.; Chalmers, B.; Jackson, K.A. Interaction Between Particles and a Solid-Liquid Interface. *J. Appl. Phys.* **1964**, *35*, 2986–2993. [CrossRef]
2. Bolling, G.F.; Cissé, J. A Theory for the Interaction of Particles with a Solidifying Front. *J. Cryst. Growth* **1971**, *10*, 56–66. [CrossRef]
3. Omenyi, S.N.; Neumann, A.W. Thermodynamic Aspect of particle Engulfment by Solidifying Melts. *J. Appl. Phys.* **1976**, *47*, 3956–3962. [CrossRef]

4. Chernov, A.A.; Temkin, D.E.; Melnikova, A.M. Theory of the Capture of Solid Inclusions during the Growth of Crystals from the Melts. *Sov. Phys.-Cryst.* **1976**, *21*, 369–373.
5. Shangguan, D.; Ahuja, S.; Stefanescu, D.M. An analytical model for the interaction between an insoluble particle and an advancing solid/liquid interface. *Met. Mater. Trans. A* **1992**, *23*, 669–680. [CrossRef]
6. Pötschke, J.; Rogge, V. On the behaviour of foreign particles at an advancing solid-liquid interface. *J. Cryst. Growth* **1989**, *94*, 726–738. [CrossRef]
7. Kao, J.C.T.; Golovin, A.A.; Davis, S.H. Particle capture in binary solidification. *J. Fluid Mech.* **2009**, *625*, 299–320. [CrossRef]
8. Chen, L.-Y.; Xu, J.-Q.; Choi, H.; Pozuelo, M.; Ma, X.; Bhowmick, S.; Yang, J.-M.; Mathaudhu, S.; Li, X.-C. Processing and properties of magnesium containing a dense uniform dispersion of nanoparticles. *Nature* **2015**, *528*, 539–543. [CrossRef]
9. Zhang, H.; Hussain, I.; Brust, M.; Butler, M.F.; Rannard, S.; Cooper, A.I. Aligned two- and three-dimensional structures by directional freezing of polymers and nanoparticles. *Nat. Mater.* **2005**, *4*, 787–793. [CrossRef] [PubMed]
10. Ishiguro, H.; Rubinsky, B. Mechanical Interactions between Ice Crystals and Red Blood Cells during Directional Solidification. *Cryobiology* **1994**, *31*, 483–500. [CrossRef]
11. Corte, A.E. Particle Sorting by Repeated Freezing and Thawing. *Science* **1963**, *142*, 499–501. [CrossRef]
12. Corte, A.E. Vertical Migration of Particles in Front of a Moving Freezing Plane. *J. Geogr. Res.* **1962**, *67*, 1085–1090. [CrossRef]
13. Kaptay, G. Discussion of "Particle Engulfment and Pushing by Solidifying Interfaces: Part II. Microgravity Experiments and Theoretical Analysis". *Metall. Mater. Trans. A* **1999**, *30*, 1887–1894. [CrossRef]
14. Mortensen, A.; Jin, I. Solidification processing of metal matrix composites. *Int. Mater. Rev.* **1992**, *37*, 101–128. [CrossRef]
15. Wilde, G.; Perepezko, J.H. Experimental study of particle incorporation during dendritic solidification. *Mater. Sci. Eng. A* **2000**, *283*, 25–37. [CrossRef]
16. Han, Q.; Lindsay, J.P.; Hunt, J.D. The Effects of Fluid Flow on Particle Pushing. *Cast Met.* **1994**, *6*, 237–239. [CrossRef]
17. Wang, M.; Han, Q. Particle Pushing during Solidification of Metals and Alloys. *Mater. Forum* **2014**, *783*, 155–160. [CrossRef]
18. Liu, Y.; Han, Q. Interaction between nucleant particles and a solid-liquid interface in Al-4.5Cu alloy. *Acta Mater.* **2021**, *213*, 116956. [CrossRef]
19. Youssef, Y.M.; Dashwood, R.J.; Lee, P.D. Effect of Clustering on Particle Pushing and Solidification Behaviour in TiB$_2$ Reinforced Aluminum PMMCs. *Compos. Part A* **2005**, *36*, 747–763. [CrossRef]
20. Han, Q.; Hunt, J.D. Redistribution of Particles during Solidification. *ISIJ Int.* **1995**, *35*, 693–699. [CrossRef]
21. Han, Q.; Hunt, J. Particle pushing: Critical flow rate required to put particles into motion. *J. Cryst. Growth* **1995**, *152*, 221–227. [CrossRef]
22. Han, Q.; Hunt, J. Particle pushing: The attachment of particles on the solid-liquid interface during fluid flow. *J. Cryst. Growth* **1994**, *140*, 406–413. [CrossRef]
23. Chang, A.; Dantzig, J.A.; Darr, B.T.; Hubel, A. Modeling the interaction of biological cells with a solidifying interface. *J. Comput. Phys.* **2007**, *226*, 1808–1829. [CrossRef]
24. Yang, Y.; Garvin, J.; Udaykumar, H. Sharp interface numerical simulation of directional solidification of binary alloy in the presence of a ceramic particle. *Int. J. Heat Mass Transf.* **2008**, *51*, 155–168. [CrossRef]
25. Sen, S.; Curreri, P.; Kaukler, W.F.; Stefanescu, D.M. Dynamics of solid/liquid interface shape evolution near an insoluble particle—An X-ray transmission microscopy investigation. *Met. Mater. Trans. A* **1997**, *28*, 2129–2135. [CrossRef]
26. Körber, C.; Rau, G.; Cosman, M.; Cravalho, E. Interaction of particles and a moving ice-liquid interface. *J. Cryst. Growth* **1985**, *72*, 649–662. [CrossRef]
27. Cissé, J.; Bolling, G. A study of the trapping and rejection of insoluble particles during the freezing of water. *J. Cryst. Growth* **1971**, *10*, 67–76. [CrossRef]
28. Sen, S.; Dhindaw, B.K.; Stefanescu, D.M.; Catalina, A.; Curreri, P. Melt convection effects on the critical velocity of particle engulfment. *J. Cryst. Growth* **1997**, *173*, 574–584. [CrossRef]
29. Juretzko, F.R.; Dhindaw, B.K.; Stefanescu, D.M.; Sen, S.; Curreri, P.A. Particle Engulfment and Pushing by Solidifying Interfac-es: Part I. Ground Experiments. *Metall. Mater. Trans. A* **1998**, *29*, 1691–1696. [CrossRef]
30. Stefanescu, D.M.; Juretzko, F.R.; Catalina, A.; Dhindaw, B.K.; Sen, S.; Curreri, P.A. Particle engulfment and pushing by solidifying interfaces: Part II. Microgravity experiments and theoretical analysis. *Met. Mater. Trans. A* **1998**, *29*, 1697–1706. [CrossRef]
31. Surappa, M.K.; Rohatgi, P.K. Heat diffusivity criterion for the entrapment of particles by a moving solid-liquid interface. *J. Mater. Sci.* **1981**, *16*, 562–564. [CrossRef]
32. Gilpin, R. Theoretical studies of particle engulfment. *J. Colloid Interface Sci.* **1980**, *74*, 44–63. [CrossRef]

Article

Wear Characterization of Laser Cladded Ti-Nb-Ta Alloy for Biomedical Applications

Raj Soni [1], Sarang Pande [1,*], Santosh Kumar [2], Sachin Salunkhe [3], Harshad Natu [4] and Hussein Mohammed Abdel Moneam Hussein [5,6]

1. Department of Mechanical Engineering, Marwadi University, Rajkot 360003, India
2. Department of Mechanical Engineering, Indian Institute of Technology, Varanasi 221005, India
3. Department of Mechanical Engineering, Vel Tech Rangarajan Dr. Sagunthala R & D Institute of Science and Technology, Chennai 600062, India
4. Magod Fusion Technologies Pvt. Ltd., Pune 411026, India
5. Mechanical Engineering Department, Faculty of Engineering and Technology, Future University in Egypt, New Cairo 11835, Egypt
6. Mechanical Engineering Department, Faculty of Engineering, Helwan University, Cairo 11732, Egypt
* Correspondence: sarang.pande@marwadieducation.edu.in

Abstract: Additive manufacturing (AM) has started to unfold diverse fields of applications by providing unique solutions to manufacturing. Laser cladding is one of the prominent AM technologies that can be used to fulfill the needs of custom implants. In this study, the wear resistance of the laser cladded titanium alloy, Ti-17Nb-6Ta, has been evaluated under varied loads in Ringer's solution. Microstructural evaluation of the alloy was performed by SEM and EDX, followed by phase analysis through XRD. The wear testing and analysis have been carried out with a tribometer under varied loads of 10, 15, and 20 N while keeping other parameters constant. Abrasion was observed to be the predominant mechanism majorly responsible for the wearing of the alloy at the interface. The average wear rate and coefficient of friction values were 0.016 mm^3/Nm and 0.22, respectively. The observed values indicated that the developed alloy exhibited excellent wear resistance, which is deemed an essential property for developing biomedical materials for human body implants such as artificial hip and knee joints.

Keywords: Ti alloys; β-phase; laser cladding; wear characterization; biomedical applications

1. Introduction

Over the years, there has been a growing need for hard-tissue replacements in the human body like hip/knee joints, teeth, etc., [1]. Some of the key reasons are attributable to the increasing age of the population and increased life expectancy due to better health care in the past decades [2]. A high number of accidents also increases the demand [3]. Due to this, interest has surged in exploring the possibilities for the materials that fulfil the replacement needs. Metallic materials are outstanding hard-tissue replacement materials because of their versatile properties like high mechanical strength, toughness, and promising biocompatibility [4]. Titanium (Ti) alloys possess the required properties like high strength, outstanding biocompatibility, low density, enhanced resistance to corrosion, etc., making them a perfect candidate for use in the human body [5,6]. One of the critical reasons for Ti and its alloys' outstanding biological and physical properties is the presence of an oxide film (TiO$_2$), which is generated naturally on their surface [7]. However, the low hardness and wear resistance affect their full flange use [8]. The ASTM standardized Ti alloy Ti-6Al-4V has high resistance to corrosion, excellent machinability, and tensile strength [9,10]. However, due to a mismatch in Young's modulus (~110 GPa) compared to the human bone (~30 GPa) and reported claims of creating severe health problems such as adverse tissue reactions and neurological disorders [11,12], there has been renewed interest in the search for its alternatives for use in biomedical applications.

Titanium is an allotropic element present in various phases like α, α + β, and β. α-stabilizers and β-stabilizers are added to enhance the α- and β-phases, respectively [13]. Over the past years, numerous studies have been conducted in search of β-Ti alloys, which potentially possess desired mechanical properties and resistance to corrosion. Due to the drawback of Al- and V- ions in Ti-6Al-4V, past studies have tried replacing those elements with other β-stabilizers such as Ta, Sn, Mo, Nb, and Zr to make β-Ti alloys [14,15]. Wei et al. [16] examined Ti-Ta-Nb alloys with 15, 23, and 30 Ta mass% for evaluating the microstructural, biocompatibility, and mechanical properties. Ti-30Ta-10Nb was the optimum alloy, with a Young's modulus of around 60 GPa, a strength of around 1250 MPa, and a hardness of around 3.1 GPa. A Ti-23Ta-10Nb alloy, was also suitable with a Young's modulus of around 75 GPa, hardness of around 3.0 GPa, overall strength of around 1300 MPa, and favorable biocompatibility. The porous foams of this alloy exhibited strength and Young's modulus values of around 180 MPa and 10 GPa, respectively, which were observed to be close to human bone properties.

2. Wear in Ti Alloys

Ti alloys possess low thermal conductivity and low resistance to plastic shearing. Hence, the frictional heat keeps the mating materials attached [17]. Wear resistance is an essential property for implants, and Ti alloys are often used in mechanical joints. When these alloys are used for implant applications such as hip or knee joint replacements, their efficiency may be lower because of the sliding or slight amplitude movement generated from frictional wear [18]. This can create wear debris that is abrasive and is responsible for producing metallosis and metallic allergies [18]. Past studies have tried to upgrade the wear characteristics of Ti alloys by using surface improvement techniques such as PVD, CVD, ion implantation, thermal spraying, etc., [19–22]. However, it was observed that most processes were not suitable due to insufficient adhesion and layer thickness requirements, though some experimentation was done to generate an irregular surface. Irregular surfaces generate oxide layers, which increase the surface area by about 25–35%, due to which there is a noticeable improvement in osseointegration and bonding between the implant and the bone [23]. Barfeie et al. [24] reported that rough surfaces are supposed to give better osseointegration. However, the amount of roughness required for better results is still unknown.

In their study, Khan and Nisar [25] discussed the abrasive wear of a zinc-aluminum alloy. They used pin-on-disc equipment with a 1 m/s sliding speed to obtain the necessary wear characteristics. Loads of 5, 10, 15, and 20 N and 125, 250, 375, and 500 m sliding distances were chosen for the tests. Each sample was finished with up to #1200 grade SiC paper and cleaned with acetone. Increased wear with deeper grooves was noted when the applied loads were increased from 5 to 20 N.

Another study by Jing et al. [26] prepared a bio-functional gradient coating (BFGC) of hydroxyapatite (HA) on Ti6Al4V alloy samples using laser cladding. The four samples in the study were subjected to SEM, EDX, and XRD analyses, and the wear properties were evaluated in simulated body fluid (SBF). It was observed that the wear scars on the BFGC surface were very light because of the specimens' high hardness and wear resistance. The rate of wear of the grinding ball against the four specimens indirectly indicated the wear resistance of the four specimens. Overall, the abrasive wear mechanism was predominant in all four specimens.

Lee et al. [17] evaluated the wear mechanisms of untreated alloy Ti-Nb-Ta-Zr (TNTZ) and Ti-6Al-4V using a ball-on-disc type configuration in Ringer's solution and compared the results with those gained from air exposure. It was found that the volume losses (V_{loss}) of the TNTZ discs and balls were higher in the Ringer's solution than that of air exposure alone. Due to lower resistance to plastic shearing in the TNTZ combination, delamination wear was predominant, resulting in an excessive wear rate. On the other hand, the V_{loss} of the Ti-6Al-4V disc and ball combination decreased in Ringer's solution more than in air exposure. It may be due to the abrasive wear generated in the Ti-6Al-4V combination being effectively suppressed by Ringer's solution.

Dittrick et al. [27] evaluated the tribological behavior of laser processed Ta coatings on pure Ti and found the wear to be one order of magnitude less than pure Ti. The wear rates of pure Ti and Ta coating on Ti under 5 N load against hardened chrome steel balls were 1.39×10^{-3} mm^3/Nm and 1.89×10^{-4} mm^3/Nm, respectively. The authors claimed that the Ta coatings on Ti could be used in place of HA coatings for excellent wear resistance and bioactivity of human body implants.

Additive manufacturing (AM) helps in boosting the research towards making successful implants. At the same time, it fulfils the need for customization. This study tries to explore the possibilities beyond the currently used metallic biomaterials, which in turn leads to an increase in the durability of the implant and human comfort. The present study reports on the manufacturing of the Ti alloy Ti-17Nb-6Ta by laser cladding, which to the best of authors' knowledge is the first of its kind. Like titanium, metals like tantalum (Ta) and niobium (Nb) are also highly corrosion-resistant and inert in body fluids, which are responsible for enhancing the β phase as discussed earlier. XRD analysis was done to analyze the phase of the developed alloy, and microstructures were evaluated using SEM and EDX techniques. Wear characterization was carried out in Ringer's solution by varying the loads, and the friction coefficient was also obtained for the test samples. Detailed descriptions have been given in the upcoming sections.

3. Materials and Methods

3.1. Laser Cladding

AM refers to an integral part of the industrial revolution 4.0. It is a new-age manufacturing technology that boosts research in healthcare industries in a revolutionary way. It is a digital manufacturing process that uses a computer model to fabricate the final product. Laser cladding, a leading AM process, produces net-shaped parts in a layer-by-layer pattern by using laser radiation. The three significant benefits of using laser cladding over other AM techniques are:

(1) It uses significantly less material to fabricate a product;
(2) There is no requirement to use only spherical metallic powder for manufacturing, which leads to lower final product costs [28];
(3) Gradient fabrication is possible only with this technology [29].

For large-scale acceptability of any technology, it is required that it doesn't come with any limitations. Laser cladding can be used to deposit the material on the large parts no matter what the size is, which is not seen in other AM techniques.

The setup of the laser cladding unit used in the study is illustrated in Figure 1.

For proper melting of the alloy, the minimum energy supply needed is designated as the laser energy density (E_d). This energy is calculated from the laser power (P), scan speed (V), and laser spot diameter (d) [30]. The relationship between the variables is represented below in Equation 1:

$$E_d = \frac{P}{Vd} \quad (1)$$

For this study, titanium powder was procured from M/s. Parshvamani Metals Pvt. Ltd., Mumbai, India. Tantalum and niobium powder were procured from M/s. Aritech Chemazone Pvt. Ltd., Kurukshetra, India. The average particle size of all the metal powders was 30–70 μm. The laser cladding for this research was performed at M/s. Magod Fusion Technologies Pvt. Ltd., Pune, India, with a 4 KW diode laser. The metal powders were mixed in a mini vertical blender in pre-set proportions (77, 17, and 6% of Ti, Nb, and Ta, respectively), followed by sieving, and then poured into a laser cladding unit. Laser cladding was performed using a 1200 W laser with a scan speed of 30 mm/s, a laser spot size of 3 mm, a powder flow rate of 12 g/min, a carrier and shielding gas velocity of 15 L/min, and a pass layer distance of 1.5 mm. Argon gas was used to maintain the oxygen level at a low ppm to avoid contamination of the fabricated product. A pure Ti plate preheated to 200 °C was used as the substrate to ensure homogeneous cladding. The cladding was performed in a closed chamber with an inert argon atmosphere to eliminate oxidation. Post manufacturing, the samples were wire-cut into 20 mm × 20 mm × 5 mm dimensions for further characterization.

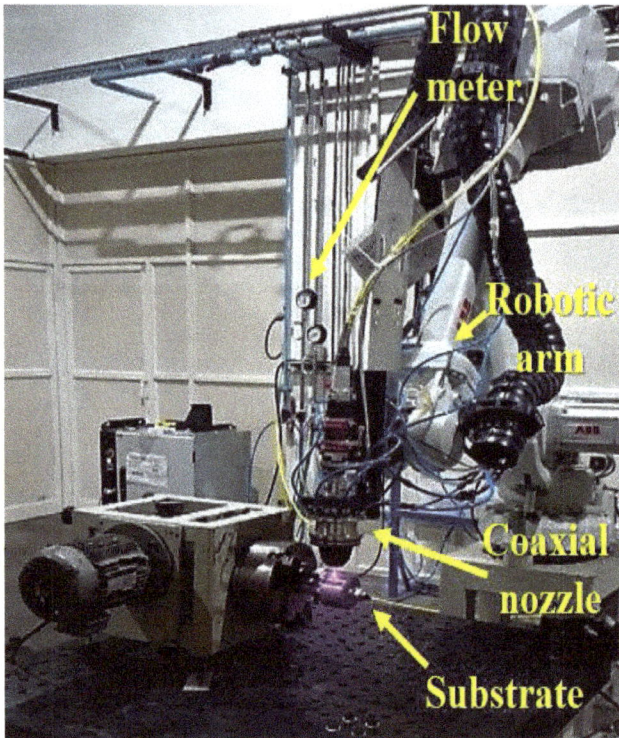

Figure 1. Setup of the laser cladding unit.

3.2. SEM, EDX, XRD, and Micro-Hardness Analyses

SEM and EDX analyses were performed to analyze the microstructure and alloy constituents of the manufactured alloy. SEM and EDX were performed on the EVO-MA15/18SEM and 51N1000-EDS System. Following SEM and EDX, XRD analysis was conducted to analyze the phase of the manufactured alloy. The XRD analysis was done using the Rigaku Miniflex 600 Desktop XRD. For metallographic analysis, the samples were wet and finished with up to #2000 grade SiC paper. Vickers micro-hardness testing was performed on the Micro-Mach micro-hardness tester MMV-M machine.

3.3. Wear Testing

A sliding wear test on the fabricated alloy was conducted using a reciprocating ball-on-disk tribometer (BioTribometer, DUCOM, Bengaluru, India) as illustrated in Figure 2. A zirconia ball of 10 mm in diameter was used. Zirconia has been used in the past in metallic parts for implant surgeries. Hence, it was chosen as a counter material for wear testing [31]. Before the testing procedure, the zirconia ball and specimens were subjected to ultrasonication for 20 min to remove any surface impurities. All the wear tests were done in a temperature-controlled environment (37 °C) using Ringer's solution as a medium. Ringer's solution was made by adding 9 g NaCl, 0.43 g KCl, 0.2 g $NaHCO_3$, and 0.24 g $CaCl_2$ to 1 L of distilled water followed by autoclaving at 121 °C for 15 min. The applied loads were set at 10, 15, and 20 N, respectively, based on past studies [18,26,32]. The maximum Hertzian contact pressures corresponding to applied loads were 682.9, 781.7, and 860.4 MPa, respectively. The track length was 10 mm, and the frequency was 2 Hz. Each run was limited to the 1800 s. The specimens were weighed pre- and post-testing in a microelectronic balance having 10^{-3} g accuracy [33]. The density was calculated with the help of Archimedes' principle.

Figure 2. The setup of the Tribometer system.

4. Results and Discussion

4.1. SEM and EDX Analyses

Ti, Nb, and Ta metal powders having irregular shapes are illustrated in Figure 3. SEM images delineate the pores in the fabricated alloy, which may either be due to the release of entrapped residual gases during manufacturing or to the partial evaporation of Ti from the alloy. Unmelted Ta can also be seen in the SEM images.

The EDX spectrum of the alloy in Figure 4 shows the peaks of Ti, Nb, and Ta in the manufactured alloy along with some Fe and Cu peaks in trace amounts. Fe and Cu do not affect the biomedical properties of the alloy. As a result, they can be ignored.

From EDX analysis, the proportion of each element obtained in the cladded layer is shown in Table 1.

Table 1. Proportion of each element in the laser cladded sample.

	Proportion (%)
Ti	75.58
Nb	15.97
Ta	5.25
Fe	2.68
Cu	0.53

Figure 3. *Cont.*

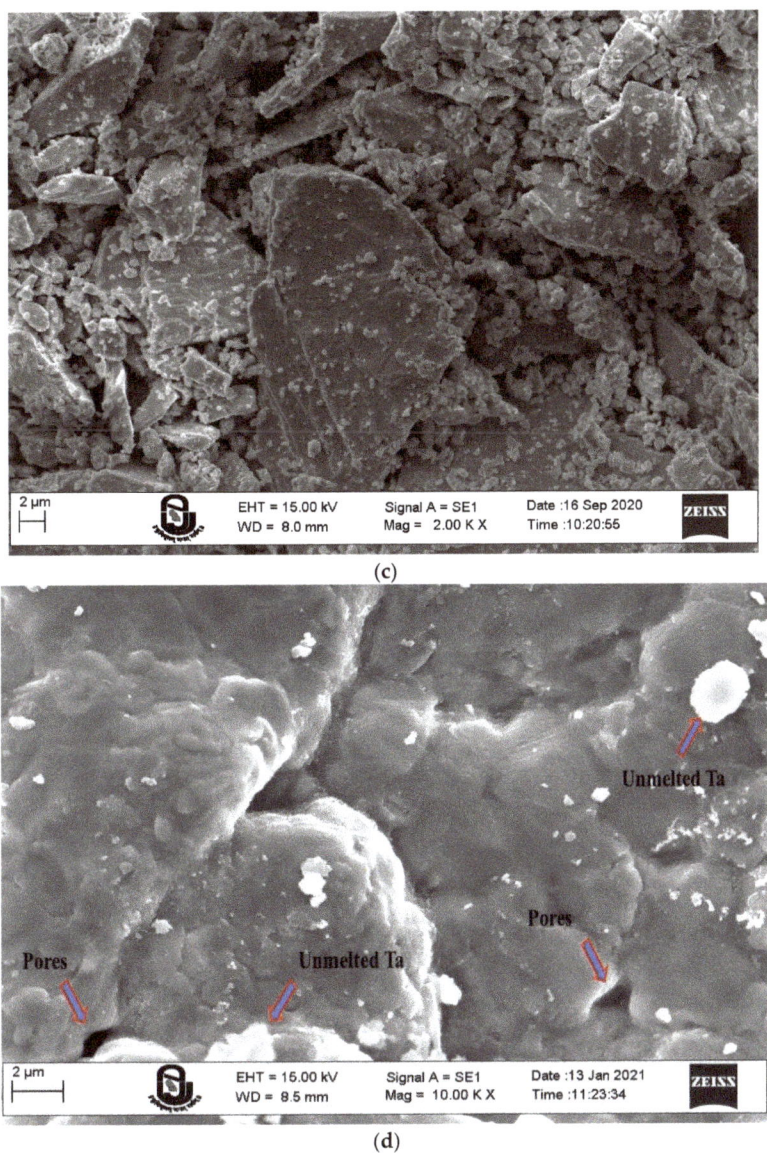

Figure 3. SEM images of (**a**) Ti, (**b**) Nb, (**c**) Ta, and (**d**) laser cladded alloy Ti-17Nb-6Ta.

The XRD spectrum of Ti-17Nb-6Ta is shown in Figure 5. The analysis was done using the X'pert Highscore and Origin Pro software. The majority of peaks belong to the BCC structure, which means that the manufactured alloy has a pure β-phase except for some small peaks which could be unrecognized due to contained impurities. The β-phase alloy possesses good wear characteristics in comparison to α or α + β alloys.

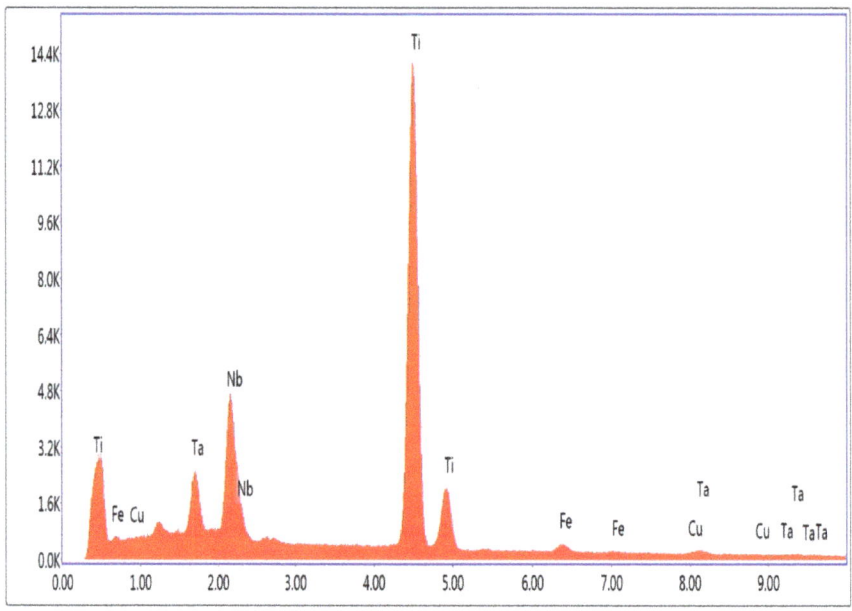

Figure 4. EDX spectrum of the alloy Ti-17Nb-6Ta.

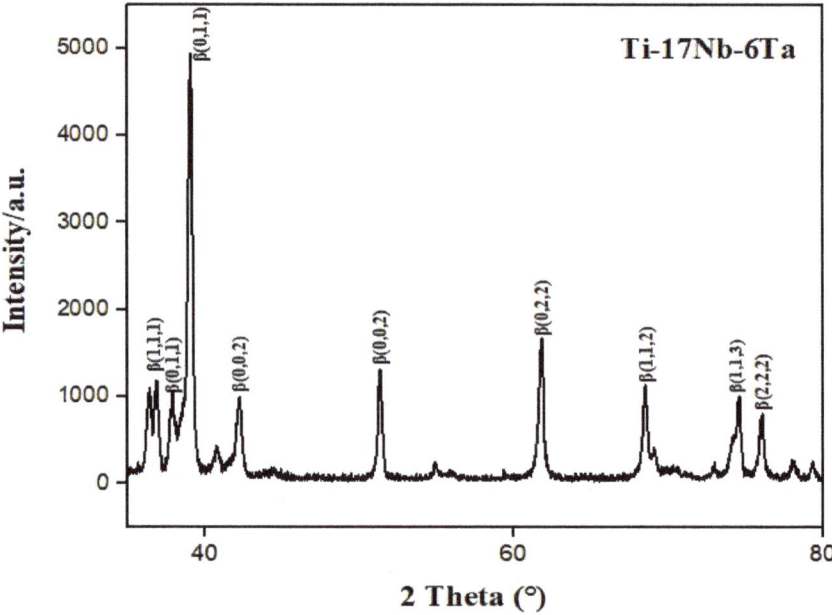

Figure 5. XRD spectrum of the alloy Ti-17Nb-6Ta.

4.2. Hardness Testing

The laser cladded Ti-17Nb-6Ta alloy showed a micro-hardness of 176 HV, which was higher than the pure Ti substrate, which showed a hardness of 165 HV.

4.3. Wear Testing

The specimens that were subjected to wear testing are demonstrated in Figure 6a–c along with the SEM images in Figure 6d,e. The worn specimens demonstrated scattered grooves in parallel lines, which were most likely caused by abrasion due to the zirconia ball in the tribometer. The hardness of the zirconia ball was sufficient to dislodge the asperities from the surface in the form of the wear debris at the time of reciprocating movement. These asperities drag against the opposite surface and create scratches. So, abrasive wear was the predominant wear phenomenon, concurrent with the findings reported earlier by Sukhpreet et al. [8].

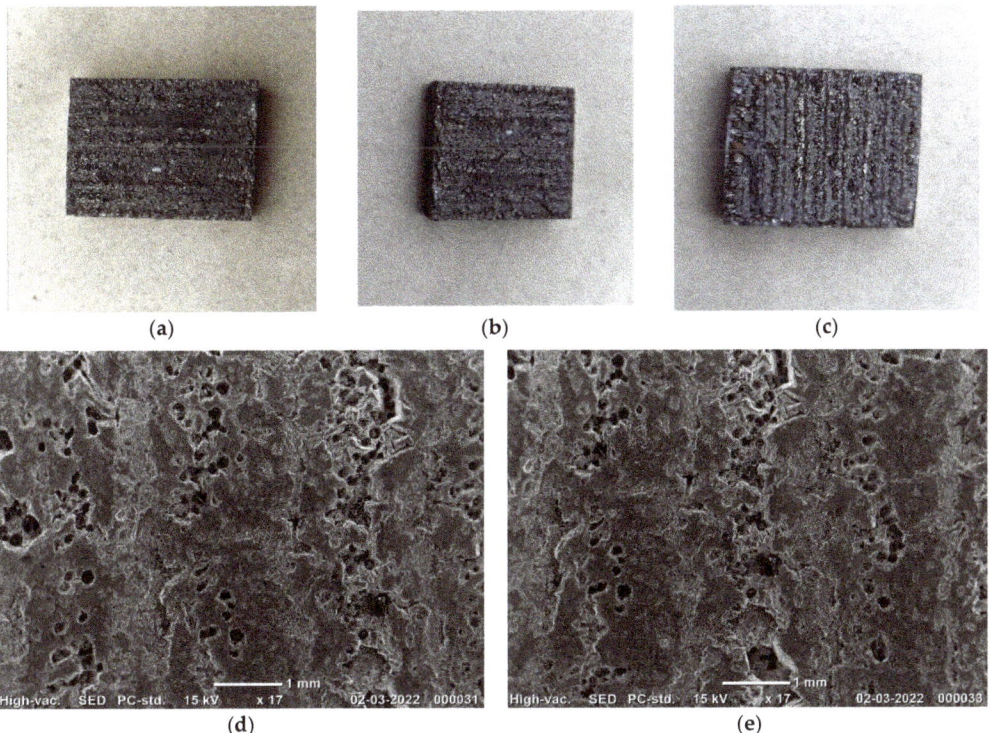

Figure 6. Wear tracks on (**a–c**) cladded specimens and (**d,e**) SEM images.

4.4. Coefficient of Friction (CoF)

The graphs illustrated in Figure 7 a,b depict the relationship of CoF vs. time and wear vs. time.

From the graphs, it can be seen that the value of CoF increased suddenly and reached a maximum of ~0.7. This may be due to the initial stages during which the zirconia ball starts rubbing against the cladded surface, followed by dislodging the surface asperities, thereby creating more friction at the interface. After 100 s, the CoF noticeably decreased for 15 and 20 N loads and settled around 0.15. For the 10 N loading, it was observed to fluctuate around 0.3. After 200 s, the steady-state phase is achieved, and very slight changes were observed except for the 10 N load, which still showed some resistance to achieving the steady-state. This can be attributed to lower loads on the ball generating frictional force, which was not enough to break off the material from the alloy specimens. Conversely, at higher loads, the friction was high enough to break off the upper layer of the specimen, and metallic debris was generated. Upon immersion in the Ringer's solution, the debris acted as a solid lubricant, thereby reducing the actual area of contact, ultimately responsible for the decrease in the CoF and hence the wear [17].

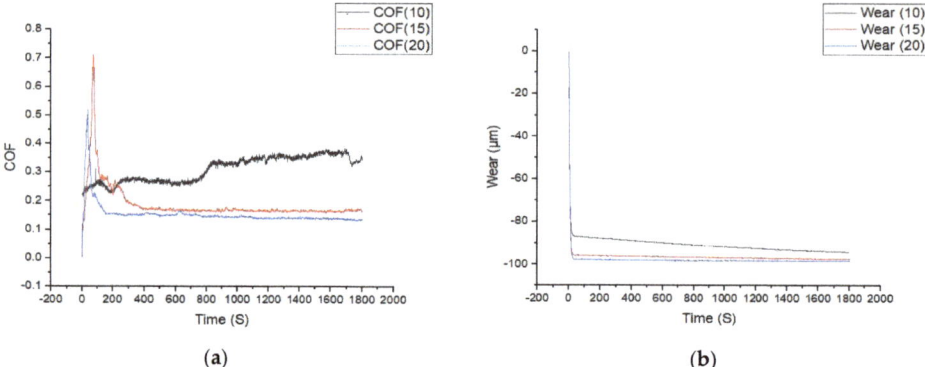

Figure 7. Graphs of (**a**) CoF vs. time and (**b**) wear vs. time.

4.5. Wear Rate

The wear rate was calculated from the following equation as described by the authors of [31]:

$$\omega = \frac{V}{N \times L} \quad (2)$$

where ω is the wear rate (mm³/Nm), N is the load (N), V is the wear volume (mm³), and L is the sliding distance (m). The value of V_{loss} was negligible in the evaluated alloy specimens. As per Archard's law, the volumetric loss of the material is inversely proportional to the hardness of the material. Materials with high hardness essentially have good wear resistance [25,33].

The calculated volume loss, wear rate, and CoF of the specimens are highlighted in Table 2.

Table 2. Volume loss (V_{loss}), wear rate, and CoF of Ti-17Nb-6Ta.

Load (N)	Initial Weight (g)	Final Weight (g)	Wloss (g)	Density (g/cm³)	Vloss (mm³)	Wear Rate (mm³/Nm)	CoF
10	8.791	8.657	0.134	5.179	25.854	0.035	0.312
15	8.793	8.740	0.053	5.179	10.214	0.009	0.155
20	8.785	8.760	0.025	5.179	4.808	0.003	0.189

5. Conclusions

This study describes the wear characteristics of the Ti alloy Ti-17Nb-6Ta. To the best of the authors' knowledge, using laser cladding to manufacture this alloy is the first time this has been performed. The following conclusions from this study are drawn:

(1) Ti-17Nb-6Ta alloy possesses the β phase attributable to the elements Nb and Ta, which was confirmed by the XRD analysis. The β phase is deemed essential for the alloy to be compatible with biomedical applications.
(2) Abrasive wear plays a significant role in the wear mechanism. The wear rate of the specimens with 15 N (0.0045 mm³/Nm) and 20 N load (0.007 mm³/Nm) was relatively lower in comparison to the specimens with 10 N load (0.035 mm³/Nm). The average wear rate is close to 0.016 mm³/Nm.
(3) CoF was higher during the initial stages due to the surface unevenness of the alloy surface, but after 100 s, a decreasing trend was observed. After 200 s of the run, the

CoF fluctuated in a narrow range. CoF for 10 N load (0.312) was higher in comparison to 15 N (0.155) and 20 N load (0.189). The average CoF is close to 0.22.

The wear results look advantageous from this characterization. The wear rate was close to what was reported in the earlier studies [25,27]. However, this was an effort only to check out the possibility of making the alloy by laser cladding and evaluate the wear behavior. Further characterizations are necessary to evaluate its full potential to be used as an artificial joint material in artificial hip and knee joint replacements.

Author Contributions: Conceptualization and methodology, R.S., S.P., S.K., S.S. and H.N.; writing—original draft preparation, R.S., S.P., S.K., S.S. and H.N.; writing—review and editing, R.S., S.P., S.K., S.S. and H.M.A.M.H. All authors have read and agreed to the published version of the manuscript.

Funding: This research received no external funding.

Institutional Review Board Statement: Not applicable.

Informed Consent Statement: Not applicable.

Data Availability Statement: Not applicable.

Acknowledgments: Authors duly acknowledge the facilities provided by CIF, IIT-BHU, Varanasi for conducting SEM, EDX, and XRD Testing. The authors gratefully acknowledge the contribution of R K Gautam for help in the wear testing.

Conflicts of Interest: The authors declare that they have no conflict of interest.

References

1. Niinomi, M. Mechanical Biocompatibilities of Titanium Alloys for Biomedical Applications. *J. Mech. Behav. Biomed. Mater.* **2008**, *1*, 30–42. [CrossRef] [PubMed]
2. Niinomi, M. Recent Research and Development in Titanium Alloys for Biomedical Applications and Healthcare Goods. *Sci. Technol. Adv. Mater.* **2003**, *4*, 445–454. [CrossRef]
3. Hussein, A.H.; Gepreel, M.A.H.; Gouda, M.K.; Hefnawy, A.M.; Kandil, S.H. Biocompatibility of New Ti-Nb-Ta Base Alloys. *Mater. Sci. Eng. C* **2016**, *61*, 574–578. [CrossRef] [PubMed]
4. Niinomi, M. Biologically and Mechanically Biocompatible Titanium Alloys. *Mater. Trans.* **2008**, *49*, 2170–2178. [CrossRef]
5. Liu, J.; Ruan, J.; Chang, L.; Yang, H.; Ruan, W. Porous Nb-Ti-Ta Alloy Scaffolds for Bone Tissue Engineering: Fabrication, Mechanical Properties and in Vitro/Vivo Biocompatibility. *Mater. Sci. Eng. C* **2017**, *78*, 503–512. [CrossRef]
6. Liu, J.; Yang, Q.; Yin, J.; Yang, H. Effects of Alloying Elements and Annealing Treatment on the Microstructure and Mechanical Properties of Nb-Ta-Ti Alloys Fabricated by Partial Diffusion for Biomedical Applications. *Mater. Sci. Eng. C* **2020**, *110*, 110542. [CrossRef]
7. Chen, Y.H.; Chuang, W.S.; Huang, J.C.; Wang, X.; Chou, H.S.; Lai, Y.J.; Lin, P.H. On the Bio-Corrosion and Biocompatibility of TiTaNb Medium Entropy Alloy Films. *Appl. Surf. Sci.* **2020**, *508*, 145307. [CrossRef]
8. Kaur, S.; Ghadirinejad, K.; Oskouei, R.H. An Overview on the Tribological Performance of Titanium Alloys with Surface Modifications for Biomedical Applications. *Lubricants* **2019**, *7*, 65. [CrossRef]
9. Kuroda, D.; Niinomi, M.; Morinaga, M.; Kato, Y.; Yashiro, T. Design and Mechanical Properties of New β Type Titanium Alloys for Implant Materials. *Mater. Sci. Eng. A* **1998**, *243*, 244–249. [CrossRef]
10. Ran, J.; Jiang, F.; Sun, X.; Chen, Z.; Tian, C.; Zhao, H. Microstructure and Mechanical Properties of Ti-6al-4v Fabricated by Electron Beam Melting. *Crystals* **2020**, *10*, 972. [CrossRef]
11. Sumitomo, N.; Noritake, K.; Hattori, T.; Morikawa, K.; Niwa, S.; Sato, K.; Niinomi, M. Experiment Study on Fracture Fixation with Low Rigidity Titanium Alloy: Plate Fixation of Tibia Fracture Model in Rabbit. *J. Mater. Sci. Mater. Med.* **2008**, *19*, 1581–1586. [CrossRef] [PubMed]
12. Choe, H.C. Nanotubular Surface and Morphology of Ti-Binary and Ti-Ternary Alloys for Biocompatibility. *Thin Solid Films* **2011**, *519*, 4652–4657. [CrossRef]
13. Geetha, M.; Singh, A.K.; Asokamani, R.; Gogia, A.K. Ti Based Biomaterials, the Ultimate Choice for Orthopaedic Implants—A Review. *Prog. Mater. Sci.* **2009**, *54*, 397–425. [CrossRef]
14. Manivasagam, G.; Dhinasekaran, D.; Rajamanickam, A. Biomedical Implants: Corrosion and Its Prevention—A Review. *Recent Patents Corros. Sci.* **2010**, *2*, 40–54. [CrossRef]
15. Zhou, Y.L.; Niinomi, M.; Akahori, T.; Fukui, H.; Toda, H. Corrosion Resistance and Biocompatibility of Ti-Ta Alloys for Biomedical Applications. *Mater. Sci. Eng. A* **2005**, *398*, 28–36. [CrossRef]
16. Wei, T.Y.; Huang, J.C.; Chao, C.Y.; Wei, L.L.; Tsai, M.T.; Chen, Y.H. Microstructure and Elastic Modulus Evolution of TiTaNb Alloys. *J. Mech. Behav. Biomed. Mater.* **2018**, *86*, 224–231. [CrossRef]

17. Lee, Y.S.; Niinomi, M.; Nakai, M.; Narita, K.; Cho, K. Differences in Wear Behaviors at Sliding Contacts for β-Type and (α + β)-Type Titanium Alloys in Ringer's Solution and Air. *Mater. Trans.* **2015**, *56*, 317–326. [CrossRef]
18. Niinomi, M.; Nakai, M.; Akahori, T. Frictional Wear Characteristics of Biomedical Ti-29Nb-13Ta-4.6Zr Alloy with Various Microstructures in Air and Simulated Body Fluid. *Biomed. Mater.* **2007**, *2*, S167. [CrossRef]
19. Cai, Z.B.; Zhang, G.A.; Zhu, Y.K.; Shen, M.X.; Wang, L.P.; Zhu, M.H. Torsional Fretting Wear of a Biomedical Ti6Al7Nb Alloy for Nitrogen Ion Implantation in Bovine Serum. *Tribol. Int.* **2013**, *59*, 312–320. [CrossRef]
20. Marin, E.; Offoiach, R.; Regis, M.; Fusi, S.; Lanzutti, A.; Fedrizzi, L. Diffusive Thermal Treatments Combined with PVD Coatings for Tribological Protection of Titanium Alloys. *Mater. Des.* **2016**, *89*, 314–322. [CrossRef]
21. Seth, S.; Jones, A.H.; Lewis, O.D. Wear Resistance Performance of Thermally Sprayed Al3Ti Alloy Measured by Three Body Micro-Scale Abrasive Wear Test. *Wear* **2013**, *302*, 972–980. [CrossRef]
22. De Souza, T.M.; Leite, N.F.; Trava-Airoldi, V.J.; Corat, E.J. Studies on CVD-Diamond on Ti6A14V Alloy Surface Using Hot Filament Assisted Technique. *Thin Solid Films* **1997**, *308–309*, 254–257. [CrossRef]
23. Ramos-Saenz, C.R.; Sundaram, P.A.; Diffoot-Carlo, N. Tribological Properties of Ti-Based Alloys in a Simulated Bone-Implant Interface with Ringer's Solution at Fretting Contacts. *J. Mech. Behav. Biomed. Mater.* **2010**, *3*, 549–558. [CrossRef]
24. Barfeie, A.; Wilson, J.; Rees, J. Implant Surface Characteristics and Their Effect on Osseointegration. *Br. Dent. J.* **2015**, *218*, E9. [CrossRef] [PubMed]
25. Khan, M.M.; Nisar, M. Effect of in Situ TiC Reinforcement and Applied Load on the High-Stress Abrasive Wear Behaviour of Zinc–Aluminum Alloy. *Wear* **2022**, *488–489*, 204082. [CrossRef]
26. Jing, Z.; Cao, Q.; Jun, H. Corrosion, Wear and Biocompatibility of Hydroxyapatite Bio-Functionally Graded Coating on Titanium Alloy Surface Prepared by Laser Cladding. *Ceram. Int.* **2021**, *47*, 24641–24651. [CrossRef]
27. Dittrick, S.; Balla, V.K.; Bose, S.; Bandyopadhyay, A. Wear Performance of Laser Processed Tantalum Coatings. *Mater. Sci. Eng. C* **2011**, *31*, 1832–1835. [CrossRef] [PubMed]
28. Amado, J.M.; Rodríguez, A.; Montero, J.N.; Tobar, M.J.; Yáñez, A. A Comparison of Laser Deposition of Commercially Pure Titanium Using Gas Atomized or Ti Sponge Powders. *Surf. Coat. Technol.* **2019**, *374*, 253–263. [CrossRef]
29. Dobbelstein, H.; Gurevich, E.L.; George, E.P.; Ostendorf, A.; Laplanche, G. Laser Metal Deposition of Compositionally Graded TiZrNbTa Refractory High-Entropy Alloys Using Elemental Powder Blends. *Addit. Manuf.* **2019**, *25*, 252–262. [CrossRef]
30. Bhardwaj, T.; Shukla, M.; Paul, C.P.; Bindra, K.S. Direct Energy Deposition-Laser Additive Manufacturing of Titanium-Molybdenum Alloy: Parametric Studies, Microstructure and Mechanical Properties. *J. Alloys Compd.* **2019**, *787*, 1238–1248. [CrossRef]
31. Singh, R.; Kurella, A.; Dahotre, N.B. Laser Surface Modification of Ti-6Al-4V: Wear and Corrosion Characterization in Simulated Biofluid. *J. Biomater. Appl.* **2006**, *21*, 49–73. [CrossRef] [PubMed]
32. Kang, S.; Tu, W.; Han, J.; Li, Z.; Cheng, Y. A Significant Improvement of the Wear Resistance of Ti6Al4V Alloy by a Combined Method of Magnetron Sputtering and Plasma Electrolytic Oxidation (PEO). *Surf. Coatings Technol.* **2019**, *358*, 879–890. [CrossRef]
33. Fellah, M.; Labaïz, M.; Assala, O.; Dekhil, L.; Taleb, A.; Rezag, H.; Iost, A. Tribological Behavior of Ti-6Al-4V and Ti-6Al-7Nb Alloys for Total Hip Prosthesis. *Adv. Tribol.* **2014**, *2014*, 451387. [CrossRef]

Article

Effect of Pre-Oxidation on a Ti PVD Coated Ferritic Steel Substrate during High-Temperature Aging

Maria-Rosa Ardigo-Besnard [1,*], Aurélien Besnard [2], Galy Nkou Bouala [2], Pascal Boulet [3], Yoann Pinot [2] and Quentin Ostorero [1]

[1] Laboratoire Interdisciplinaire Carnot de Bourgogne (ICB), UMR 6303 CNRS, Université Bourgogne Franche-Comté, BP 47870, 21078 Dijon, France
[2] Arts et Metiers Institute of Technology, LaBoMaP, Université Bourgogne Franche-Comté, HESAM Université, 71250 Cluny, France
[3] Institut Jean Lamour, UMR CNRS 7198, Université de Lorraine, 54000 Nancy, France
* Correspondence: maria-rosa.ardigo-besnard@u-bourgogne.fr; Tel.: +33-0-380396016

Citation: Ardigo-Besnard, M.-R.; Besnard, A.; Nkou Bouala, G.; Boulet, P.; Pinot, Y.; Ostorero, Q. Effect of Pre-Oxidation on a Ti PVD Coated Ferritic Steel Substrate during High-Temperature Aging. *Crystals* **2022**, *12*, 1732. https://doi.org/10.3390/cryst12121732

Academic Editors: Yang Zhang and Yuqiang Chen

Received: 11 November 2022
Accepted: 25 November 2022
Published: 1 December 2022

Publisher's Note: MDPI stays neutral with regard to jurisdictional claims in published maps and institutional affiliations.

Copyright: © 2022 by the authors. Licensee MDPI, Basel, Switzerland. This article is an open access article distributed under the terms and conditions of the Creative Commons Attribution (CC BY) license (https://creativecommons.org/licenses/by/4.0/).

Abstract: A PVD coating is often applied on the surface of metallic alloys to improve their high-temperature resistance. In the present work, a thin titanium layer (1.2 µm) was deposited by PVD on the surface of a stainless steel substrate before high-temperature exposure (800 °C in ambient air). The underlying idea is that metallic Ti converts into Ti oxide (TiO_2) during high-temperature aging at 800 °C, thereby slowing down the substrate oxidation. The stability of the coating with and without substrate pre-oxidation was investigated. Morphological, structural, and chemical characterizations were performed and completed by simulation of the film growth and measurement of the mechanical state of the film and the substrate. In the case of the sample that was not pre-oxidized, the oxidation of the steel was slowed down by the TiO_2 scale but spallation was observed. On the other hand, when the steel was pre-oxidized, TiO_2 provided more significant protection against high-temperature oxidation, and spalling or cracking did not occur. A combination of different kinds of stress could explain the two different behaviors, namely, the mechanical state of the film and the substrate before oxidation, the growing stress, and the thermal stress occurring during cooling down.

Keywords: physical vapor deposition; pre-oxidation; coating adhesion; titanium; ferritic stainless steel

1. Introduction

The application of a coating on the surface of metallic alloys, particularly stainless steels, can improve some properties, such as high-temperature (800–1000 °C) resistance [1–3]. The aim is to limit inward oxygen diffusion and outward metallic cation (mainly chromium and iron) migration. As a consequence, the coating should inhibit the growth rate of the protective Cr_2O_3 layer and avoid the formation of non-protective Fe oxides. It has also been reported in the literature that the presence of a coating can increase the adherence of the oxide scale [4]. Over the years, several kinds of materials have been developed and used as coatings for high-temperature applications [5], such as perovskites and spinel-type oxides, reactive element oxides, and MAlCrYO systems (where M represents a metal, typically Co, Mn, and/or Ti). Coatings can be deposited on stainless steels by different techniques, including magnetron sputtering [6], screen printing [3], sol–gel [7], chemical vapor deposition (CVD) [8], plasma spraying [9], electrodeposition [10], pulsed laser deposition [11], and large area filtered arc deposition [12]. Pre-oxidation is often performed on bare alloys before DC magnetron sputtering process [13–17]. The results indicate that irrespective of the chemical nature of the deposited coating and the aging conditions (atmosphere, time, and temperature), pre-oxidation is effective in inhibiting diffusion of Cr and Fe from the substrate into the coating and in decreasing the kinetic growth of the Cr_2O_3 scale during high-temperature aging. For example, Hoyt et al. [13] performed

PVD Co$_{1.5}$Mn$_{1.5}$O$_4$ coatings on commercial ferritic stainless steel 441HP samples, with pre-oxidized beforehand or not, prior to exposure at 800 °C in laboratory air. They showed that in as little as 3 h of pre-oxidation at 800 °C in laboratory air, Fe transport from the substrate toward the coating was inhibited, and the thickness of the coating did not increase during exposure at 800 °C for 100 h or less. Zhao et al. [16] studied the high-temperature behavior of the ferritic stainless steel SUS 430 coated by NiFe$_2$. The coatings were performed by magnetron sputtering on bare and pre-oxidized (100 h at 800 °C in air) steel samples. All samples were then exposed to air at 800 °C for 15 weeks. The authors observed that the oxidation resistance of the coated pre-oxidized samples was improved and that pre-oxidation reduced the diffusion of Cr from the substrate to the outer layer. More generally, it has been shown in the literature that pre-oxidation of uncoated ferritic stainless steel also has a beneficial effect on the reduction of the oxidation rate during thermal aging. Talic et al. [18] demonstrated that pre-oxidation of the ferritic stainless steel Crofer 22APU under different atmospheres (air or N$_2$-H$_2$) reduced the oxidation rate of the alloy at 800 °C in air for long exposure times (around 1000 h). Even in a dual atmosphere environment (air on one side of the sample and hydrogen or another fuel on the other side), pre-oxidation contributed to improve oxidation resistance. Goebel et al. [19] investigated the behavior of the ferritic stainless steel AISI 441. Pre-oxidation of the alloy was performed in air at 800 °C for different times (from 10 to 280 min) before discontinuous exposure to a dual atmosphere at 600 °C for 1000 h. The authors found that in dual atmosphere conditions, oxidation resistance was enhanced by longer pre-oxidation times. All these results clearly suggest that alloy pre-oxidation, followed or not by subsequent application of a coating, affects the high-temperature oxidation resistance. However, in these works, the effects of substrate pre-oxidation, especially when it is coupled with a specific coating, were not reported or not completely understood and therefore deserve to be further addressed. This is the aim of the present study.

In the present study, a thin titanium layer (1.2 μm) was deposited on the surface of ferritic stainless steel substrates by the physical vapor deposition (PVD) technique before high-temperature exposure (800 °C in ambient air). Some of the advantages of the PVD technique are fine control of the film thickness and uniform covering of the surfaces, even in the case of complex shapes [20]. Large surfaces can be coated, making this technique appropriate for industrial-scale use. In the present work, the idea is that metallic Ti converts into Ti oxide during high-temperature aging at 800 °C, thereby limiting further substrate oxidation. It has been reported in the literature that the oxidation of Ti under pure oxygen follows a parabolic law between approximately 600 and 800 °C [21]. Under air, the oxidation behavior of Ti seems to be improved, likely due to the presence of impurities (like nitrogen) in the oxidant atmosphere, and the parabolic regime starts at lower temperatures compared to a pure oxygen atmosphere [22,23]. Ti has a strong chemical affinity with oxygen. According to the Ellingham–Richardson diagram [24], the oxygen partial pressure needed for TiO$_2$ formation at 298 K is 10^{-156} bar, which is lower compared to the oxygen partial pressure of 10^{-123} bar required for Cr$_2$O$_3$ formation. Moreover, the mobility of titanium in iron is extremely low [25,26], thus reducing the risk of contamination of the Fe-based substrate. Finally, in the present study, the interest in choosing Ti as a coating is that it is a minor alloying element, thus making it possible to distinguish the contribution of the Ti deposited by PVD from the Ti contained in the alloy to oxide scale formation during high-temperature exposure. The effect of short pre-oxidation in the same conditions as the final oxidation (1 h under ambient air at 800 °C) performed before the application of the coating was studied. This ensured that only the thickness of the oxide scale was smaller, with the nature of the oxide remaining the same. The stability of the coating with and without substrate pre-oxidation was also investigated based on the simulation of the film growth and on the mechanical state of the substrate and the film. In addition, the different oxidation behaviors are presented and discussed as well as the possible oxide scale growth mechanisms. When the coating was deposited on the substrate that was not pre-oxidized, the TiO$_2$ formed from the Ti film slowed down the oxidation of the steel but suffered severe spallation. On the contrary, when pre-oxidation of the substrate was performed before the application of the coating, the TiO_2 scale formed from the Ti coating provided more effective protection against oxidation compared to the steel that was not pre-oxidized, and spalling or cracking was not observed. In the present work, it was found that substrate pre-oxidation promoted the stability of the coating during subsequent high-temperature oxidation. To explain this effect, particular attention was given to the mechanical state of the film and the substrate before long-term oxidation. The different behaviors observed between the substrates that were pre-oxidized or not can be explained as the result of the association of residual stress, growing stress developed during isothermal treatment, and thermal stress that occurred during cooling down.

2. Materials and Methods

The alloy chosen and tested in this study was a commercial ferritic stainless steel (type AISI 441) with the following composition (wt.%): Fe: 18, Cr: 0.58, Si: 0.25, Mn: 0.64 Ti + Nb. Samples with a size of 10 mm × 10 mm and thickness of 1.5 mm were polished using SiC papers (down to 1200 grit) and diamond paste (down to 3 µm) and then rinsed in an ultrasonic bath with ethanol.

Ti coatings were performed in an industrial DC magnetron PVD system KS40V (Kenosistec, Binasco, Italy). The substrate holder oscillated in front of the titanium target (99.95 at.% purity, 406 × 127 × 6 mm in dimension) with a solid angle of 30° and a minimal target-to-substrate distance of 95 mm. The nominal speed was 0.7 rpm, and 80 sccm of pure argon was injected to obtain a working pressure of 0.43 Pa. An electrical power of 1500 W was applied to the target (U = 340 V, I = 4.41 A) during the deposition. The deposition length was 22 min, corresponding to 160 scans (a scan is a half oscillation). A coating thickness of about 1.2 µm (±0.1) was obtained.

Pre-oxidation and oxidation tests were carried out at 800 °C for 1 and 100 h, respectively, under ambient air in a horizontal furnace (Carbolite, Carbolite Gero, Neuhausen, Germany). The experimental approach of the study is represented in Figure 1.

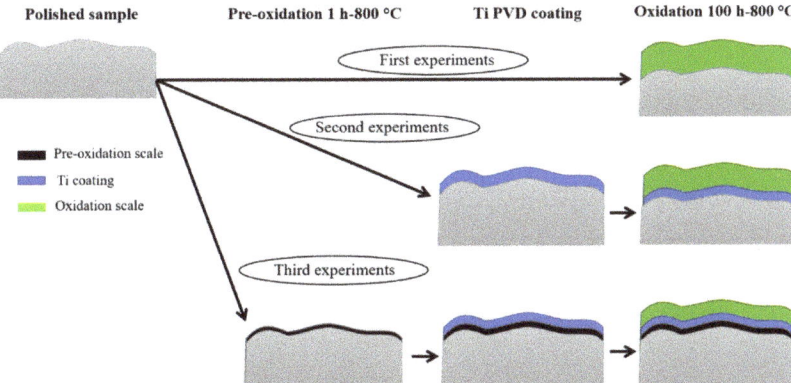

Figure 1. Schematic approach of the experimental study.

In the first experiment, the polished steel sample was directly oxidized. In the second experiment, a Ti coating was applied before oxidation. In the third experiment, pre-oxidation was performed before the application of the coating, and the coated pre-oxidized sample was then oxidized.

The sample's surfaces and cross sections were characterized by a scanning electron microscope equipped with a field emission gun (SEM-FEG JSM-7600F, JEOL, Tokyo, Japan) and coupled with an energy-dispersive X-ray spectrometer (EDX). Chemical phase identification was performed by X-ray diffraction (XRD) using a Bruker D8-A25 diffractometer (Bruker-AXS, Karlsruhe, Germany)) with Cu Kα (λ = 0.154056 nm).

The mechanical state of the coated substrates with and without 1 h of pre-oxidation at 800 °C under ambient air was evaluated by the $\sin^2\psi$ method [27]. A D8 Discover Bruker diffractometer with Co Kα radiation (λ_{Co} = 1.79026 Å) was used. The measurements were performed at the Bragg angle 2θ = 99.7°, corresponding to the (2,1,1) peak of the ferritic substrate, in the direction ϕ = 90° at 14 ψ–positions, from ψ = −71.57° to ψ = 71.57°. Assuming a biaxial stress state, residual stress corresponds to the slope of the regression line obtained by plotting the deformation $\varepsilon_{\phi\psi}$ versus $\sin^2\psi$, which is equal to ((1 + ν)/E) σ_ϕ. The Young's modulus (E) and the Poisson ratio (ν) used for the calculations were 220,264 MPa and 0.28, respectively.

The film's residual stress was obtained using the curvature method and the Stoney formula [28] on a silicon (100) substrate. The substrate thickness, Young's modulus (E), and Poisson ratio (ν) used for the calculations were 380 μm, 130,000 MPa, and 0.28, respectively.

The characteristic of the Ti flux was calculated with SiMTRA [29] using the experimental working pressure and system geometry. The initial angular and energy distribution was obtained by SRIM [30] using the experimental ion energy. First, 10^5 and 5×10^8 particles were simulated in SRIM and SiMTRA, respectively. The film growth was then simulated with NASCAM (v4.8.1, UNamur, Namur, Belgium) [31]. The profile of the substrate was extracted from SEM images and converted into a NASCAM substrate using the plug-in "Make substrate" with dimensions of 5 × 1241 particles. In these simulations, a particle represents a volume with a side length of 10 nm. A total of 8×10^5 particles were deposited. Quantitative pore analyses were performed using the plug-in "Porosity" with a probe size of 2.

3. Results and Discussion

3.1. Oxidation of Polished Sample

Figure 2 shows the SEM micrographs of the surface of the polished sample after 100 h at 800 °C in ambient air. The mirror-polished surface before high-temperature oxidation is displayed in the inset of Figure 2a for comparison. The oxide scale was homogeneous with two kinds of crystals: big pyramidal grains and platelets covering a continuous lower layer formed by small grains. EDX analysis revealed that platelets and small grains were enriched in Cr, while pyramidal grains contained high amounts of Cr and Mn.

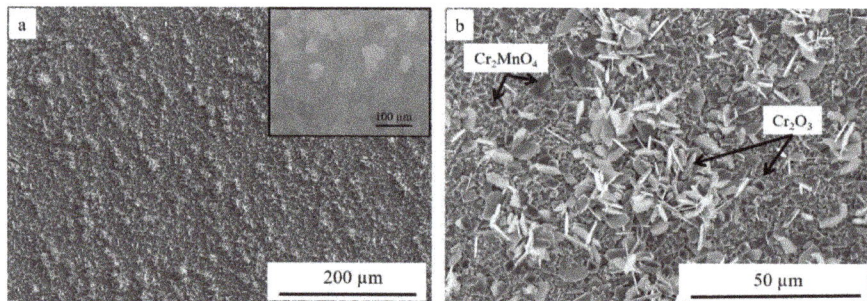

Figure 2. Secondary electron images of the surface of the polished sample after 100 h aging at 800 °C under ambient air: (**a**) global view and (**b**) magnification. Inset in (**a**): global view of the mirror-polished surface before high-temperature oxidation.

XRD analyses performed in Bragg–Brentano conditions (see Supplementary Material S1) confirmed the formation of chromia Cr_2O_3, corresponding to platelets and small grains covering the substrate, and a Cr_2MnO_4 spinel phase, corresponding to the pyramidal crystals. The substrate was also detected. This result is in agreement with the literature, where it has been reported that chromia and Cr-Mn spinel oxides are the typical phases that form at the surface of ferritic steels after aging at 800 °C under air [32–34]. As revealed by the backscattered image of the cross section of the oxidized sample, the thickness of the oxide layer ranged from about 4 to almost 8 µm considering the top of the platelets and the pyramidal grains (Figure 3). Laves phases enriched in Nb and Si decorated the grain boundaries [32].

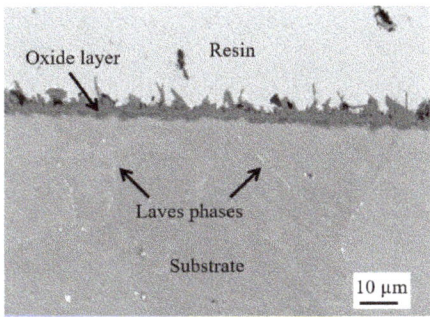

Figure 3. Backscattered electron image of the cross section of the polished sample after 100 h aging at 800 °C under laboratory air.

3.2. Oxidation of Ti-Coated Sample

The morphology of the as-deposited Ti PVD coating (surface and cross section) is presented in Figure 4.

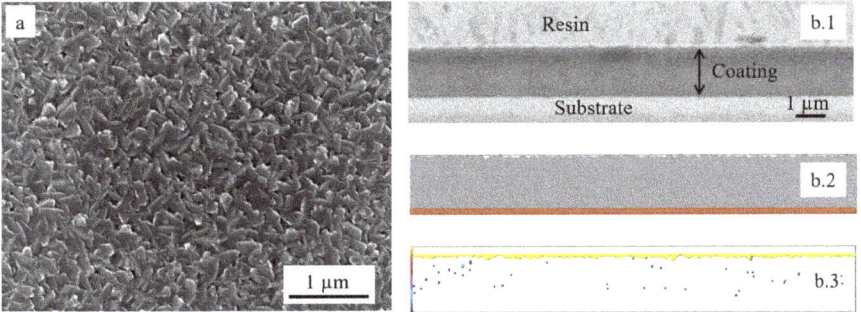

Figure 4. Morphologies of the as-deposited Ti PVD coating: (**a**) surface and (**b.1**) cross section, (**b.2**) simulated film growth using NASCAM, and (**b.3**) porosity of the simulated film.

The coating had a columnar-type structure (Figure 4a). The SEM cross section observation (Figure 4b.1) indicated that the coating was dense, well adherent to the substrate, and had a homogeneous thickness of about 1.2 µm. XRD analyses performed in Bragg–Brentano conditions (see Supplementary Material S2) confirm that the film was only composed of the hexagonal alpha Ti phase with a (002) texture. According to calculation using the Laue–Sherrer's method, the coherent domain size was 16.5 ± 0.5 nm. EDX analyses revealed that the film was composed of Ti with an expected oxygen contamination of about 5 to 7 at.%. Figure 4b.2 shows the simulated film cross section (where gray is the film and brown is the substrate), which confirmed the SEM observations. The coating was dense and smooth and had a homogeneous thickness. Figure 4b.3 presents the coating porosity, where white represents the dense matter (film and substrate), yellow represents the air-connected pores (here, only the roughness), and blue represents the occluded pores. These pores were randomly distributed in the film and represented about 0.1% of the volume. The roughness of the bare substrate was about 13 nm, while the roughness after coating was slightly increased to 19 nm.

After 100 h of aging at 800 °C under ambient air, a significant surface spalling could be observed (Figure 5a).

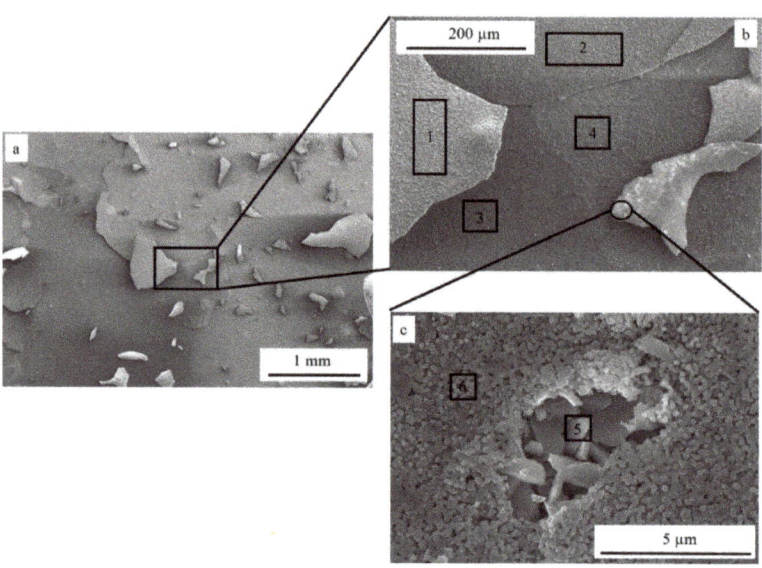

Figure 5. Secondary electron images of the surface of the Ti-coated sample after 100 h aging at 800 °C under laboratory air: (**a**) global view, (**b**) magnification of a spalled zone, and (**c**) magnification of spalled TiO_2 scale upside down.

EDX analyses (Figure 5b and Table 1) revealed that the spalled layer was the initial Ti coating transformed into TiO_2 during aging (points 1 and 2). Points 3 and 4 corresponded to areas located beneath the TiO_2 layer. These zones contained oxides formed by substrate elements and a particularly high amount of Si. Figure 5c shows a spalled TiO_2 scale upside down. Some zones of wrenching could be observed. Inside this area, big grains with a morphology similar to Cr_2O_3 platelets formed at the surface of the uncoated sample after 100 h of oxidation at 800 °C (Figure 2b) could be distinguished and was confirmed by EDX characterizations (point 5). Finally, small grains surrounding the big Cr_2O_3 platelets presented a stoichiometry close to $(Cr, Ti)_2O_3$ (point 6). The lack of chromia platelets and spinel crystals at the surface of the steel was an indication that the spalling occurred at the end of the oxidation step or during the cooling.

Table 1. EDX results (at.%) of areas indicated in Figure 5b,c.

Element	O	Si	Ti	Cr	Mn	Fe	Nb
Point 1	64.7	0.1	32.1	2.6	0.3	0.2	-
Point 2	60.7	-	33.5	4.9	0.4	0.5	-
Point 3	31.1	14.5	0.6	9.6	-	43.6	0.6
Point 4	27.1	6.5	0.5	17.4	0.5	48.0	-
Point 5	61.3	-	1.6	35.1	-	2.0	-
Point 6	58.9	0.5	7.3	25.3	-	1.8	6.2

The backscattered electron micrograph of the cross section taken in an area where the coating did not spall is displayed in Figure 6, together with the EDX elementary maps. The results confirmed the surface observations that a Cr_2O_3 layer had formed beneath TiO_2. The thicknesses of both scales were almost the same of between 2 and 4 µm. As a comparison, in the case of the uncoated steel, Cr_2O_3 thickness ranged between 4 and 8 µm. On the other hand, the initial Ti coating was 1.2 µm thick, almost 2–3 times lower than the thickness measured for TiO_2. This can be partially, but not only, explained by the presence of O atoms in the lattice, which were responsible for a volume expansion that led to an increased thickness when passing from metallic Ti to the oxide phase. A very small amount of Cr and Fe seemed to have diffused toward the external coating, likely forming a mixed Ti-Cr-Fe oxide between the Cr_2O_3 and TiO_2 layers. Mn signal was slightly more pronounced inside the chromia layer, but elementary maps did not show evidence of the formation of Cr-Mn spinel oxides as in the case of the uncoated sample. Moreover, O and Si enrichment, likely SiO_2, could be observed at the substrate/oxide scale interface as well as inside the substrate. The formation of this phase is thermodynamically possible as SiO_2 is stable at low oxygen partial pressure (10^{-33} bar at 800 °C [24]). However, as reported by other authors [5], due to the non-miscibility between Cr_2O_3 and SiO_2, spalling between these two phases can occur, especially during long-term oxidation.

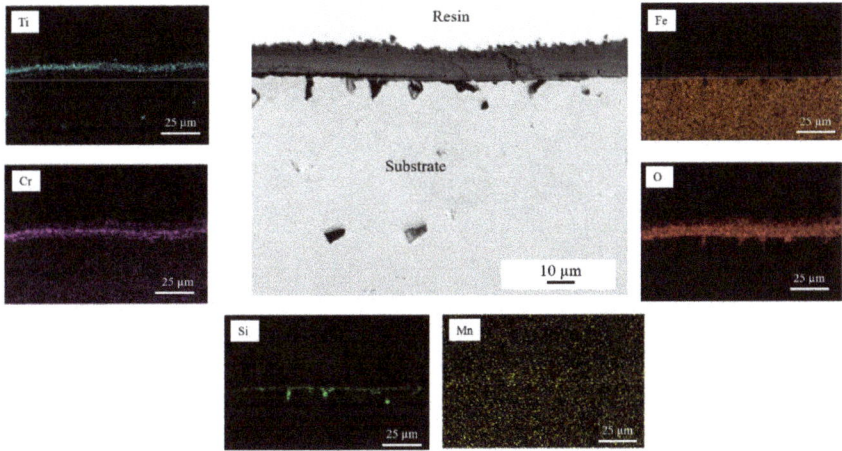

Figure 6. Backscattered electron image and EDX elementary maps of the cross section of the Ti-coated sample after 100 h aging at 800 °C under ambient air.

These observations suggest that three zones could be distinguished inside the oxide scale (Figure 7): an inner Cr_2O_3 layer, an outer TiO_2 oxide, and a transition zone between them, probably a mixed Cr-Fe-Ti oxide (Figure 7). It could be noticed that the TiO_2 scale appeared quite dense, and some porosities were present at the interface between TiO_2 and the mixed Cr-Fe-Ti oxide.

If it is assumed that TiO_2 represents the starting interface, it is possible to claim that Cr_2O_3 is formed by inward oxygen diffusion and the mixed Cr-Fe-Ti by mixed inward oxygen and outward Cr-Fe diffusion through the Cr_2O_3 oxide layer.

3.3. Oxidation of Ti-Coated Sample after Pre-Oxidation

3.3.1. Ti Deposition after Pre-Oxidation

The polished steel substrate was pre-oxidized for 1 h at 800 °C in ambient air. Figure 8a shows that the surface was formed by a layer of small grains enriched in chromium and that big pyramidal crystals containing Cr and Mn, previously identified as Cr-Mn spinel oxides on the surface of the polished oxidized steel, were dispersed on the surface. The cross-sectional observation (Figure 8b) revealed that the oxide layer was very thin (a few hundred nm) and continuous. XRD performed at an incidence angle of 2° (Supplementary Material S3) confirmed that the oxide scale consisted of a chromium oxide Cr_2O_3. The Cr-Mn spinel phase was not detected. The crystals had quite a big size but were too dispersed.

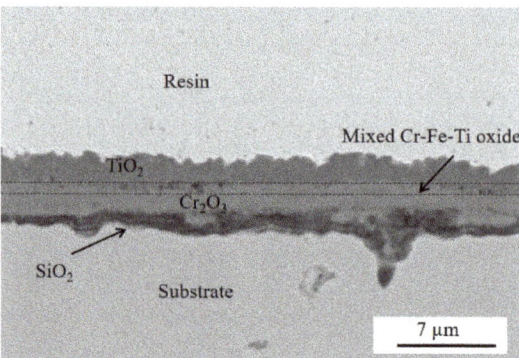

Figure 7. Magnification of the oxide layer formed in the case of the Ti-coated sample after 100 h aging at 800 °C under laboratory air.

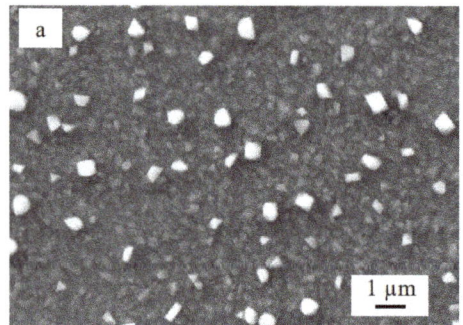 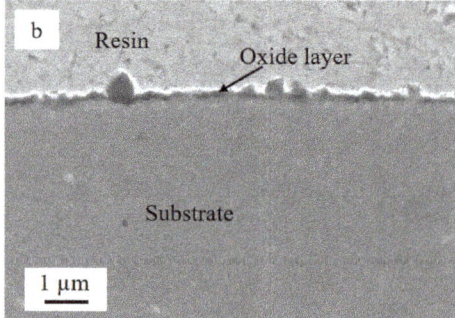

Figure 8. Backscattered electron image of the steel substrate after 1 h of pre-oxidation at 800 °C in ambient air: (**a**) surface and (**b**) cross section.

Ti film was then deposited on the pre-oxidized steel substrate. The morphology of the coating (surface and cross section) is displayed in Figure 9.

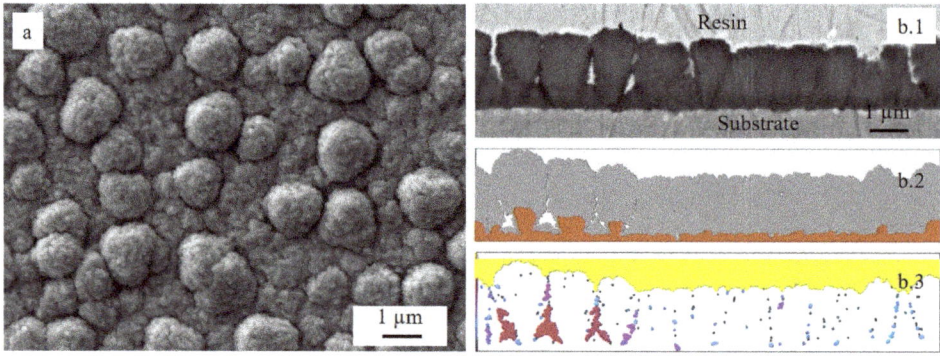

Figure 9. Morphology of the Ti PVD coating deposited on the pre-oxidized steel: (**a**) surface and (**b.1**) cross section, (**b.2**) simulated film growth using NASCAM, and (**b.3**) porosity of the simulated film.

The columnar-type structure was still present as observed without pre-oxidation. However, the surface was covered by a significant number of Ti nodules (Figure 9a). As reported in the literature [16], the morphology of the coating is directly influenced by the surface conditions of the substrate, notably roughness or irregularities, suggesting that the growth of Ti nodules was promoted by the presence of the Cr-Mn spinel oxide grains formed after 1 h of pre-oxidation at 800 °C. XRD and EDX analyses (texture, coherent domain size, and composition) presented exactly the same results as those for the film on the sample that was not pre-oxidized. The SEM cross-sectional micrography (Figure 9b.1) revealed that the coating was not fully dense over the total length due to the presence of the Ti nodules. Figure 9b.2 shows the simulated film cross section (gray is the film and brown is the substrate: steel and oxide together), which confirmed the SEM observations. The film nodules were due to the presence of the big oxide crystals. The discontinuities in the local topography induced shadowing and consequently voids in the film. Figure 9b.3 presents the coating porosity, where white represents the dense matter (film and substrate), yellow represents the air-connected pores (here, only the roughness), and blue to red represents the occluded pores (blue is for small pores and red is for big ones). These pores were vertically aligned and homogeneously distributed in the film. The porosity in this film represented about 3% of the volume, which was 30 times higher than that without substrate pre-oxidation. The roughness of the pre-oxidized substrate was about 118 nm, while the roughness after coating was slightly increased to 166 nm, which was the same proportion as for the substrate that was not pre-oxidized.

EDX elementary maps (Figure 10) confirmed that the average thickness of the Ti coating was 1.2 µm and showed evidence of the presence of a thin chromia layer and Cr-Mn spinel crystals between the substrate and the coating, which were formed during the pre-oxidation step.

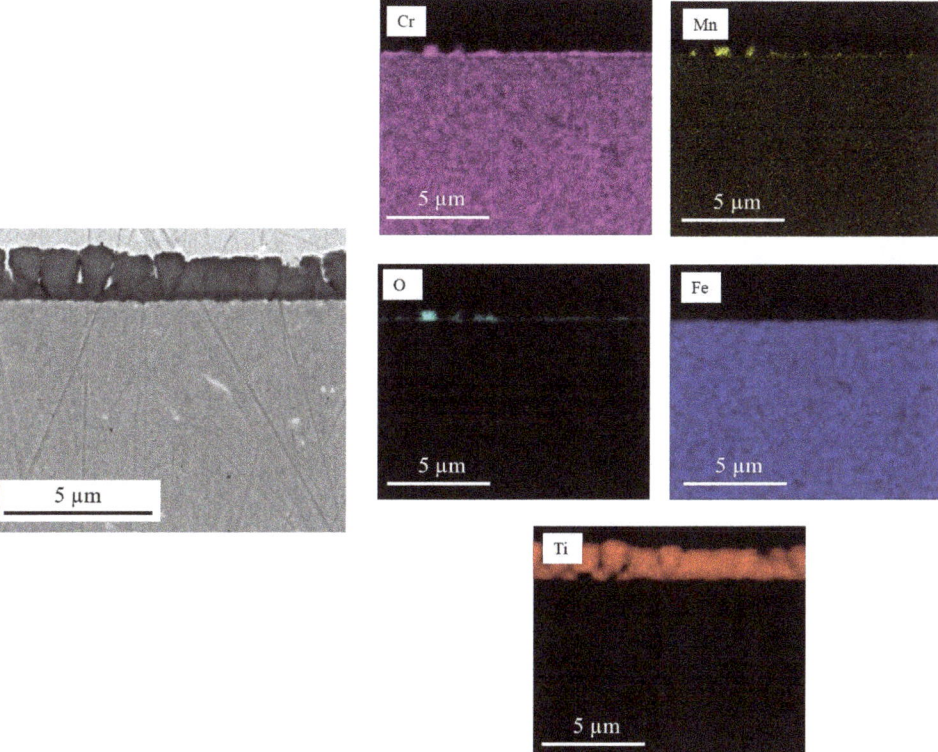

Figure 10. Backscattered electron image and EDX elementary maps of the cross section of the Ti coating deposited on the pre-oxidized steel.

3.3.2. Oxidation of Coated Pre-Oxidized Sample

The surface morphology of the pre-oxidized coated steel after 100 h of oxidation at 800 °C in ambient air is presented in Figure 11.

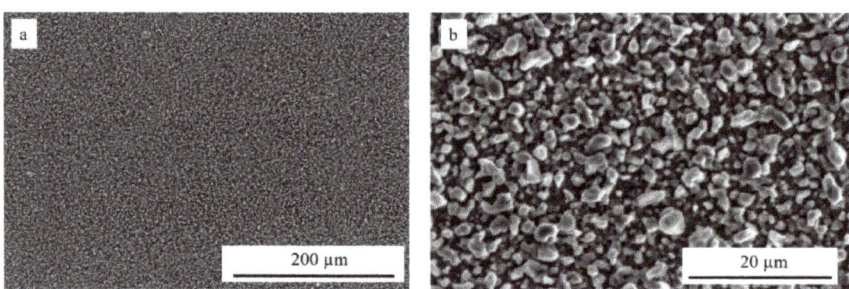

Figure 11. Secondary electron image of the surface of the polished sample after 100 h aging at 800 °C under laboratory air: (**a**) global view and (**b**) magnification.

It is worth noting that spalling or cracking did not occur. The oxide scale was quite homogeneous, with big and rather spherical grains covering a lower layer formed by smaller grains. These big grains had the same distribution as the Cr-Mn spinel crystals formed after pre-oxidation (Figure 8a) and the Ti nodules formed after PVD deposition (Figure 9a). EDX analyses indicated that the scale was mainly composed of titanium oxide; a small amount of Cr was also detected. XRD performed at an incidence angle of 2° confirmed the formation of TiO_2 as the majority phase, Ti_5O_9 (Magnéli phase), and the presence of a Cr-Mn spinel oxide (see Supplementary Material S4). Magnéli phases were substoichiometric Ti oxides of general formula Ti_nO_{2n-1} (n = 4–9) [35], which form for oxygen contents ranging between 60 and 66.67 at.% (corresponding to TiO_2 formation) [36].

These results were confirmed by EDX elementary maps (Figure 12), showing the formation of titanium oxide and a duplex $Cr_2O_3/(Cr,Mn)_3O_4$ scale below.

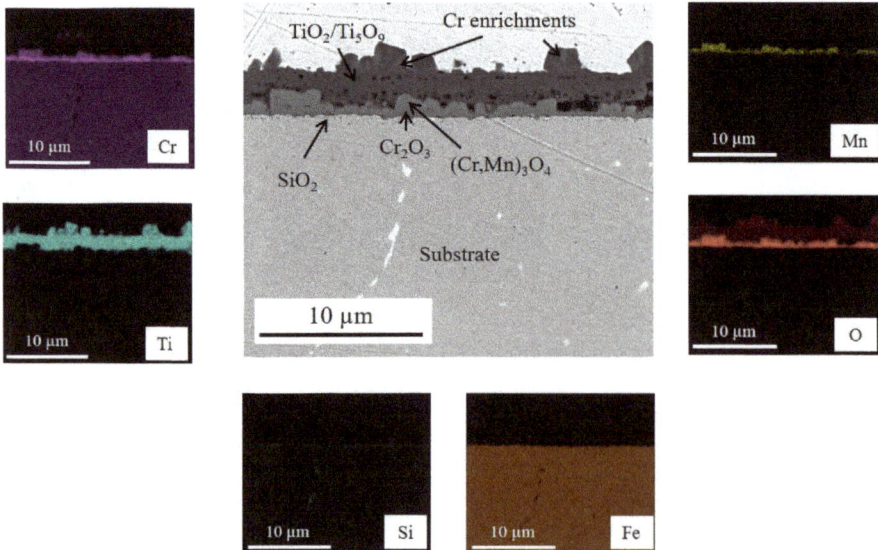

Figure 12. Backscattered electron image and EDX elementary maps of the cross section of the Ti-coated pre-oxidized sample after 100 h aging at 800 °C under laboratory air.

The presence of SiO$_2$ at the metal/oxide interface could also be observed. The thickness of the titanium oxide ranged between 1.2 and 3.3 μm considering the nodules. This value was lower than the one measured for the sample without pre-oxidation (between 2 and 4 μm). The thickness of the duplex Cr$_2$O$_3$/(Cr,Mn)$_3$O$_4$ layer varied from about 500 nm to 1.2 μm, i.e., more than 4 times lower than in the case of the steel that was not pre-oxidized. Several porosities were present inside the titanium oxide layer and at the TiO$_2$/(Cr,Mn)$_3$O$_4$ interface. Their size and number were more important than in the case of the coated sample without pre-oxidation, in agreement with other works [16,37]. EDX maps revealed that a small amount of Cr diffused toward the external coating, but the diffusion of Fe was not observed. This was due to the Cr$_2$O$_3$ layer that was formed during pre-oxidation, which acted as a barrier and prevented the outward diffusion of Fe from the substrate. This phenomenon, in turn, promoted the formation of many pores inside the scale [16,37]. Concerning the reduction of the oxidation rate, some hypotheses can be suggested. During aging after pre-oxidation, the coating remained adherent to the Cr$_2$O$_3$ pre-formed scale, forming a barrier that limited potential oxygen diffusion. In parallel, to react with oxygen, Cr cations had to diffuse through the thermally grown Cr$_2$O$_3$ scale formed during pre-oxidation. Moreover, the formation of quite a continuous Cr-Mn spinel oxide layer above Cr$_2$O$_3$ (not evidenced after aging of the not coated steel that was not pre-oxidized) also acted as a barrier, thus reducing outward diffusion of Cr [38]. It should be noted that the formation of the (Cr,Mn)$_3$O$_4$ phase on top of the Cr$_2$O$_3$ layer was due to the higher diffusivity of Mn^{2+} in the Cr$_2$O$_3$ scale compared to Cr^{3+} (about 100 times higher) [39].

3.4. Discussion about Adhesion

The question that arises at this stage of the study relates to why the spalling of the oxide layer occurred in the case of the coated substrate that was not pre-oxidized.

The first consideration concerns the mechanical state of the substrate and the film before 100 h of oxidation. The film residual stress, obtained with the curvature method, was tensile and about 192 ± 40 MPa. As it was measured on a mirror-polished silicon substrate, it is assumed that this value also corresponds to the stress of the coating on the simply polished steel. As the coating on the pre-oxidized steel had about 3% of porosity, one can assume that the stress will be released. The substrate residual stress was determined for both samples (with and without pre-oxidation and after deposition) by X-ray diffraction using the sin$^2\psi$ method. In both cases, the stresses were compressive, but the values obtained for the coated sample that was not pre-oxidized (−542.5 ± 55.7 MPa) were about five times higher than those measured for the pre-oxidized one (−108.6 ± 50.6 MPa). These findings indicate the existence of an important difference in the mechanical state between the two substrates and the two films before undergoing high-temperature oxidation. If stress is generated during oxidation, the sample without pre-oxidation will have less ability to resist than the pre-oxidized one.

The second consideration is about the growth of the Cr$_2$O$_3$ scale underneath the TiO$_2$ layer. This growth, which occurs during isothermal aging at high-temperature, generates growing stress. Indeed, the size of the Cr$_2$O$_3$ grains was much bigger than the TiO$_2$ ones as revealed by SEM characterizations (Figure 5c). Once the coating cracks, the substrate is in direct contact with the oxidizing atmosphere, leading to an increase in the oxidation rate. This failure process is well known in the literature as mechanically induced chemical failure (MICF) [40].

The third consideration concerns the thermal stress, which occurs during cooling down and due to the differences between the thermal expansion coefficients (TEC) of the substrate and the different oxide scales. The TEC of the ferritic substrate tested in the present study was equal to 11–12 × 10^{-6} °C^{-1} between 100 and 800 °C, while the TEC of TiO$_2$ was about 7–8 × 10^{-6} °C^{-1} between 25 and 1000 °C [41]. Cr$_2$O$_3$ has a TEC of 9.6 × 10^{-6} °C^{-1} between 25 and 1000 °C [41], which is an intermediate value between TiO$_2$ and the ferritic substrate. It is possible to suggest that the presence of a continuous, albeit thin, Cr$_2$O$_3$

layer covering the substrate (Figure 8b) before the application of coating, i.e., from the first exposure instant of the Ti coating to the oxidant atmosphere, had a beneficial effect on the adhesion of the TiO$_2$ scale forming during high-temperature exposure. Furthermore, the use of an appropriate interlayer between the PVD coating and the substrate has often been successfully reported in the literature to reduce the mismatch between the coating and the substrate and to limit the development of thermal stress during high-temperature exposure [42–44]. It is also noteworthy that in the case of the substrate that was not pre-oxidized, a thick and continuous SiO$_2$ layer was detected. The main disadvantage of this oxide was its low TEC (0.5 × 10^{-6} °C^{-1}) [41], which promoted spalling of the oxide scale.

In the present study, all these considerations play a role in the different behaviors observed for the two samples. To understand the mechanisms of scale spallation, in-situ SEM observations during oxidation are necessary.

4. Conclusions

In this study, a thin titanium layer (1.2 µm) was deposited on the surface of a ferritic stainless steel substrate by PVD technique before high-temperature exposure (100 h at 800 °C in ambient air). The effects of short pre-oxidation of the substrate (1 h at 800 °C) on the adhesion of the coating were investigated.

The major findings can be summarized as follows:

- Metallic Ti converted into Ti oxide (TiO$_2$) during high-temperature aging at 800 °C. TiO$_2$ scale formed on the steel that was not pre-oxidized slowed down the oxidation of the substrate. Indeed, the thickness of the chromia layer ranged from about 4 to almost 8 µm in the case of the uncoated sample and between 2 and 4 µm for the sample that was not pre-oxidized. However, significant spallation of the TiO$_2$ scale occurred during cooling down.
- The TiO$_2$ scale grown on the pre-oxidized steel effectively protected the substrate against oxidation. The thickness of the duplex Cr$_2$O$_3$/(Cr,Mn)$_3$O$_4$ layer varied from about 500 nm to 1.2 µm and was more than four times lower than in the case of the steel that was not pre-oxidized.
- These different behaviors were the result of the combination of different kinds of stress, namely, residual, growing, and thermal stress.
- The film residual stress on the polished substrate before 100 h of oxidation was tensile. On the other hand, the coating on the pre-oxidized steel had about 3% of porosity. This value was 30 times higher than the one measured in the case of the substrate that was not pre-oxidized. The substrate residual stress determined for both steel samples (with and without pre-oxidation) before undergoing high-temperature oxidation was compressive, but it was five times higher in the case of the steel that was not pre-oxidized (-542.5 ± 55.7 and -108.6 ± 50.6 MPa, respectively). If stress is generated during oxidation, the sample without pre-oxidation will have less ability to resist than the pre-oxidized one.
- In the case of the substrate that was not pre-oxidized, the growing stress was due to the growth of the Cr$_2$O$_3$ scale underneath the TiO$_2$ layer during isothermal oxidation, which led to the cracking of TiO$_2$.
- The thermal stress occurred during cooling down and was due to differences between the thermal expansion coefficients (TEC) of the substrate and the different oxide scales. Cr$_2$O$_3$ has a TEC intermediate between TiO$_2$ and the ferritic steel substrate. In the case of the pre-oxidized steel, the presence of a continuous, albeit thin, Cr$_2$O$_3$ layer covering the substrate before the application of the coating, i.e., from the first exposure instant of the Ti coating to the oxidant atmosphere, had a beneficial effect on the adhesion of the TiO$_2$ scale forming during high-temperature exposure.

To better understand the mechanism of the TiO$_2$ scale spallation and to provide further responses, in-situ SEM-EDX characterizations are necessary to observe what occurs during isothermal oxidation.

Supplementary Materials: The following supporting information can be downloaded at: https://www.mdpi.com/article/10.3390/cryst12121732/s1. Figure S1: XRD patterns (Bragg-Brentano conditions) of the ferritic steel oxidized for 100 h at 800 °C in ambient air. Figure S2: XRD patterns (Bragg-Brentano conditions) of the Ti coating on Si substrate; Figure S3: XRD patterns (incidence angle of 2°) of the ferritic steel oxidized for 1 h at 800 °C in ambient air. Figure S4: XRD patterns (incidence angle of 2°) of the ferritic steel pre-oxidized for 1 h at 800 °C in ambient air, coated with Ti and then oxidized for 100 h at 800 °C in ambient air.

Author Contributions: Conceptualization, M.-R.A.-B. and A.B.; methodology, M.-R.A.-B., A.B. and G.N.B.; validation, M.-R.A.-B., A.B. and G.N.B.; investigation, M.-R.A.-B., A.B., Q.O., Y.P. and P.B.; writing—original draft preparation, M.-R.A.-B.; writing—review and editing, M.-R.A.-B., A.B. and G.N.B.; visualization, M.-R.A.-B.; supervision, M.-R.A.-B.; project administration, M.-R.A.-B. and A.B. All authors have read and agreed to the published version of the manuscript.

Funding: This research received no external funding.

Institutional Review Board Statement: Not applicable.

Conflicts of Interest: The authors declare no conflict of interest.

References

1. Ardigo-Besnard, M.R.; Popa, I.; Chevalier, S. Spinel and perovskite coatings effect on long term oxidation of a ferritic stainless steel in H_2/H_2O atmosphere. *Corros. Sci.* **2019**, *148*, 251–263. [CrossRef]
2. Ardigo, M.R.; Popa, I.; Chevalier, S.; Girardon, P.; Perry, F.; Laucournet, R.; Brevet, A.; Desgranges, C. Effect of coatings on long term behaviour of a commercial stainless steel for solid oxide electrolyser cell interconnect application in H_2/H_2O atmosphere. *Int. J. Hydrogen Energy* **2014**, *39*, 21673–21677. [CrossRef]
3. Ardigo, M.R.; Popa, I.; Chevalier, S.; Parry, V.; Galerie, A.; Girardon, P.; Perry, F.; Laucournet, R.; Brevet, A.; Rigal, E. Coated interconnects development for high temperature water vapour electrolysis: Study in anode atmosphere. *Int. J. Hydrogen Energy* **2013**, *38*, 15910–15916. [CrossRef]
4. Petric, A.; Ling, H. Electrical conductivity and thermal expansion of spinels at elevated temperatures. *J. Am. Ceram. Soc.* **2007**, *90*, 1515–1520. [CrossRef]
5. Shaigan, N.; Qu, W.; Ivey, D.G.; Chen, W. A review of recent progress in coatings, surface modifications and alloy developments for solid oxide fuel cell ferritic stainless steel interconnects. *J. Power Sources* **2010**, *195*, 1529–1542. [CrossRef]
6. Mardare, C.C.; Asteman, H.; Spiegel, M.; Savan, A.; Ludwig, A. Investigation of thermally oxidised Mn–Co thin films for application in SOFC metallic interconnects. *Appl. Surf. Sci.* **2008**, *255*, 1850–1859. [CrossRef]
7. Yoon, J.S.; Lee, J.; Hwang, H.J.; Whang, C.M.; Moon, J.M.; Kim, D.H. Lanthanum oxide-coated stainless steel for bipolar plates in solid oxide fuel cells (SOFCs). *J. Power Sources* **2008**, *181*, 281–286. [CrossRef]
8. Cabouro, G.; Caboche, G.; Chevalier, S.; Piccardo, P. Opportunity of metallic interconnects for ITSOFC: Reactivity and electrical property. *J. Power Sources* **2006**, *156*, 39–44. [CrossRef]
9. Lim, D.P.; Lim, D.S.; Oh, J.S.; Lyo, I.W. Influence of post-treatments on the contact resistance of plasma-sprayed $La_{0.8}Sr_{0.2}MnO_3$ coating on SOFC metallic interconnector. *Surf. Coat. Technol.* **2005**, *200*, 1248–1251. [CrossRef]
10. Wu, J.; Johnson, C.D.; Jiang, Y.; Gemmen, R.S.; Liu, X. Pulse plating of Mn–Co alloys for SOFC interconnect applications. *Electrochim. Acta* **2008**, *54*, 793–800. [CrossRef]
11. Mikkelsen, L.; Chen, M.; Hendriksen, P.V.; Persson, A.; Pryds, N.; Rodrigo, K. Deposition of $La_{0.8}Sr_{0.2}Cr_{0.97}V_{0.03}O_3$ and $MnCr_2O_4$ thin films on ferritic alloy for solid oxide fuel cell application. *Surf. Coat. Technol.* **2007**, *202*, 1262–1266. [CrossRef]
12. Gorokhovsky, V.I.; Gannon, P.E.; Deibert, M.C.; Smith, R.J.; Kayani, A.; Kopczyk, M.; VanVorous, D.; Yang, Z.G.; Stevenson, J.W.; Visco, S.; et al. Deposition and Evaluation of Protective PVD Coatings on Ferritic Stainless Steel SOFC Interconnects. *J. Electrochem. Soc.* **2006**, *153*, A1886–A1893. [CrossRef]
13. Hoyt, K.O.; Gannon, P.E.; White, P.; Tortop, R.; Ellingwood, B.J.; Khoshuei, H. Oxidation behavior of $(Co,Mn)_3O_4$ coatings on preoxidized stainless steel for solid oxide fuel cell interconnects. *Int. J. Hydrogen Energy* **2012**, *37*, 518–529. [CrossRef]
14. Pan, Y.; Geng, S.; Chen, G.; Wang, F. Effect of pre-oxidation on surface scale microstructure and electrical property of Cu-Fe coated steel interconnect. *Corros. Sci.* **2020**, *170*, 108680. [CrossRef]
15. Yang, P.; Liu, C.-K.; Wu, J.-Y.; Shong, W.-J.; Lee, R.-Y.; Sung, C.-C. Effects of pre-oxidation on the microstructural and electrical properties of $La_{0.67}Sr_{0.33}MnO_{3-\delta}$ coated ferritic stainless steels. *J. Power Sources* **2012**, *213*, 63–68. [CrossRef]
16. Zhao, Q.; Geng, S.; Chen, G.; Wang, F. Influence of preoxidation on high temperature behavior of $NiFe_2$ coated SOFC interconnect steel. *Int. J. Hydrogen Energy* **2019**, *44*, 13744–13756. [CrossRef]
17. Amendola, R.; Gannon, P.; Ellingwood, B.; Hoyt, K.; Piccardo, P.; Genocchio, P. Oxidation behavior of coated and preoxidized ferritic steel in single and dual atmosphere exposures at 800 °C. *Surf. Coat. Technol.* **2012**, *206*, 2173–2180. [CrossRef]
18. Talic, B.; Molin, S.; Hendriksen, P.V.; Lein, H.L. Effect of pre-oxidation on the oxidation resistance of Crofer 22 APU. *Corros. Sci.* **2018**, *138*, 189–199.

19. Goebel, C.; Alnegren, P.; Faust, R.; Svensson, J.-E.; Froitzheim, J. The effect of pre-oxidation parameters on the corrosion behavior of AISI 441 in dual atmosphere. *Int. J. Hydrogen Energy* **2018**, *43*, 14655–14674. [CrossRef]
20. Evrard, M.; Besnard, A.; Lucas, S. Study of the influence of the pressure and rotational motion of 3D substrates processed by magnetron sputtering: A comparative study between Monte Carlo modelling and experiments. *Surf. Coat. Tech.* **2019**, *378*, 125070. [CrossRef]
21. Kofstad, P. *High Temperature Corrosion*; Elsevier Applied Science: London, UK; Elsevier Applied Science: New York, NY, USA, 1988.
22. Coddet, C.; Chaze, A.M.; Beranger, G. Measurements of the adhesion of thermal oxide film: Application to the oxidation of titanium. *J. Mater. Sci.* **1987**, *22*, 2969–2974. [CrossRef]
23. Chaze, A.M.; Coddet, C. The role of nitrogen in the oxidation behavior of titanium and some binary alloys. *J. Less Common Metals* **1986**, *124*, 73–84. [CrossRef]
24. Hasegawa, M. Chapter 3.3—Ellingham Diagram. In *Treatise on Process Metallurgy Volume 1: Process Fundamentals*; Seetharaman, S., McLean, A., Guthrie, R., Sridhar, S., Eds.; Elsevier Ldt.: Oxford, UK, 2014; pp. 507–516.
25. Nakajima, H.; Koiv, M. Diffusion in Titanium. *ISIJ Int.* **1991**, *31*, 757–766. [CrossRef]
26. Shapovalov, V.P.; Kurasov, A.N. Diffusion of titanium in iron. *Metalloved. Termich. Obrab. Metall.* **1975**, *9*, 71–73. [CrossRef]
27. Noyan, I.C.; Cohen, J.B. *Residual Stress, Measurement by Diffraction and Interpretation*; Springer: Berlin/Heidelberg, Germany, 1987.
28. Besnard, A.; Ardigo, M.R.; Imhoff, L.; Jacquet, P. Curvature radius measurement by optical profiler and determination of the residual stress in thin films. *Appl. Surf. Sci.* **2019**, *487*, 356–361. [CrossRef]
29. Depla, D.; Leroy, W.P. Magnetron sputter deposition as visualized by Monte Carlo modeling. *Thin Solid Films* **2012**, *520*, 6337–6354. [CrossRef]
30. Ziegler, J.F.; Biersack, J.P. The stopping and range of ions in matter. In *Treatise Heavy-Ion Science*; Bromley, D.A., Ed.; Springer: Boston, MA, USA, 1985; Volume 6, Astrophys. Chem. Condens. Matter; pp. 93–129.
31. Moskovkin, P.; Lucas, S. Computer simulations of the early stage growth of Ge clusters at elevated temperatures, on patterned Si substrate using the kinetic Monte Carlo method. *Thin Solids Film* **2013**, *536*, 313–317. [CrossRef]
32. Niewolak, L.; Young, D.J.; Hattendorf, H.; Singheiser, L.; Quadakkers, W.J. Mechanisms of oxide scale formation on ferritic steel in simulated low and high pO2 service environments of solid oxide fuel cells. *Oxid. Met.* **2014**, *82*, 123–143. [CrossRef]
33. Fontana, S.; Amendola, R.; Chevalier, S.; Piccardo, P.; Caboche, G.; Viviani, M.; Molins, R.; Sennour, M. Metallic interconnects for SOFC: Characterisation of corrosion resistance and conductivity evaluation at operating temperature of differently coated alloys. *J. Power Sources* **2007**, *171*, 652–662. [CrossRef]
34. Bednarz, M.; Molin, S.; Bobruk, M.; Stygar, M.; Długoń, E.; Sitarz, M.; Brylewski, T. High-temperature oxidation of the Crofer 22 H ferritic steel with $Mn_{1.45}Co_{1.45}Fe_{0.1}O_4$ and $Mn_{1.5}Co_{1.5}O_4$ spinel coatings under thermal cycling conditions and its properties. *Mater. Chem. Phys.* **2019**, *225*, 227–238. [CrossRef]
35. Arif, A.F.; Balgis, R.; Ogi, T.; Iskandar, F.; Kinoshita, A.; Nakamura, K.; Okuyama, K. Highly conductive nano-sized Magnéli phases titanium oxide (TiO_x). *Sci. Rep.* **2017**, *7*, 3646. [CrossRef] [PubMed]
36. Waldner, P.; Eriksson, G. Thermodynamic modelling of the system titanium-oxygen. *Calphad* **1999**, *23*, 189–218. [CrossRef]
37. Geng, S.; Zhao, Q.; Li, Y.; Mu, J.; Chen, G.; Wang, F.; Zhu, S. Sputtered MnCu metallic coating on ferritic stainless steel for solid oxide fuel cell interconnects application. *Int. J. Hydrogen Energy* **2017**, *42*, 10298–10307. [CrossRef]
38. Yang, Z.; Xia, G.G.; Singh, P.; Stevenson, J.W. Electrical contacts between cathodes and metallic interconnects in solid oxide fuel cells. *J. Power Sources* **2006**, *155*, 246–252. [CrossRef]
39. Lobnig, R.E.; Schmidt, H.P.; Hennesen, K.; Grabke, H.G. Diffusion of cations in chromia layers grown on iron-base alloys. *Oxid. Met.* **1992**, *37*, 81–93. [CrossRef]
40. Evans, H.E.; Donaldson, A.T.; Gilmour, T.C. Mechanisms of breakaway oxidation and application to a chromia-forming steel. *Oxid. Met.* **1999**, *54*, 379–402. [CrossRef]
41. Zhu, W.Z.; Deevi, S.C. Development of interconnect materials for solid oxide fuel cells. *Mater. Sci. Eng.* **2003**, *A348*, 227–243. [CrossRef]
42. Li, S.; Xiao, L.; Liu, S.; Zhang, Y.; Xu, J.; Zhou, X.; Zhao, G.; Cai, Z.; Zhao, X. Ultra-high temperature oxidation resistance of a novel (Mo, Hf, W, Ti)Si_2 ceramic coating with Nb interlayer on Ta substrate. *J. Eur. Ceram. Soc.* **2022**, *42*, 4866–4880. [CrossRef]
43. Anton, R.; Hüning, S.; Laska, N.; Weber, M.; Schellert, S.; Gorr, B.; Christ, H.J.; Schulz, U. Graded PVD Mo-Si interlayer between Si coating and Mo-Si-B alloys: Investigation of oxidation behavior. *Corros. Sci.* **2021**, *192*, 109843. [CrossRef]
44. Lima, C.R.C.; Cinca, N.; Guilemany, J.M. Study of the high temperature oxidation performance of Thermal Barrier Coatings with HVOF sprayed bond coat and incorporating a PVD ceramic interlayer. *Ceram. Int.* **2012**, *38*, 6423–6429. [CrossRef]

Article

Wetting of Refractory Ceramics with High-Manganese and Structural Steel and Description of Interfacial Interaction

Vlastimil Novák [1,*], Lenka Řeháčková [1], Silvie Rosypalová [1] and Dalibor Matýsek [2]

[1] Faculty of Materials Science and Technology, VŠB-Technical University of Ostrava, 17. listopadu 15, Poruba, 708 00 Ostrava, Czech Republic
[2] Faculty of Mining and Geology, VŠB-Technical University of Ostrava, 17. listopadu 15, Poruba, 708 00 Ostrava, Czech Republic
* Correspondence: vlastimil.novak@vsb.cz

Abstract: This work aims to describe the interfacial interaction at the interface between refractory material and high-manganese (XT 720) and structural (11 523) steel using a wetting test up to 1600 °C. The contact angles were determined through the sessile drop method, and the results were put into context through degradation testing and the characterization of the interfacial interface by Energy Dispersive X-Ray (EDX), X-Ray Diffraction (XRD) analyses, and Scanning Electron Microscopy (SEM). The lowest resistance to molten steel was observed for chamotte materials, while the highest was observed for materials based on electrofused corundum. High-manganese steel was strongly erosive to the materials tested, with the wetting angle decreasing significantly from 10 to 103° with decreasing Al_2O_3 content (an increase of 2.4 to 59.4% corundum) in the refractories. Structural steel showed wetting angles from 103 to 127° for identical refractories. These results were consistent with the average erosion depth for Mn steel (0.2–7.8 mm) and structural steel (0–2.4 mm).

Keywords: high-manganese steel; structural steel; phase interface; reactive wetting; contact angle; refractory material

1. Introduction

High-alloy steels, such as manganese steel with the designation XT 720, are in high demand in technical practice, primarily because of their ability to harden when exposed to sufficiently strong impact loads and pressures. This gives the steel a high resistance to abrasive wear, making it suitable for the construction of excavators and other mining equipment. It is also widely applied in the national defense, automobile, and petrochemical industries due to its excellent strength, plasticity, and low-temperature flexibility [1,2]. However, some alloying elements contained in these steels have an adverse effect on refractory ceramics, whose participation in steel casting is essential. Refractories are necessary components not only in casting steels in the form of ingots but also in melting non-ferrous alloys. They serve as linings for thermal aggregates and furnaces and casting routes for transferring molten steel to the mold. They are also used to produce melting pots for the recovery of non-ferrous intermetallic alloys [3–5]. The steel and foundry industry is continuously focusing on developing and improving products, mainly in terms of their physical and mechanical properties. In this context, it is crucial to study the effect of refractory material on the steels with which it is in contact to maintain steel chemical composition within the required range, to achieve the desired cleanliness concerning the amount and character of non-metallic inclusions, and to prevent defects on the steel surface [6–8].

Chamotte and high-alumina materials are commonly used in the refractory industry to design casting routes. The latter are high-quality ceramic materials possessing high strength, heat resistance, low porosity, and high resistance to the erosive action of the melt. These properties are of great importance, as poor quality casting routes can cause the release of unwanted elements into the melt due to ongoing reactions between the ceramic and the melt, which affects the purity of the steel and its microstructure, closely related to its final performance [9–11]. In secondary steel production, one of the main challenges is to ensure high steel cleanliness, which is influenced, among other things, by the resistance of the refractory material during the production process. In particular, non-metallic inclusions are a source of considerable concern as they can act as stress enhancers in cast steels. For this reason, one of the indicators of steel quality is the amount of non-metallic inclusions present [12–16].

One method of assessing the degree of interaction between molten steel and an arbitrary substrate is through a high-temperature wetting test [17]. Wettability is the ability of a liquid to spread on a substrate. It is measured as the contact angle between a tangent drawn at the triple point and the substrate surface [18]. The spreading of a liquid on a substrate without reaction/absorption of the liquid by the substrate material is termed non-reactive wetting, while a wetting process affected by a reaction between the spreading liquid and the substrate material is termed reactive wetting. In most publications, the equilibrium contact angle is given respecting Young's equation, which puts the interfacial tensions into relation. However, it should be noted that the latter was derived assuming the spreading of a non-reactive liquid on a physically and chemically inert, smooth, and homogeneous substrate [19,20]. The wetting of a solid by a liquid is a complex phenomenon depending on many factors. In the case of non-reactive wetting, it is controlled by the physical properties of the spreading liquid and wetting system, chemical heterogeneity causing wetting hysteresis, surface roughness, and gaseous environment [21]. Conversely, the reactive wetting process is influenced by several other factors such as the use of flux added to reduce the barrier effects of the oxide layers, the addition of trace impurities, etc. [19,22,23].

The interaction between refractories and manganese steel has been studied by several researchers [24–28]. Wang et al. investigated the interfacial reactions between high-Mn and Al steel, MgO-C refractory material, and CaO-SiO_2-Al_2O_3-MgO refining slag concerning the effect of complex reactions on steel purity [24]. Alibeigi et al. investigated the reactive wetting of several steels containing manganese up to approximately 5 wt% during annealing, finding that the strength of the wetting in a given arrangement is a function of the partial pressure of oxygen and is directly related to the thickness of the outer MnO film formed and not to the Mn content of the alloy [25]. Articles on similar topics can be found elsewhere [26,27]. Kong et al. observed the formation of interfacial layers of (Mn, Mg)O and (Mn, Mg)O·Al_2O_3 as a result of the interaction between medium-manganese steel and the refractory material MgO [28].

The objective of this paper was to describe the interaction at the interface of high-manganese steel and structural steel with a refractory ceramic material through a wetting test and to find a relationship between wetting and the erosive action of the melt. The phase interface after high-temperature tests was also characterized by SEM, EDX, and XRD analyses. Knowledge of interfacial behavior improves an understanding of the processes and reactions occurring at the interphase at elevated temperatures. Based on such tests, it is possible to optimize refractory ceramics in terms of their composition or physicochemical properties, which is vital for enhancing the casting process's quality and minimizing the ceramic material's negative influence on the final product.

2. Materials and Methods

2.1. Specification and Preparation of Samples

Six types of refractory ceramic materials produced by SEEIF Ceramic a.s. were selected for the wetting tests, differing in their physicomechanical properties, chemical and mineralogical composition, firing temperature, application, and cost of the final product. A detailed characterization of the physical properties of the studied ceramic materials determined according to the standard EN 993-1:2018 (Methods of test for dense shaped refractory products—Part 1: Determination of bulk density, apparent porosity and true porosity) is given in Table 1.

Table 1. Physical properties of ceramic materials.

Sample	Firing Temperature [°C]	Volumetric Weight [g·cm^{-3}]	Water Absorption [%]	Apparent Porosity [%]
F 36	1250	2.03	12.48	25.4
ZR 50	1280	2.97	6.29	18.7
B 70	1250	2.36	8.95	21.1
ML 65	1250	2.37	8.83	20.9
M 70	1380	2.45	8.83	21.6
MK 82	1380	2.61	8.75	22.9

A Bruker-AXS D8 powder X-ray diffractometer was used to determine the mineralogical composition of the refractory specimens (Table 2).

Table 2. Mineralogical composition of ceramic materials [wt%].

Sample	Corundum	Mullite	Cristobalite	Quartz	Zircon	Baddeleyite	Amorphous phase
F 36	2.4	45.6	0.9	6.9	—	—	44.2
ZR 50	14.4	26.4	—	—	54.1	5.1	—
B 70	23.7	35.1	3.2	4.0	—	—	34.0
ML 65	17.9	46.7	0.2	0.5	—	—	34.7
M 70	26.2	45.3	0.3	1.0	—	—	27.2
MK 82	59.4	24.4	—	0.1	—	—	16.1

The ceramic materials were prepared from pure components by accurately batching them into a wheel mixer according to process guidelines (SEEIF Ceramic, a.s.). The mixture was moistened with water (the water content was between 10 and 15 wt%) and thoroughly mixed to ensure homogeneity. The molding was carried out using a press operating at a specific pressure of up to 10 MPa. From the materials thus prepared, plates with dimensions of 25 × 25 × 5 mm were made. The chemical composition of the prepared refractories is listed in Table 3.

Table 3. Chemical composition of ceramic materials [wt%].

Sample	Al_2O_3	SiO_2	TiO_2	Fe_2O_3	CaO	MgO	K_2O	Na_2O	ZrO_2
F 36	40.3	54.5	1.6	1.8	0.3	0.3	1.0	0.2	—
ZR 50	23.1	35.4	0.7	0.8	0.2	0.2	0.6	0.3	38.7
B 70	70.5	23.7	2.5	1.9	0.3	0.2	0.8	0.1	—
ML 65	64.1	30.8	1.7	1.7	0.4	0.2	0.9	0.2	—
M 70	72.6	25.2	0.5	0.6	0.1	0.1	0.7	0.2	—
MK 82	83.1	14.8	0.4	0.6	0.1	0.1	0.6	0.3	—

Contact angles (wetting angles) were determined at the steel/ceramic interface. Two steels were selected for the wetting tests, XT 720 high manganese abrasion resistant steel and 11 523 low alloy structural steel, whose chemical composition was obtained using a SPECTRUMA GDA-750 HP glow discharge optical emission spectrometer, while the content of oxygen, sulfur, and carbon was measured with an ELTRA 2000 ONH and 2000 CS combustion analyzer. The results of the analyses are presented in Table 4.

Cubes with an edge length of 4.5 mm were manufactured from these steels by electro-spark machining. Before the experiment, the surface of the metal samples was first cleaned mechanically and then cleaned with acetone to remove oxides. Likewise, the surface of the ceramic materials was cleaned with acetone immediately prior to the wetting test.

Table 4. Chemical composition of XT 720 and 11 523 steels [wt%].

Steel	C	Mn	Si	P	S	Cu	Ni	Cr	Mo	V	Ti	Al	N
XT 720	1.29	19.33	0.73	0.04	0.004	0.11	0.10	1.99	0.06	0.02	0.01	0.04	0.03
11 523	0.22	0.67	0.40	0.01	0.003	0.09	0.11	0.14	0.03	—	—	—	0.01
					The remainder up to 100 wt% was iron.								

2.2. Determination of Wetting Angles

The wetting angles were determined according to the sessile drop method in a CLASIC high-temperature observation furnace within the temperature range from the melting point of the steel sample to 1600 °C (Figure 1). The prepared samples (steel/ceramic substrate) were placed in the furnace tube, which was then hermetically sealed, evacuated to 0.1 Pa, and purged with argon of high purity (>99.9999%). The last two steps were repeated. The temperature program consisted of a 2 °C·min^{-1} temperature ramp rate ending at a maximum temperature of 1600 °C. A Pt-13% Rh/Pt thermocouple, which was placed near the sample, registered the temperature. To prevent oxidation of the sample, all measurements were carried out in an inert atmosphere of argon. A CANON EOS 550D high-resolution camera took images of drop silhouettes during the heating ramp. The wetting angles were evaluated using the ADSA (axisymmetric drop shape analysis) method based on fitting the drop profiles to a Laplacian curve using a nonlinear regression procedure [29].

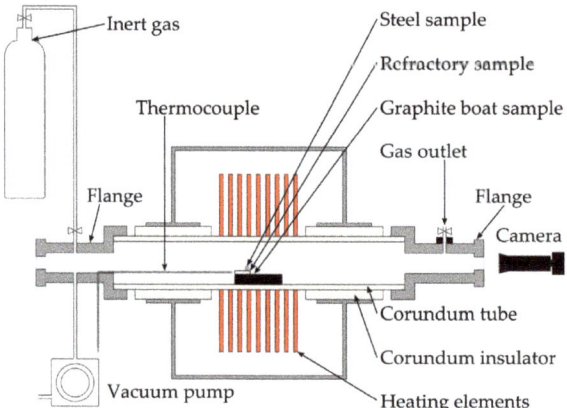

Figure 1. Schematic of the apparatus for the high-temperature wetting tests.

2.3. Degradation Testing of Refractory Materials

Degradation tests provide information on the resistance of the refractory material to molten steel. The tests were carried out in a TERMEL ITEP001 tilt induction furnace with a maximum power of 15 kW and a frequency range of 10–12 kHz. The furnace consists of a rectifier, a filtering capacity, an intermediate circuit, an inverter, a SIEMENS control system, and a cooling system. The maximum attainable temperature is 1700 °C. The steel melting was carried out in ceramic crucibles with dimensions of 125 × 85 × 45 mm made of the refractory materials under test. Cylinders with a diameter of 40 mm and a length of 68 mm were made from the steel specimens. The melting of the steel took 30 min, starting from the moment the steel was melted. During the melting process, the slag forming on the surface was removed. The temperature of the melt was measured with a thermocouple. After cooling, the crucible was cut along the central axis and prepared for evaluation of the ceramic's erosion losses due to the steel's action.

2.4. SEM, EDX, and XRD Methods

After the experiments, the interaction between the steel samples and ceramic crucibles was assessed by a JEOL 6490LV scanning electron microscope (SEM) equipped with an INCA EDX (Energy Dispersive X-ray Spectroscopy) analyzer enabling X-ray analysis of microsized particles and the determination of the chemical composition. The settings were as follows: thermos-emission cathode LaB6 and voltage 20 kV. The specimen chamber was kept at a high vacuum.

The phase composition of the ceramics crucibles after interaction with the steels was measured by a Bruker AXS D8 Advance X-ray diffractometer equipped with a LynxEye position-sensitive silicon strip detector under the following conditions: CuKα/Ni-filtered radiation, voltage 40 kV, current 40 mA, step mode with a step of 0.014° 2θ, total time 25 s per step, and 2θ range 5–80°. The data were processed by Bruker AXS Diffrac and Bruker EVA software. The PDF-2 database (International Centre for Diffraction Data) was used for phase identification.

3. Results and Discussion

3.1. Wetting Tests

The temperature dependencies of the contact angles at the XT 720 steel/ceramic interface are shown in Figure 2.

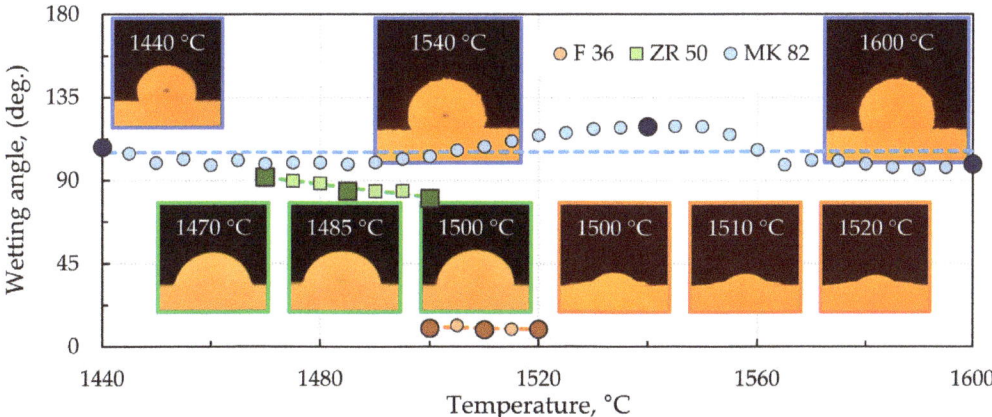

Figure 2. Temperature dependencies of the contact angles at the XT 720 steel/ceramic interface.

Since high-alloy manganese steel showed a severe interaction at the interface with the investigated ceramic material, only three temperature dependencies of the contact angles are selected in Figure 2, namely, with the basic fireclay material F36, the zirconia material ZR50, and the high corundum material MK82. The most pronounced interaction is observed at the interface with the material designated F36, with an apparent wetting angle of only about 10° at temperatures around 1500 °C for this material with manganese steel. In the case of zirconia material ZR 50, the apparent wetting angle at the same temperature is 80°, and that of MK 82 is 103°. Though the temperature dependencies of the wetting angle were investigated up to a maximum temperature of 1600 °C, not all the angles could be adequately evaluated. Therefore, only the evaluable sections are shown in Figure 2, including images of the recorded droplets at the specified temperatures. A discussion regarding the interaction at these phase interfaces is given in Section 3.3.

Figure 3 shows the temperature dependencies of the wetting angles at the 11 523 steel/ceramic substrate interface. The low-alloy 11 523 structural steel showed considerably weaker interaction than the high-alloy Mn steel on the refractory surfaces investigated. The most significant interaction occurred at the interface with the conventional F 36 fireclay material.

In the temperature range of 1540–1600 °C, the contact angles ranged from 105° to 102°, while the lowest interaction was observed for the high corundum material MK82, where the contact angles in the same temperature range were 130°–123°. The values of the contact angles decrease slightly with increasing temperature as the sulfur content is less than 60 ppm [30,31].

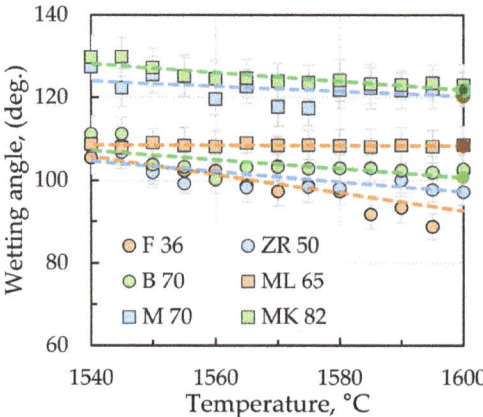

Figure 3. Temperature dependencies of contact angles at the 11 523 steel/ceramic interface (error lines parallel to the y-axis).

Images of the steel droplets observed on the ceramic material, including contact angle values at 1500 °C for XT 720 steel and at 1550 °C for 11 523 steel, are shown in Figures 4 and 5, respectively.

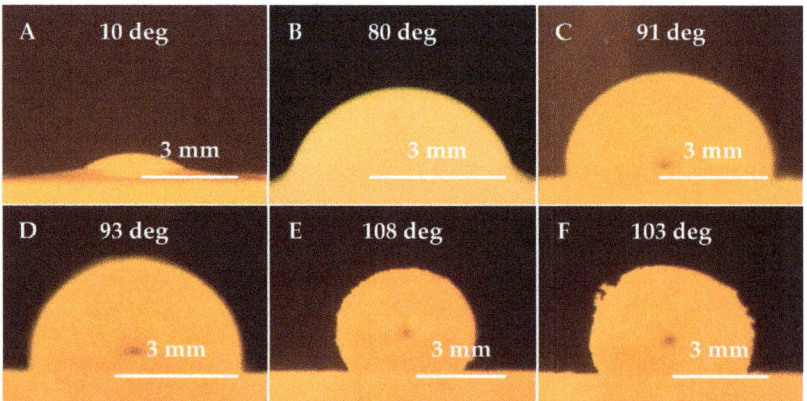

Figure 4. Images of XT 720 steel droplets wetting refractory ceramic substrates at 1500 °C; (**A**) F 36 ceramics, (**B**) ZR 50 ceramics, (**C**) B 70 ceramics, (**D**) ML 65 ceramics, (**E**) M 70 ceramics, and (**F**) MK 82 ceramics.

In general, when the wetting angle $\Theta \geq 90°$, the liquid (melt) exhibits non-wetting behavior towards the substrate. If the wetting angle is $\Theta \leq 90°$, the liquid (melt) exhibits wetting behavior. The interaction of the phases in contact (molten steel and ceramic substrate) becomes more likely as the wetting behavior increases. Judging by the drop silhouettes of the tested steels, it can be concluded that the interaction of Mn (XT 720) steel with the tested ceramic materials is much more pronounced than that of 11 523 steel.

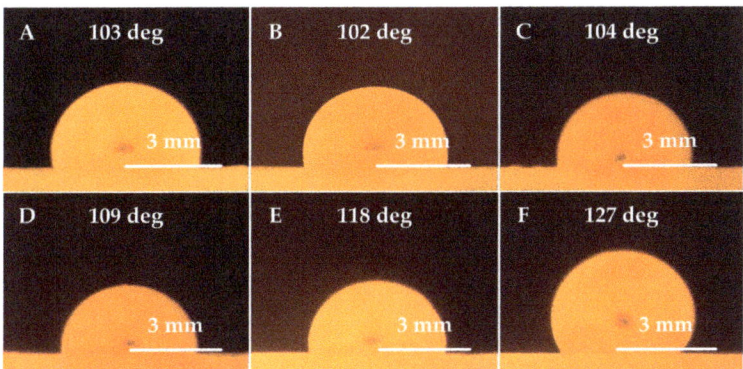

Figure 5. Images of 11 523 steel droplets wetting refractory ceramic substrates at 1550 °C; (**A**) F 36 ceramics, (**B**) ZR 50 ceramics, (**C**) B 70 ceramics, (**D**) ML 65 ceramics, (**E**) M 70 ceramics, and (**F**) MK 82 ceramics.

3.2. Degradation Testing

Figure 6 shows the degradation test results, i.e., losses of ceramics caused by high-manganese steel and low-alloy structural steel.

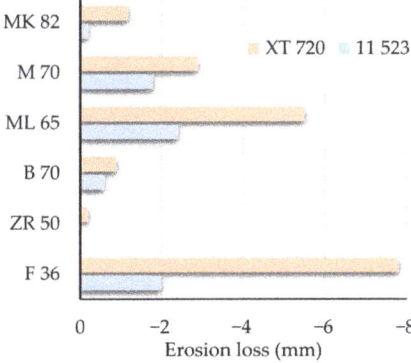

Figure 6. Degradation testing of high-manganese steel and low-alloy structural steel interactions with ceramics.

The type of steel used was found to greatly influence the degradation results of the refractories. Compared to conventional low-alloy structural steel, high-alloy manganese steel caused more significant wear to the materials tested. For this corrosive steel, the use of conventional fireclay materials proved to be entirely unsuitable. The most pronounced erosion loss interacting with this steel was observed for material F36, where even the crucible melted during the degradation tests (see Figure 7A). A more detailed discussion of the interaction is given in Section 3.3. The interaction rate of the steels with the refractory material decreases with increasing corundum content and decreasing silica and cristobalite content in the ceramics. The final state of the crucibles made from materials F36 and MK82 after degradation tests in interaction with XT 720 and 11 523 steels is shown in Figures 7A,B and 8A,B.

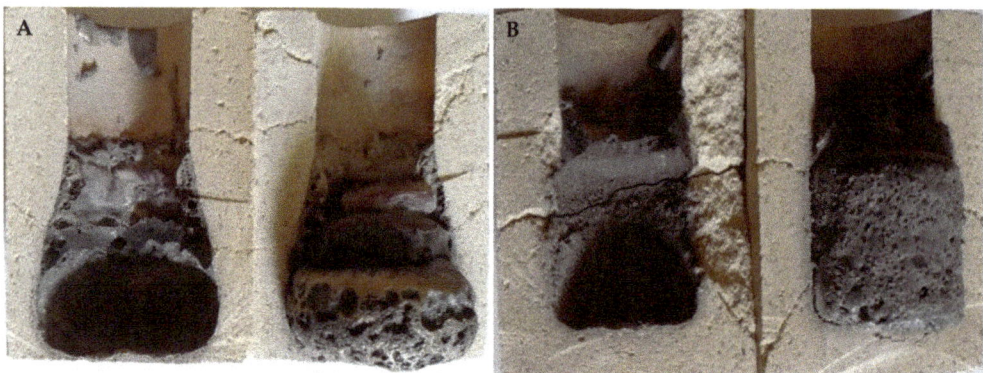

Figure 7. Degradation of refractory material F 36 after high-temperature testing; (**A**) steel XT 720, (**B**) steel 11 523.

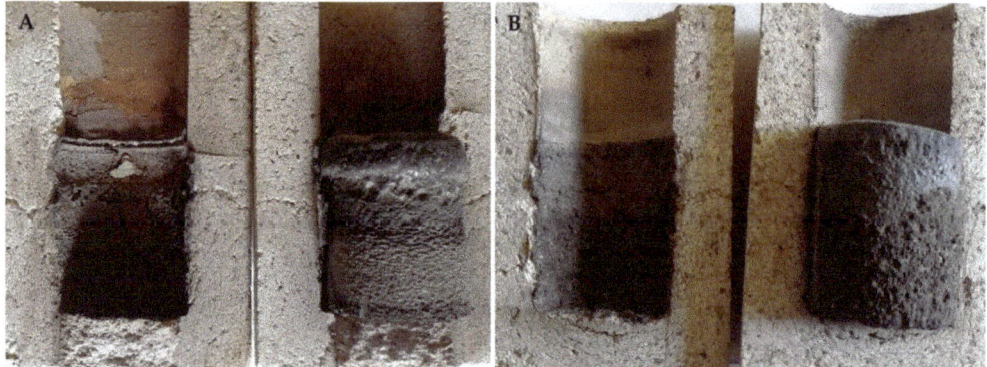

Figure 8. Degradation of refractory material MK 82 after high-temperature testing; (**A**) steel XT 720, (**B**) steel 11 523.

3.3. Results of SEM, EDX, and XRD Analyses

After the high-temperature wetting tests, the phase interfaces between the tested steels and the corresponding refractories were subjected to SEM, EDX, and XRD analyses. It was confirmed that the high-manganese XT 720 steel interacts more strongly with the tested refractories. The results of XRD analyses for the interfacial interfaces of this steel with the conventional chamotte material F 36, the material ZR 50 containing zirconia, and the material MK 82 with the highest corundum content, are compared below and in Figure 9. The XRD analyses confirmed the formation of new phases at the interfaces. In particular, galaxite-spinel and sillimanite phases were newly identified at the interface of the F 36 material. In the case of the ZR 50 material, galaxite-spinel ($MnAl_2O_4$) and tazheranite ((Zr, Ti, Ca)O_2) phases were newly formed. Finally, for the MK 82 material, galaxite-spinel, sillimanite ($Al_2(SiO_4)O$), and tephroite (Mn_2SiO_4) phases were determined. It is worth noting that there is a high degree of isomorphic substitution in the case of spinels but in our case, it is not easy to deduce the degree of isomorphism [32]. However, qualitative evaluation points to the formation of galaxite but the formation of (Mn, Fe, Mg) (Al, Fe, Cr)$_2$O$_4$ ferrites cannot be excluded. The crystalline fractions for the refractory material F 36, Zr 50, and MK 82 were 31.9 wt%, 43.2 wt%, and 73.9 wt%, respectively. Thus, sample F 36 contained the most amorphous phase.

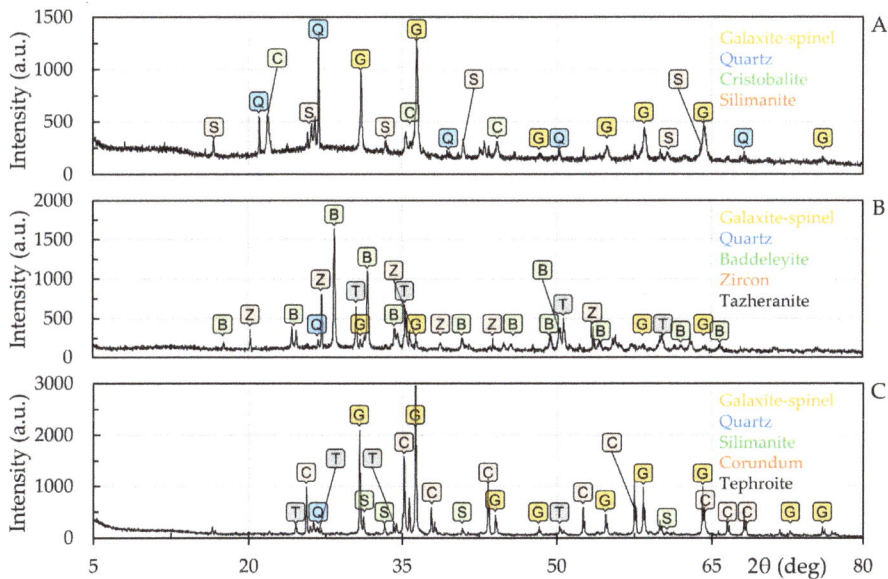

Figure 9. Diffractograms of refractory materials after high-temperature tests; (**A**) F 36 ceramics, (**B**) ZR 50 ceramics, and (**C**) MK 82 ceramics.

Based on SEM/EDX analyses (Figure 10), it can be argued that, in the case of the refractory material F 36, the main newly formed phase is the glassy phase with manganese oxide particles on its surface. Pores were also present to a higher degree in this sample, which was probably due to bubbles trapped in the rapidly cooling layer. In the case of ZR 50 ceramics, the glassy phase has also formed, and additionally, particles of baddeleyite (ZrO_2) and iron, manganese, and chromium oxide particles were identified. Ceramics MK 82 contained the least glassy phase after the high-temperature test. The crystalline phase was dominated by corundum, the crystals of which had aluminum partially substituted by manganese on their surface. In addition, aluminosilicates with manganese admixture were present, forming a dendritic structure. This structure was directed towards the hotter surface region where the crystal nucleation rate is lower in contrast to the low-temperature region where they originate. Some dendrites showed a tip-split of the primary dendritic arm, and, in addition, tertiary arm formation was observed [33]. The corresponding results of the EDX point analysis are summarized in Table 5.

Table 5. Results of semi-quantitative EDX microanalyses on the interaction of refractory samples with XT 720 steel.

Point	Caption	Mn	Si	Al	O	C	P	Fe	Zr	Zn	Cr
						(wt%)					
1	Aluminosilicate with Mn oxide particles	36.1	20.0	10.1	33.8	—	—	—	—	—	—
2	Manganese oxide	54.8	9.5	7.4	28.3	—	—	—	—	—	—
3	Zirconium silicate	—	13.4	—	48.3	—	—	—	38.3	—	—
4	Probably Schreibersite and Fe, Mn. and Cr oxide particles	21.9	2.7	1.0	30.7	5.1	2.7	27.5	—	1.9	6.5
5	Baddeleyite	1.9	—	—	31.3	—	—	—	66.8	—	—
6	Aluminosilicate	23.9	21.5	12.8	41.8	—	—	—	—	—	—
7	Corundum (Al partially substituted by Mn)	26.2	1.3	32.1	40.4	—	—	—	—	—	—
8	Aluminosilicate with Mn oxide	42.5	18.9	8.3	30.3	—	—	—	—	—	—

In the EDX point analysis, the influence of the surroundings must be considered for small particles and thin layers.

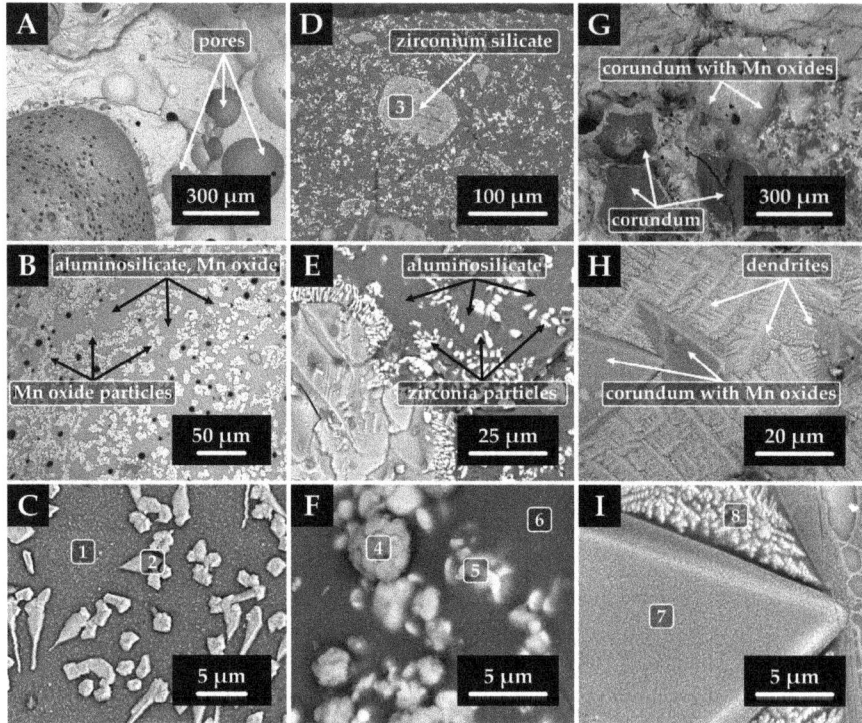

Figure 10. SEM images of refractory F 36 (**A–C**), refractory ZR 50 (**D–F**), and refractory MK 82 (**G–I**) after high-temperature test with steel XT 720; numbers 1–8 refer to EDX point analyses.

4. Conclusions

The interaction at the interface of high-manganese steel, or low-alloy construction steel, and refractory ceramics was characterized by wetting tests performed up to 1600 °C. The findings were supported by degradation tests and SEM, EDX, and XRD analyses. The findings can be summarized as:

- The intensity of the interaction at a given interface is influenced by the type of steel involved. High-manganese steel causes much more wear on refractory materials than conventional construction steel.
- Apparent wetting angles (contact angles) were significantly lower for the XT 720 steel than for the 11 523 steel. For high-manganese steel, the contact angles ranged from 10° to 103° depending on the substrate, in contrast to structural steel where non-wetting behavior was observed, and contact angles ranged from 103° to 127°. The decrease in contact angles correlated with the fraction of corundum in the refractories tested, i.e., they decreased with decreasing corundum. In addition, the contact angles decreased slightly with increasing temperature.
- The degradation tests confirmed the aggressive behavior of the XT 720 steel. The most significant erosion occurred in the case of the conventional chamotte material F 36 with a low corundum content and a higher SiO_2 and cristobalite content. Conversely, almost no erosion was observed for steel 11 523 in contact with the zirconia material ZR 50 or the high corundum material MK 82.
- The results of SEM, EDX, and XRD analyses confirmed the presence of newly formed phases at the interface of XT 720 steel with the tested refractories. In the case of the refractory material F 36, where the most significant interaction was observed, a

glassy phase with manganese-oxide particles was formed at the interface with the high manganese steel. Of the newly detected phases, galaxite-spinel and sillimanite phases were identified.

The results of this study suggest that, as the corundum content of the refractories increases, their resistance to erosion by molten steel also increases. Materials containing a higher content of fluxes (chamotte, bauxite, sintered mullite) have an increased content of the amorphous phase and show a weaker resistance to erosion. For high-alloy steels such as XT 720 steel, the use of conventional chamotte products is entirely inappropriate. For this steel, materials made of electrofused corundum can be recommended.

Author Contributions: Conceptualization, L.Ř. and V.N.; methodology, L.Ř., V.N. and S.R.; software, L.Ř., V.N. and S.R.; formal analysis, L.Ř., V.N. and S.R.; investigation, L.Ř., V.N. and D.M.; resources, L.Ř.; writing—original draft preparation, L.Ř. and V.N.; writing—review and editing, L.Ř., V.N. and S.R.; visualization, L.Ř. and V.N.; supervision, L.Ř.; project administration, S.R.; funding acquisition, L.Ř. All authors have read and agreed to the published version of the manuscript.

Funding: This work was supported by the project No.CZ.02.1.01/0.0/0.0/17_049/0008399—EU and CR financial funds—provided by the Operational Programme Research, Development and Education, Call 02_17_049 Long-Term Intersectoral Cooperation for ITI, Managing Authority: Czech Republic—Ministry of Education, Youth and Sports, and student project SP2022/39.

Institutional Review Board Statement: Not applicable.

Informed Consent Statement: Not applicable.

Data Availability Statement: The data presented in this study are available on request from the corresponding author.

Acknowledgments: We also thank the employees of SEEIF, a.s. Ing. Libor Bravanský and Ing. Tadeáš Franek for their help and comments on the manuscript.

Conflicts of Interest: The authors declare no conflict of interest.

Abbreviations

ADSA	Axisymmetric Drop Shape Analysis
EDX	Energy Dispersive X-ray Spectroscopy
GDOES	Glow Discharge Optical Emission Spectrometry
SEM	Scanning Electron Microscopy
XRD	X-Ray Powder Diffraction

References

1. Grässel, O.; Krüger, L.; Frommeyer, G.; Meyer, L.W. High strength Fe-Mn-(Al, Si) TRIP/TWIP steels development-properties-application. *Int. J. Plast.* **2000**, *16*, 1391–1409. [CrossRef]
2. Krawczyk, J.; Bembenek, M.; Pawlik, J. The role of chemical composition of high-manganese cast steels on wear of excavating chain in railway shoulder bed ballast cleaning machine. *Materials* **2021**, *14*, 7794. [CrossRef] [PubMed]
3. Dana, K.; Sinhamahapatra, S.; Tripathi, H.S.; Ghosh, A. Refractories of alumina-silica system. *Trans. Indian Ceram. Soc.* **2014**, *73*, 1–13. [CrossRef]
4. Fu, L.; Gu, H.; Huang, A.; Or, S.W.; Zou, Y.; Zou, Y.; Zhang, M. Design, fabrication and properties of lightweight wear lining refractories: A review. *J. Eur. Ceram. Soc.* **2022**, *42*, 744–763. [CrossRef]
5. Horckmans, L.; Nielsen, P.; Dierckx, P.; Ducastel, A. Recycling of refractory bricks used in basic steelmaking: A review. *Resour. Conserv. Recycl.* **2019**, *140*, 297–304. [CrossRef]
6. Poirier, J. A review: Influence of refractories on steel quality. *Metall. Res. Technol.* **2015**, *112*, 410. [CrossRef]
7. Wei, X.; Dudczig, S.; Storti, E.; Ilatovskaia, M.; Endo, R.; Aneziris, C.G.; Volkova, O. Interaction of molten Armco iron with various ceramic substrates at 1600 °C. *J. Eur. Ceram. Soc.* **2022**, *42*, 2535–2544. [CrossRef]
8. Wang, Y.; Huang, A.; Wu, M.; Gu, H. Corrosion of alumina-magnesia castable by high manganese steel with respect to steel cleanness. *Ceram. Int.* **2019**, *45*, 9884–9890. [CrossRef]
9. Ren, X.-M.; Ma, B.-Y.; Li, S.-M.; Li, H.-X.; Liu, G.-Q.; Yang, W.-G.; Qian, F.; Zhao, S.-X.; Yu, J.-K. Comparison study of slag corrosion resistance of MgO–MgAl$_2$O$_4$, MgO–CaO and MgO–C refractories under electromagnetic field. *J. Iron Steel Res. Int.* **2021**, *28*, 38–45. [CrossRef]

10. Alhussein, A.; Yang, W. Mechanism of interface reactions between Fe-2%Al alloy and high-silica tundish refractory. *Trans. Indian Inst. Met.* **2019**, *72*, 591–602. [CrossRef]
11. Alhussein, A.; Yang, W.; Zhang, L. Effect of interactions between Fe–Al alloy and MgO-based refractory on the generation of MgO·Al$_2$O$_3$ spinel. *Ironmak. Steelmak.* **2020**, *47*, 424–431. [CrossRef]
12. Jiang, M.; Wang, X.; Chen, B.; Wang, W. Formation of MgO·Al$_2$O$_3$ inclusions in high strength alloyed structural steel refined by CaO-SiO$_2$-Al$_2$O$_3$-MgO slag. *ISIJ Int.* **2008**, *48*, 885–890. [CrossRef]
13. Ehara, Y.; Yokoyama, S.; Kawakami, M. Formation mechanism of inclusions containing MgO·Al$_2$O$_3$ spinel in type 304 stainless steel. *Tetsu Hagane* **2007**, *93*, 208–214. [CrossRef]
14. Park, J.H.; Lee, S.-B.; Gaye, H.R. Thermodynamics of the formation of MgO-Al$_2$O$_3$-TiO$_x$ inclusions in Ti-stabilized 11Cr ferritic stainless steel. *Metall. Mater. Trans. B* **2008**, *39*, 853–861. [CrossRef]
15. Zhang, L.; Thomas, B.G. State of the art in the control of inclusions during steel ingot casting. *Metall. Mater. Trans. B* **2006**, *37*, 733–761. [CrossRef]
16. Park, J.H.; Zhang, L. Kinetic modeling of nonmetallic inclusions behavior in molten steel: A review. *Metall. Mater. Trans. B* **2020**, *51*, 2453–2482. [CrossRef]
17. Polkowski, W.; Sobczak, N.; Polkowska, A.; Nowak, R.; Kudyba, A.; Bruzda, G.; Giuranno, D.; Generosi, A.; Paci, B.; Trucchi, D.M. Ultra-high temperature interaction between h-BN-based composite and molten silicon. *Metall. Mater. Trans. A* **2019**, *50*, 997–1008. [CrossRef]
18. López, V.H.; Kennedy, A.R. Flux-assisted wetting and spreading of Al on TiC. *J. Colloid Interface Sci.* **2006**, *298*, 356–362. [CrossRef]
19. Kumar, G.; Prabhu, K.N. Review of non-reactive and reactive wetting of liquids on surfaces. *Adv. Colloid Interface* **2007**, *133*, 61–89. [CrossRef]
20. Shibata, H.; Jiang, X.; Valdez, M.; Cramb, A.W. The contact angle between liquid iron and a single-crystal magnesium oxide substrate at 1873 K. *Metall. Mater. Trans. B* **2004**, *35*, 179–181. [CrossRef]
21. Eustathopoulos, N.; Voytovych, R. The role of reactivity in wetting by liquid metals: A review. *J. Mater. Sci.* **2016**, *51*, 425–437. [CrossRef]
22. Sobczak, N.; Singh, M.; Asthana, R. High-temperature wettability measurements in metal/ceramic systems—Some methodological issues. *Curr. Opin. Solid State Mater. Sci.* **2005**, *9*, 241–253. [CrossRef]
23. Zhou, X.B.; De Hosson, J.T.M. Reactive wetting of liquid metals on ceramic substrates. *Acta Mater.* **1996**, *44*, 421–426. [CrossRef]
24. Wang, L.; Zhu, H.; Zhao, J.; Song, M.; Xue, Z. Steel/refractory/slag interfacial reaction and its effect on inclusions in high-Mn high-Al steel. *Ceram. Int.* **2022**, *48*, 1090–1097. [CrossRef]
25. Alibeigi, S.; Kavitha, R.; Meguerian, R.J.; McDermid, J.R. Reactive wetting of high Mn steels during continuous hot-dip galvanizing. *Acta Mater.* **2011**, *59*, 3537–3549. [CrossRef]
26. Pourmajidian, M.; McDermid, J.R. On the reactive wetting of a medium-Mn advanced high-strength steel during continuous galvanizing. *Surf. Coat. Technol.* **2019**, *357*, 418–426. [CrossRef]
27. Yang, T.; He, Y.; Chen, Z.; Zheng, W.; Wang, H.; Li, L. Effect of dew point and alloy composition on reactive wetting of hot dip galvanized medium manganese lightweight steel. *Coatings* **2020**, *10*, 37. [CrossRef]
28. Kong, L.; Deng, Z.; Zhu, M. Reaction behaviors of Al-killed medium-manganese steel with different refractories. *Metall. Mater. Trans. B* **2018**, *49*, 1444–1452. [CrossRef]
29. Saad, S.M.I.; Neumann, A.W. Axisymmetric drop shape analysis (ADSA): An outline. *Adv. Colloid. Interface* **2016**, *238*, 62–87. [CrossRef]
30. Brooks, R.F.; Quested, P.N. The surface tension of steels. *J. Mater. Sci.* **2005**, *40*, 2233–2238. [CrossRef]
31. Dubberstein, T.; Heller, H.-P.; Klostermann, J.; Schwarze, R.; Brillo, J. Surface tension and density data for Fe–Cr–Mo, Fe–Cr–Ni, and Fe–Cr–Mn–Ni steels. *J. Mater. Sci.* **2015**, *50*, 7227–7237. [CrossRef]
32. Lu, Z.; Wang, Z.; Wang, S.; Zhao, H.; Cai, Z.; Wang, Y.; Ma, H.-A.; Chen, L.; Jia, X. Orientational dependences of diamonds grown in the NiMnCo-Silicate-H$_2$O-C system under HPHT conditions and implications to natural diamonds. *ACS Earth Space Chem.* **2022**, *6*, 987–998. [CrossRef]
33. Bhagurkar, A.G.; Qin, R. The microstructure formation in slag solidification at continuous casting mold. *Metals* **2022**, *12*, 617. [CrossRef]

MDPI
St. Alban-Anlage 66
4052 Basel
Switzerland
Tel. +41 61 683 77 34
Fax +41 61 302 89 18
www.mdpi.com

Crystals Editorial Office
E-mail: crystals@mdpi.com
www.mdpi.com/journal/crystals

www.ingramcontent.com/pod-product-compliance
Lightning Source LLC
LaVergne TN
LVHW070431100526
838202LV00014B/1575